The Notion of Equality

The International Research Library of Philosophy
Series Editor: John Skorupski

Metaphysics and Epistemology

Identity *Harold Noonan*
Personal Identity *Harold Noonan*
Scepticism *Michael Williams*
Infinity *A.W. Moore*
Theories of Truth *Paul Horwich*
The Existence of God *Richard M. Gale, Alexander Pruss*
Knowledge and Justification, Vols I & II *Ernest Sosa*
Spacetime *Jeremy Butterfield, Mark Hogarth, Gordon Belot*
A Priori Knowledge *Albert Casullo*
Vagueness *Timothy Williamson, Delia Graff*
Supervenience *Jaegwon Kim*

The Philosophy of Mathematics and Science

The Ontology of Science *John Worrall*
Mathematical Objects and Mathematical Knowledge *Michael D. Resnik*
Theory, Evidence and Explanation *Peter Lipton*

The Philosophy of Logic, Language and Mind

Understanding and Sense, Vols I and II *Christopher Peacocke*
The Limits of Logic *Stewart Shapiro*
Events *Robert Casati, Achille C. Varzi*
Consciousness *Frank Jackson*
Reason, Emotion and Will *R. Jay Wallace*

The Philosophy of Value

Consequentialism *Philip Pettit*
Punishment *Antony Duff*
Meta-ethics *Michael Smith*
Human Sexuality *Igor Primoratz*
The Ethics of the Environment *Andrew Brennan*
Medical Ethics *R.S. Downie*
Feminist Ethics *Moira J. Gatens*
Public Reason *Fred D'Agostino, Gerald F. Gaus*

The Notion of Equality

Edited by

Mane Hajdin

Dominican University of California, USA

Routledge
Taylor & Francis Group
LONDON AND NEW YORK

First published 2001 by Ashgate Publishing

Reissued 2018 by Routledge
2 Park Square, Milton Park, Abingdon, Oxon, OX14 4RN
711 Third Avenue, New York, NY 10017, USA

Routledge is an imprint of the Taylor & Francis Group, an informa business

A Library of Congress record exists under LC control number: 00038100

ISBN 13: 978-1-138-71126-6 (hbk)
ISBN 13: 978-1-138-71124-2 (pbk)
ISBN 13: 978-1-315-19979-5 (ebk)

Contents

Acknowledgements

The editor and publishers wish to thank the following for permission to use copyright material.

Blackwell Publishers for the essays: Janet Radcliffe Richards (1997), 'Equality of Opportunity', *Ratio (new series)*, **X**, 253–79, Copyright © 1997 Janet Radcliffe Richards; Derek Parfit (1997), 'Equality and Priority', *Ratio (new series)*, **X**, pp. 202–21, Copyright © 1997 Blackwell Publishers Ltd.

Cambridge University Press for the essay: Allen Buchanan (1995), 'Equal Opportunity and Genetic Intervention', *Social Philosophy & Policy*, **12** (2), pp. 105–35, Copyright © 1995 Social Philosophy and Policy Foundation, published by Cambridge University Press.

Ronald Dworkin (1981), 'What Is Equality? Part 1: Equality of Welfare', *Philosophy & Public Affairs*, **10**, pp. 185–246, Copyright © 1981 Ronald Dworkin; Ronald Dworkin (1981), 'What Is Equality? Part 2: Equality of Resources', *Philosophy & Public Affairs*, **10**, pp. 283–345, Copyright © 1981 Ronald Dworkin.

Edinburgh University Press for the essay: Klemens Kappel (1997), 'Equality, Priority, and Time', *Utilitas*, **9**, pp. 203–25, Copyright © 1997 Edinburgh University Press.

R.M. Hare (1978), 'Justice and Equality', in John Arthur and William H. Shaw (eds), *Justice and Economic Distribution*, Englewood Cliffs, NJ: Prentice-Hall, pp. 116–31, Copyright © 1978 R.M. Hare.

Johns Hopkins University Press for the essay: Madison Powers (1996), 'Forget About Equality', *Kennedy Institute of Ethics Journal*, **6**, pp. 129–44, Copyright © 1996 Johns Hopkins University Press.

Kluwer Academic Publishers for the essay: Richard J. Arneson (1989), 'Equality and Equal Opportunity for Welfare', *Philosophical Studies*, **56**, pp. 77–93, Copyright © 1989, with kind permission from Kluwer Academic Publishers.

Lieber Atherton, Inc. for the essay: Richard E. Flathman (1967), 'Equality and Generalization, A Formal Analysis', in J. Roland Pennock and John W. Chapman (eds), *Nomos 9: Equality*, New York: Atherton Press, pp. 38–60. Reprinted by permission of the publishers, Lieber-Atherton, Inc.

Oxford University Press for the essays: J. Raz (1978), 'Principles of Equality', *Mind*, **LXXXVII**, pp. 321–42; D.A. Lloyd Thomas (1977), 'Competitive Equality of Opportunity', *Mind*, **LXXXVI**, pp. 388–404; Derek Browne (1978), 'Nonegalitarian Justice', *Australasian Journal of Philosophy*, **56**, pp. 48–60.

Princeton University Press for the essay: Larry S. Temkin (1986), 'Inequality', *Philosophy & Public Affairs*, **15**, pp. 99–121, Copyright © 1986 PAPA. Reprinted by permission of Princeton University Press.

The University of Calgary Press for the essay: Bruce M. Landesman (1983), 'Egalitarianism', *Canadian Journal of Philosophy*, **XIII**, pp. 27–56.

The University of Chicago Press for the essays: Douglas Rae (1981), 'Two Contradictory Ideas of (Political) Equality', *Ethics*, **91**, pp. 451–56, Copyright © 1981 University of Chicago; Peter Westen (1985), 'The Concept of Equal Opportunity', *Ethics*, **95**, pp. 837–50, Copyright © 1985 University of Chicago; Harry Frankfurt (1987), 'Equality as a Moral Ideal', *Ethics*, **98**, pp. 21–43, Copyright © 1987 University of Chicago; Dennis McKerlie (1989), 'Equality and Time', *Ethics*, **99**, pp. 475–91, Copyright © 1989 The University of Chicago; Dennis McKerlie (1996), 'Equality', *Ethics*, **106**, pp. 274–96, Copyright © 1996 The University of Chicago.

University of Pittsburgh for the essay: Felix E. Oppenheim (1970), 'Egalitarianism as a Descriptive Concept', *American Philosophical Quarterly*, **7**, pp. 143–52.

Every effort has been made to trace all the copyright holders, but if any have been inadvertently overlooked the publishers will be pleased to make the necessary arrangement at the first opportunity.

Series Preface

The International Research Library of Philosophy collects in book form a wide range of important and influential essays in philosophy, drawn predominantly from English-language journals. Each volume in the Library deals with a field of inquiry which has received significant attention in philosophy in the last 25 years, and is edited by a philosopher noted in that field.

No particular philosophical method or approach is favoured or excluded. The Library will constitute a representative sampling of the best work in contemporary English-language philosophy, providing researchers and scholars throughout the world with comprehensive coverage of currently important topics and approaches.

The Library is divided into four series of volumes which reflect the broad divisions of contemporary philosophical inquiry:

- Metaphysics and Epistemology
- The Philosophy of Mathematics and Science
- The Philosophy of Logic, Language and Mind
- The Philosophy of Value

I am most grateful to all the volume editors, who have unstintingly contributed scarce time and effort to this project. The authority and usefulness of the series rests firmly on their hard work and scholarly judgement. I must also express my thanks to John Irwin of the Ashgate Publishing Company, from whom the idea of the Library originally came, and who brought it to fruition; and also to his colleagues in the Editorial Office, whose care and attention to detail are so vital in ensuring that the Library provides a handsome and reliable aid to philosophical inquirers.

JOHN SKORUPSKI
General Editor
University of St. Andrews
Scotland

Introduction

Equality: The Conceptual Problem

The notion of equality has been featured prominently in a multitude of day-to-day moral and political debates. In almost every discussion on such topics as welfare reform, taxation policies, voting systems or affirmative action, the notion of equality will probably be invoked in the argument of at least one of the sides. The political developments of recent decades have also brought the notion into the debates on such topics as abortion, pornography, sexual harassment and the proper content of school curricula. Often, two opposed positions on a controversial issue will both be defended by appeals to the ideal of equality: the critics of affirmative action thus typically insist that the ideal requires the abolition of affirmative action, while its defenders claim that affirmative action is the only way to achieve the ideal.

The arguments centred around the notion of equality that are mounted by one side to such a dispute often fail to convince the other side: people often continue to disagree on such issues, even after thorough discussion. Such an impasse is frequently due not simply to stubbornness or to an irresolvable clash of vastly different normative views, but also to conceptual confusion. This anthology is a representative selection of recent philosophical attempts to expose at least some of the roots of such confusions and to thus help clear them up. In the essays that follow, contemporary philosophers examine the conceptual underpinnings of appeals to the ideal of equality, and try to illuminate the ways in which the notion of equality operates in moral and political contexts. While the specific arguments presented in these essays are controversial, taken together they certainly do reveal that the seemingly simple and clear notion of equality turns out to be quite problematic once it is brought into moral and political contexts.

The essays in this volume thus show that there are quite a few basic problems that need to be tackled before anyone can meaningfully either embrace or reject the ideal of equality, and before anyone can appeal to that ideal in a discussion of a specific moral or a political issue.

Equality in General

Egalitarianism and the Core Uses of the Notion of Equality

The first step in appreciating the problems that surround the use of the notion of equality in moral and political contexts is to note that, although the notion has been used in these contexts for centuries, this is not the notion's original 'home'. The primary, core, uses of that notion are in statements like

(1) $5 + 7 = 12$

In such contexts, equality is a relationship between numbers. Any attempt to elucidate how the notion functions in another context must explain how the use of the notion in that context relates to its core uses.

In some cases, that is easy. For example, the use of the notion of equality in a statement such as

(2) The length of this table and the length of that table are equal.

although different from its use in '5 + 7 = 12', is only one step removed from it. The difference is that, in the statement about the lengths of the tables, equality does not, on the face of it, appear to be a relationship between numbers, but between lengths. However, it does not require great analytical skills to see that that statement is equivalent to, say, 'the number that expresses the length of this table in some units of length (such as feet or metres) is equal to the number that expresses the length of that table in the same units'. In that last statement, equality is again a relationship between numbers, just as it is in '5 + 7 = 12'. The statements about lengths, weights, voltages and other quantities can thus always be reformulated as statements about numerically expressed results of measurements, and that establishes a close connection between the uses of the notion of equality exemplified by (2) and its uses exemplified by (1).

One more step away from statements such as '5 + 7 = 12' is the use of the notion of equality in statements such as

(3) This table and that table are equal in length.

Although the syntax of that statement makes equality a relationship, neither between numbers nor between lengths, but between tables, it is fairly easy to see that the statement is equivalent to (2).

One thing to realize about the uses of the notion of equality exemplified by (3) is that, if we want to treat equality as a relationship between such things as tables – that is, between things that are not themselves quantities (but which instantiate different quantities) – then we need to specify the relevant quantity elsewhere in the statement. Note that (3) said that the tables were 'equal *in length*' – that is, it not only specified what the two objects were that the equality was asserted between, but also the dimension along which the relevant, numerically expressible, measurements have been made. It was only in virtue of its having specified the dimension that our understanding of it was unproblematic. The form exemplified by (3) is thus '___ and ___ are equal in ___', where the third blank must be filled by the specification of some quantity. Only by virtue of that specification being provided can we relate statements such as (3) to statements such as (2) and, through them, to those like (1).

This means that a statement such as

(4) This table and that table are equal

is, on its own, incomplete. Tables are not themselves quantities, so (4) cannot be treated as being of the same type as (2). Tables do instantiate quantities, but each of them instantiates many quantities: length, height, weight, age, distance from the North Pole, the age of the carpenter who made it . . . . Until we know which of these is meant, we cannot relate (4) to the core uses of the notion of quality, and indeed cannot make any sense of it. Of course, when a statement like (4) is used in actual communication, the wider context often makes clear that the relevant dimension is, say, length, and we can then treat (4) as an abbreviation for something like (3), and thus relate it to the core uses of the notion of equality.

To fully appreciate that (4), on its own, says nothing, we should realize that, for any two things, such as tables, it is always possible to find some ways of measuring them that will give equal results – and, also some ways of measuring them that will give unequal results. Suppose we have one table which is 0.5 metres long and another one which is 1.5 metres long. We can say that they are equal in how much their lengths differ from 1 metre. On the other hand, suppose that we have two tables that are equal in length, width, weight and so on. They will still be unequal in their distance from the tip of my nose. (If they happen to be equal in that respect as well, I only need to select some other reference point.) Analogous manoeuvres will always be available.

The examples (2) through (4) above were examples of statements as to what *is* the case and not of claims as to what *ought* to be the case, but what they illustrate about the notion of equality does not depend on that. Everything that has been said above about (2) through (4) can also be said about (2*) through (4*):

(2*) The length of this table and the length of that table ought to be equal.

(3*) This table and that table ought to be equal in length.

(4*) This table and that table ought to be equal.

(We could easily imagine a realistic context in which a carpenter says such things in giving instructions to an apprentice.)

The popular political slogans that purport to express egalitarianism usually say something like 'Everyone ought to be treated equally'. What that presumably amounts to, is that, for any two persons:

(5) The treatment given to this person and the treatment given to that person ought to be equal.

What we immediately see is that (5) has the same form as (4*), and is thus, so far as the use of the notion of equality is concerned, in the same category as (4). The popular slogans that purport to be egalitarian thus seem to say nothing whatsoever on their own. If egalitarianism is to be a meaningful political view, something further needs to be said, so that we get something like:

(6) The treatment given to this person and the treatment given to that person ought to be equal in respect of the amounts of income that they generate for them.

So far as the way in which the notion of equality is used, (6) is in the same category as (3) and is therefore clearly meaningful. But so are:

(7) The treatment given to this person and the treatment given to that person ought to be equal in respect of the ratios between the income that each person gets and the number of hours the person has worked.

(8) The treatment given to this person and the treatment given to that person ought to be equal in respect of the ratios between the income that each person gets and the person's height.

(9) The treatment given to this person and the treatment given to that person ought to be equal in respect of the weight of the portions of chocolate-chip ice cream that each person gets.

(10) The treatment given to this person and the treatment given to that person ought to be equal in respect of the ratios between the number of years that each person spends in prison and the number allocated to the person by a throw of dice.

Infinitely many meaningful claims that embody the same structure could be formulated – some plausible, some crazily implausible. But if, say, both (7) and (10) embody the notion of equality in exactly the same way, then we seem to have no good reason for labelling one of them egalitarian and not the other. However, if the label 'egalitarianism' were to be applied to such a wide range of different views, it would not be a useful label at all.

Moreover, it seems that whatever makes some of these claims more plausible than the others cannot have anything to do with the way in which the notion of equality features in them, because it plays exactly the same role in all of them. It thus seems to follow that the notion of equality is incapable of playing a significant role in discussions about how people should be treated. In other words, consideration of examples such as (6) through (10) shows that equality *per se* cannot be morally valuable, because there is no such thing as equality *per se*. The moral value of an action or policy cannot lie in the fact that it treats people equally, but rather in the precise way in which it treats them equally.

'Equality' as an Emphasizing Device

Political rhetoric sometimes uses 'equal', 'equally', and related words in a way that is significantly different from anything that has been mentioned above. Thus, both the abolition of slavery and, more recently, of racial segregation are often characterized as moves that have advanced equality. However, being, say, a slave is not a *quantity* of anything, nor is it in any obvious way related to a quantity of anything. This means that although the claim

(11) This person and that person ought to be treated equally in that neither of them ought to be a slave

exhibits a superficial similarity to (6) through (10) and thus to (3), it cannot be analysed in the same way. The fact that (11) in no way involves anything that is expressible numerically makes it impossible to regard it as equivalent to any claim in which the notion of equality would be used in the way exhibited by (2). Claims such as (11) are thus difficult to relate to the core uses of the notion of equality.

The content of a claim such as (11) could have been expressed without using the word 'equally'. Consider:

(12) Neither this person nor that person ought to be a slave.

If we compare (11) with (12) we see that the presence of the word 'equally' in (11) doesn't really add anything significant to what is expressed in (12). If the word is playing any role in (11) at all, it is playing the role of an emphasizing device, similar to the word 'indeed'.

There is nothing wrong with using the word 'equally' as a mere emphasizing device, but we should not confound that way of using it with the ways described in the previous section. It needs to be realized that, once we accept that way of using the word as legitimate, the word can be thrown into the formulation of any conceivable moral or political view. As long as we treat people in accordance with some rule or other, it can be said that we treat them equally in that we treat them all in accordance with that rule. Even the Nazis could have said that they were treating everyone equally, in that they were treating everyone in accordance with rules such as 'All Jews ought to be sent to gas chambers'.

In order to fully understand the point that has just been made, one needs to bear in mind the currently standard understanding of the logical form of universal claims. According to that understanding, for example, the claim 'All giraffes have long necks', really says that, for every *x*, if *x* is a giraffe, then *x* has a long neck. Thus although that claim may, on the surface, seem to be only about giraffes, in a way it applies to both giraffes and non-giraffes. In that way, it applies to me: it implies that *if* I am a giraffe, *then* I have a long neck. That is a perfectly true description of me: what makes it true is the falsity of the antecedent. In the same way, the Nazis could have said that the rule 'All Jews ought to be sent to gas chambers' applied to both Jews and non-Jews, which could have then made it possible for them to say that, in following that rule, they were treating everyone, both Jews and non-Jews, in accordance with it. If they wanted to use the word 'equally' as an emphasizing device, they could then have proceeded to say that they were treating everyone equally in that they were treating everyone in accordance with the rule.

Thus, although the use of the word 'equally' exemplified by (11) is legitimate, the fact that a particular moral or political view can be expressed by using the word in that way does not enhance our understanding of that view, nor does it provide us with good ground for classifying the view as egalitarian, because any view whatsoever, including obviously outrageous ones, can be so expressed.

Is 'Egalitarianism' a Useful Concept?

The roots of the arguments outlined so far could be traced as far back as Aristotle's remark that 'All men think justice to be a sort of equality . . . But there still remains a question – equality . . . of what? Here is a difficulty which the political philosopher has to resolve' (*Politics*, 1282b). The arguments along these lines have, however, achieved particularly wide currency among English-speaking political philosophers of the mid-twentieth century.

When one acknowledges that these arguments are *prima facie* convincing, one faces the question of what one's response to them would be. Different contemporary theoreticians have given different answers to that question. Some have taken the path of endorsing and elaborating arguments of the kind outlined above, and are happy to accept their implications. A clear example of that approach is the essay by Richard Flathman, reprinted here as Chapter 1, which is a classical defence of the view that the notion of equality is, as a matter of logic, incapable of playing the role that is usually allocated to it in moral and political debates. If the argument of Flathman's essay is accepted, then the notion of equality can never play the crucial role in establishing that people ought to be treated in a particular way: we can justifiably say that two people ought to be treated equally only *after* we have established that the first one ought to be treated in a particular way and that the second one also ought to be treated in that way.

Other philosophers have taken the arguments of the kind outlined in the preceding sections as a challenge to produce an analysis of the notion of political equality that will give it a richer content than is allowed by the arguments, and to thus make 'egalitarianism' a notion of a distinct substantive ideal (or, at least, a distinct group of substantive ideals). These philosophers try to rehabilitate our pretheoretical intuition – which the above arguments attack – that moral and political views can be usefully classified into egalitarian and inegalitarian ones. An early example of such an attempt was Bernard Williams's 1962 essay, 'The Idea of Equality', which has given considerable impetus to the discussion of this topic among analytical philosophers. That essay is not reprinted in this volume, but it has already been reprinted in numerous other anthologies.

A more elaborate attempt to rehabilitate the intuition that egalitarianism is a distinct substantive ideal is made by Felix Oppenheim in 'Egalitarianism as a Descriptive Concept' (Chapter 2). According to Oppenheim's analysis, whether a particular policy is egalitarian cannot be determined by examining that policy in isolation, nor by examining solely its end-results. Whether a policy is egalitarian depends, Oppenheim claims, on how it *changes* the status quo. In other words, it is a feature of the concept of egalitarianism that it is applied against the background of pre-existing allocations; egalitarian policies are those that *re*distribute these allocations in a particular way. Oppenheim's criterion of egalitarianism implies that being egalitarian is a matter of degree: it allows one to say that a particular policy is egalitarian, but less so than another policy.[1] Before presenting his own analysis of the concept of egalitarianism, Oppenheim devotes a great deal of space to a very systematic examination of a number of alternative analyses, and even those readers who reject his conclusions are likely to find his criticism of these analyses illuminating.

Perhaps the most influential philosophical articulation of the idea that egalitarianism is a distinct type of a normative position was offered by Joseph Raz in his 1978 essay 'Principles of Equality' (Chapter 3). (A slightly different version of the same argument appears in Chapter 9 of Raz's book, *The Morality of Freedom* (1986).) In this essay Raz acknowledges that 'equality' and related words are often used to express moral claims that could be more perspicaciously expressed without them, but argues that there is nevertheless a distinct class of moral principles that can be usefully classified as strictly egalitarian. These are the principles that, in their contents, involve ineliminable comparisons: they require us to ensure that people have something *if* other people have it. In following such a principle, Raz claims, one *aims* directly at equality. Following a principle that is not strictly egalitarian may also produce equality between people, but in that case the equality is a by-product of treating each of the people involved in a way that is independently morally right.

Part I of this volume ends with an unjustly neglected essay by Derek Browne, published in the same year as Raz's essay. Like Flathman, Browne implies that we are never justified in saying that people ought to be treated equally unless we have already established, on some other ground, how they ought to be treated. However, he explicitly allows that a genuinely egalitarian moral position is logically possible – it's just that, according to him, it is, on substantive moral grounds, highly implausible. Browne's criterion of what would constitute such a genuinely egalitarian position is similar to that of Raz, although not developed at the same length.

It should be emphasized here that one's position on this conceptual (or, as is sometimes said, meta-ethical) question of whether the concept of egalitarianism is a useful concept does not in any straightforward way determine one's substantive views about day-to-day moral and political issues (although the two are certainly related). The position that no notion of egalitarianism can be articulated, that would make egalitarianism a distinct substantive ideal, does not automatically entail that various policies that are popularly referred to as egalitarian ought to be rejected; they may well be worthy of support on some ground that does not involve the notion of equality. There are philosophers who hold that the notion of equality is incapable of carrying the weight that is usually assigned to it in moral and political discussion, and who still remain committed to the political causes that are popularly referred to as egalitarian (see, for example, Westen, 1990, p. 287). On the other hand, thinking that there are ideals that can be usefully classified as egalitarian in no way entails that these ideals are worthy of endorsement: Raz, for example, after offering his analysis of the concept, makes some rather critical remarks about the views that, on that analysis, turn out to be 'strictly egalitarian'.

Equality of Resources versus Equality of Welfare

Some philosophers have accepted the arguments showing that the concepts of equality and egalitarianism allow a wide range of different views to be called egalitarian, but have still resisted the conclusion that the concepts are entirely useless in moral and political discourse. These philosophers have produced classifications of different 'egalitarianisms' – that is, of different substantive views that can all, so far as the concept is concerned, be legitimately called egalitarian. The choice of one form of egalitarianism from such a classification as the correct one is, according to these philosophers, to be made on substantive moral grounds (rather than by conceptual analysis). Thus, each such classification immediately leads to the formulation of the moral issue as to which of the egalitarianisms classified is the one we should embrace, if we embrace any.

The most influential classification of that kind is the one formulated and discussed in Ronald Dworkin's 1981 two-part essay 'What Is Equality?' (Chapters 6 and 7). Dworkin classifies versions of egalitarianism into those that advocate equality of welfare and those that advocate equality of resources. 'Welfare' is a technical term in this context, and applies to whatever matters to people *fundamentally*, as opposed to what matters to them only as means to something else. Plausible candidates for welfare are, for example, happiness, desire-satisfaction, and preference-satisfaction. 'Resources', on the other hand, is a general term for the 'stuff', ranging from peanuts to airplanes, that one can use as means to obtain welfare. Thus, treating people in such a way that they all achieve the same levels of preference-satisfaction is a form of equality of welfare, while treating them in such a way that they all obtain the same amounts of wealth would be an example of equality of resources. The most prominent versions of the ideal of equality of resources involve equalization of people's resources in their commercial value, rather than in their other dimensions; these versions are sometimes referred to as economic egalitarianism. Once the issue is formulated as to whether equality of welfare or equality of resources is morally preferable, the answer that initially seems obvious to most people is: equality of welfare. However, Dworkin argues at considerable length that a particular form of equality of resources is preferable to equality of welfare.

Dworkin's two-part essay is preceded in this volume by Douglas Rae's essay (Chapter 5) about the conflict between 'person-regarding equality' and 'lot-regarding equality', which is akin to Dworkin's conflict between equality of welfare and equality of resources. This very brief essay is an excellent introduction to the conflict. (In this essay Rae is particularly interested in how the conflict manifests itself in discussions about equality in voting, but the insight into the nature of the conflict presented by the essay applies to almost every other moral and political context in which the notion of equality is invoked.) The essay also gives a taste of the kind of conceptual analysis that Rae and his associates pursue in much greater detail in their book *Equalities* (Rae *et al.*, 1981).

The issue of which of these forms of equality is morally preferable (couched in slightly different terminology) is also at the centre of Bruce Landesman's essay 'Egalitarianism' (Chapter 8) which, despite its later publication date, was in fact written before the appearance of Dworkin's essay. Landesman argues for the version of egalitarianism that aims at equalizing what he calls well-being, but he interprets the notion of well-being broadly, so that it covers much more than just the satisfaction of one's preferences. In particular, having responsibility for autonomously coordinating and adjusting one's preferences is also an element of well-being according to Landesman. Once the ideal of equalizing well-being is understood that way, its political implications will often coincide with those of the ideal of the equality of resources.

Dworkin's 1981 two-part essay has stimulated a great deal of further philosophical exploration of the distinction between equality of welfare and equality of resources: the body of philosophical literature dealing with that distinction is now considerable. Richard Arneson's essay 'Equality and Equal Opportunity for Welfare', which belongs to that body of literature is reprinted here as Chapter 9 because it provides an important connection between the topic of Dworkin's and Landesman's essays and the topic of Part III of this volume. Arneson combines the distinction between equality of welfare and equality of resources with the distinction between 'straight equality' (also called equality of outcomes or equality of results) and equal opportunity. Given that the two distinctions are logically independent, they, taken together, generate a fourfold classification of possible egalitarian ideals:

1 straight equality of welfare;
2 straight equality of resources;
3 equal opportunity for welfare; and
4 equal opportunity for resources.

Arneson's essay gives support to the third of these ideals – that of equal opportunity for welfare. Indeed, it can be argued that many of the concerns that motivate Dworkin's defence of equality of resources can be better accommodated within the version of egalitarianism that takes equal opportunity for welfare as its aim.

Equality of Opportunity

Equality of opportunity has itself been the subject matter of a great deal of philosophical examination, some of which is presented in Part III of this volume. Most of that examination revolves around the paradox that besets much of the everyday political debate that invokes equality of opportunity. The paradox is that political movements which express their aims in terms of equality of opportunity get much of their initial appeal from the ways in which their aim differs from straight equality, but that the dynamics of political battle tends to lead them closer and closer to the defence of straight equality. Such a movement may, for example, begin by fighting against laws that explicitly prevent certain groups of people from pursuing some benefits. Once that initial aim is achieved and the laws are repealed, it is likely to be pointed out that the opportunities are still not equal, because there remain informal prejudices that discourage these groups from pursuing the benefits, and the aim of the movement becomes the eradication of such prejudices. Once that is achieved, it will be said that the opportunities are still not equal, because some people are not adequately prepared to pursue the benefits, and efforts will be made to boost their preparation. These efforts may be unsuccessful because some of the relevant people are not motivated to profit from them, and it will then be said that something should be done to motivate them, and so forth. While each step in this process may well seem justified, the process as a whole has a tendency to transform the ideal of equal opportunity to that of straight equality. As D.A. Lloyd Thomas says in articulating this paradox in Chapter 10, a competition in which everything that influences the contestants' chances of success is equalized is bound to be a tie.

Thomas proceeds to argue that, if egalitarians of opportunity managed to somehow prevent this slide towards straight equality, they would have to face another problem. As long as the ideal of equal opportunity is understood so that it permits unequal outcomes, these unequal outcomes will preclude equality of opportunity in the future. An egalitarian of opportunity who avoids the slide towards straight equality may thus be paradoxically committed to saying that a particular inequality is both just and unjust: just because it is an outcome of a competition in which the opportunities were equal, and unjust because it creates an inequality of opportunities for future competitions.

In Chapter 11 Peter Westen offers an analysis of the notion of an opportunity that, if accepted, could eliminate such paradoxes. It is a fairly obvious feature of the notion of an opportunity that an opportunity is always someone's opportunity *for* something. Westen, however, argues that that is not all there is to the notion of an opportunity. According to him, every claim about an opportunity needs to be interpreted as referring to a particular obstacle that the opportunity in question removes. If the obstacle is neither specified nor implied by the context, then all talk about opportunities is meaningless. His analysis thus implies that one cannot say, for example, that one is simply in favour of all adult citizens having equal opportunity for employment: one has to be prepared to specify whether the obstacle to be removed is a 'Jim Crow' law, or lack of childcare facilities, or harassment in hiring, or lack of education, or something entirely different. Each of these different specifications of the relevant obstacle gives us a different political goal, which needs to be defended by different substantive arguments. If the distinctness of these goals were fully appreciated in day-to-day political debates, then the paradoxes outlined above would be likely to disappear.

Allen Buchanan's and Janet Radcliffe Richards' essays (Chapters 12 and 13) both offer analyses of the tendency of the ideals of equal opportunity to keep transforming themselves in the way that keeps bringing them closer and closer to straight equality, and they show that these transformations are not inevitable. According to both of them, the concerns that make political ideals expressed in terms of equal opportunity plausible can be expressed without using the notion of equal opportunity, and are much clearer when so expressed. Once the ideals are reinterpreted in that way, the paradoxes to which they seem to lead are avoided.

The Ideal of Equality for Its Own Sake versus Other Ideals with Similar Implications

Suppose that one is persuaded that the typical uses of the notion of equality in moral and political context are hopelessly confused, but that one still feels deep commitment to upholding political measures, such as large-scale welfare programmes, that have been traditionally defended by invoking that notion. Is there a way of harmonizing such ground-level political commitments with the rejection of egalitarianism as a basic moral principle? The essays in Part IV of this volume offer some possible answers to that question.

In Chapter 14 R.M. Hare shows that many measures defended by egalitarians can also be defended on utilitarian grounds. A crucial role in his argument is played by what is known as the law of diminishing marginal utility: other things being equal, the same resources tend to produce more welfare when given to those who have little of such resources than when given to those who already have plenty of them. An egalitarian distribution of resources consequently tends to produce more welfare than other distributions of the same resources.

It needs to be noted, however, that the law of diminishing marginal utility is qualified by an 'other things being equal' clause, and other things are often not equal. A utilitarian (unlike someone who is committed to equality as a basic, underived, moral ideal) will thus often allow the conclusion of the above argument to be outweighed by other considerations. It should also be noted that, while this utilitarian reasoning gives some limited support to political measures that equalize resources, it gives no support to measures that aim at equalizing welfare.[2]

Harry Frankfurt (Chapter 15) rejects the law of diminishing marginal utility, as well as any commitment to equality as a basic moral ideal. He argues that many day-to-day political proposals that are defended by appeals to equality can be more plausibly defended by what he calls 'the doctrine of sufficiency'. The crucial difference between the two ways of defending such proposals is that the former is essentially comparative, while the latter is not. When we are concerned about the plight of the poor, the focus of our concern, according to Frankfurt, is not on the fact that they have less than somebody else, but rather on the fact that they don't have enough.

Frankfurt's argument is, however, open to the objection that, for at least some kinds of goods, it may be impossible to produce a completely non-comparative (and, at the same time, intuitively satisfying) standard of what it is to have enough of them. This objection is at the core of Madison Powers' attempt, in 'Forget About Equality' (Chapter 16) to find a compromise between Frankfurt's doctrine of sufficiency and egalitarianism proper.

New Distinctions within Egalitarianism

In his 1989 essay, 'Equality and Time' (Chapter 17), Dennis McKerlie opened philosophical discussion of a hitherto unnoticed set of problems related to egalitarianism. McKerlie provides a classification of different forms of egalitarianism that is largely independent of all the others. Regardless of what it is that egalitarians wish to equalize, they can argue that the quantity of it in each person's life, taken as a whole, should be made equal to the quantity of it in every other person's life, taken as a whole (the view that McKerlie calls 'complete lives egalitarianism'), *or* that the quantity of it in the simultaneous segments of every person's life should be equalized (the position that McKerlie calls the 'simultaneous segments view'), *or* perhaps that the quantity of it should be equalized among the segments of people's lives that are matched in some other way. McKerlie's essay shows that determining which of these versions of egalitarianism is the most plausible one is by no means a straightforward matter. He adduces a number of arguments in favour of the 'simultaneous segments view' (which may, at first glance, seem less plausible than the 'complete lives egalitarianism') but even those who disagree with these arguments cannot deny the significance of McKerlie's formulation of the problem.

In 'Equality and Priority' (Chapter 18), Derek Parfit draws two further distinctions that anybody who is tempted by egalitarianism needs to keep in mind. The first is the distinction between 'telic' and 'deontic' egalitarianism. According to the former, a state of affairs that involves inequality is bad in itself, quite apart from whether the inequality is due to human actions or natural sources, and quite apart from whether it is inequality between people living within the same society or in different, perhaps completely unrelated, societies. The latter form of egalitarianism applies, in the first instance, to actions rather than states of affairs: according to it, it is *creation* of inequality that is bad (or, more specifically, unjust).

The second distinction that Parfit makes is between egalitarianism proper (including both telic and deontic versions of it) and what he calls the 'Priority View'. According to the Priority View, 'benefiting people matters more the worse off these people are'. He suggests that the Priority View may be able to avoid the problems that make it difficult to choose between telic and deontic egalitarianism since it will, in most day-to-day situations, have implications that are similar to those of egalitarianism, but does not take equality as desirable for its own sake. The crucial difference between the Priority View and egalitarianism proper is that the latter essentially involves comparisons between people, while the former does not.

The Priority View should not be confused with the law of diminishing marginal utility, although it is indeed structurally similar to it. While the law of diminishing marginal utility is about diminishing marginal utility of *resources*, the Priority View is about diminishing marginal moral significance of *utility* (welfare) itself. The Priority View also has a certain affinity with, but is not identical to, Frankfurt's doctrine of sufficiency (unlike the doctrine of sufficiency, the Priority View does not imply the existence of the threshold of sufficiency).

It should be mentioned here that a position that is effectively the same as what Parfit calls the Priority View was (without being given a particular name) also articulated and distinguished from the view that equality is important for its own sake in McKerlie's 1984 essay 'Egalitarianism'. Something like the Priority View also seems to be hinted at, at the end of Browne's 'Nonegalitarian Justice' (see Chapter 4).

Klemens Kappel's essay (Chapter 19) combines the themes of the two preceding essays. It first discusses and casts doubt on what McKerlie calls the 'simultaneous segments view' and then proceeds to claim that the arguments that seem to support that view really lead us in the direction of the Priority View and away from egalitarianism proper.

McKerlie's second essay in this volume (Chapter 20) deals with the set of issues that are raised in Parfit's essay.[3] McKerlie tries to show that the arguments that seem to render telic egalitarianism problematic (as compared with deontic egalitarianism and the Priority View), are, in fact, not as persuasive as they seem to be. A particularly noteworthy feature of his discussion is his attempt to respond to the frequently made objection that egalitarianism is obsessively preoccupied with relationships *between* people, while it is only what is *within* people's lives that has the ultimate moral significance. (This objection is, for example, a significant source of plausibility for the Priority View and for Frankfurt's doctrine of sufficiency.) McKerlie argues that when people live in the circumstances of inequality, then some of them have unfulfilled claims to their fair shares. The aim of egalitarianism, according to McKerlie, should be understood primarily as fulfilment of these unfulfilled *individual* claims, and not simply as creation of a particular relation between people.[4]

Larry Temkin's essay, which concludes this volume, shows that there is yet another issue that must be tackled by anyone who wishes to be an egalitarian – an issue that is, again, independent of the preceding ones. Regardless of how we specify our ideal of equality, we have to take into account the fact that, in the world as it is, it probably won't be achieved perfectly. Usually the best that an egalitarian can hope for is to bring the world closer to the ideal. In order to do that, the egalitarian has to have an account of what it is for one of two states of affairs that both fall short of the ideal to be *closer* to it than the other. Temkin shows that we have intuitions on these questions that pull us in different directions. In his 1993 book, *Inequality*, Temkin continued the line of investigation he began in this essay.

* * * * * * * * *

It may be useful to add to the above account of what this volume includes, a few remarks on what it does not include, on what its limits are. First, this volume is limited to what is known as the 'analytic' tradition in philosophy; it does not include the writings of philosophers who work within, or are dominantly inspired by, the 'continental' tradition. (This limitation is not unnatural in view of the volume's focus on conceptual issues.) Second, although several of the philosophers represented in this volume are influenced by the work of economists, no essay that presupposes that its readers are trained in economics is included.

Finally, the aim of this volume was to gather together the writings that have made important contributions to our understanding of what the ideal of equality is. This includes the writings that explore how equality is related to the formal features of moral discourse, the writings that distinguish the ideal of equality from similar moral and political ideals, and the writings that provide illuminating classifications of different forms of egalitarianism. The volume does not include any writings that presuppose that the ideal itself is already clear enough, and are principally devoted to *normative* arguments for or against it. The essays that follow do contain normative arguments, but the ground for their inclusion in this volume is always their contribution to *clarifying* what the ideal of equality amounts to.

Notes

1 While saying that being egalitarian is a matter of degree is intuitively plausible, it opens up quite a few further problems. Some of these problems are addressed in Larry Temkin's work (see below).

2 Some people may characterize such utilitarianism-based endorsement of rough equality as a *version* of egalitarianism, while others would classify it as a *non*-egalitarian position that just happens to lead

to similar results as egalitarianism. I think that the latter way of labelling it is more illuminating (it is more useful to reserve the word 'egalitarianism' for the endorsement of equality as a basic, underived, ideal), but the former way of speaking about it can be acceptable as long as it is made clear what is meant.

3 While this essay of McKerlie's appeared before the publication of Parfit's essay (reprinted as Chapter 18), it was written after Parfit's ideas, as expressed in that essay, had already been circulating for some time in an unpublished form among the scholars in this field.

4 Temkin, in Chapter 21, individualizes the moral concern behind egalitarianism in a similar way. See, however, note 39 of McKerlie's essay.

References

Aristotle's *Politics* (1905), trans. Benjamin Jowett, Oxford: Clarendon Press.

McKerlie, Dennis (1984), 'Egalitarianism', *Dialogue*, **XXIII**, pp. 223–37.

Rae, Douglas *et al.* (1981), *Equalities*, Cambridge, MA: Harvard University Press.

Raz, Joseph (1986), *The Morality of Freedom*, Oxford: Clarendon Press.

Temkin, Larry S. (1993), *Inequality*, New York: Oxford University Press.

Westen, Peter (1990), *Speaking of Equality: An Analysis of the Rhetorical Force of 'Equality' in Moral and Legal Discourse*, Princeton, NJ: Princeton University Press.

Williams, Bernard (1962), 'The Idea of Equality', in Peter Laslett and W.G. Runciman (eds), *Philosophy, Politics and Society*, Second Series, Oxford: Basil Blackwell, pp. 110–31.

Part I
Equality in General

[1]

EQUALITY AND GENERALIZATION, A FORMAL ANALYSIS

RICHARD E. FLATHMAN

To treat people equally is to treat them in the same way. To treat people in the same way is to treat them according to a rule. "Equally" is defined in the *Oxford English Dictionary* to mean "According to one and the same rule." Philosophers, aware of this relationship, have attempted to explicate and refine the notion of equality through the notion of general rule or generalization. Historically, the salient names in this connection are Rousseau and Kant. Many contemporary philosophers have concerned themselves with generalization, but few have applied their conclusions to "equality." The purpose of the present paper is to analyze the concept of equality in the light of recent discussions of generalization. I will argue that the concept of equality can be explicated in terms of generalization, and that to do so shows that as normative concepts both equality and generalization are of derivative significance. These conclusions will lead to suggestions concerning the importance of utilitarian considerations in morals.

I

Recent work in ethics or meta-ethics has been concerned to identify and provide formalized statements of the principles and rules that operate, if only in a concealed manner, in moral discourse. The concept of generalization has been the object of substantial attention in these respects, and we now have several analyses of what has been called the Generalization or Universalizability Principle (hereafter GP). Professor Marcus Singer, who has made the most detailed study of the topic, states GP in the form we will adopt for the purposes of this paper: "What is right (or wrong) for one person must be right (or wrong) for every relevantly similar person in relevantly similar circumstances."[1] According to this formulation, if X is right for A, it must be right for B, C, D, . . . N unless A or his circumstances is different from B, C, D, . . . N or their circumstances in a manner justifying making an exception of A. There is, it would appear, a presumption in favor of equal treatment of all persons and any departure from that rule must be justified.

One feature of GP, its "formal" or "neutral" character, requires immediate attention. Before the universal principle that GP expresses can be applied to a problem, two particular premises must be established. The first is of the form: "This X is right for this A"; the second: "This B is, vis-à-vis this A, a relevantly similar person in relevantly similar circumstances." GP states a relationship between these premises; it does not tell us how to establish them in any instance. Until they have been established, GP, although valid as an abstract principle, has no

[1] Marcus Singer, *Generalization in Ethics,* New York: Alfred A. Knopf, 1961, pp. 19–20; alternative formulations are presented on pp. 5 and 31. There is a very extensive critical literature concerning Singer's book, and I have profited substantially from it. It would be impracticable to cite that literature here, but mention should be made of Alan Gewirth's "The Generalization Principle," *Philosophical Review,* Vol. 73 (April, 1964), p. 229, in which the reader will find arguments similar in important respects to some of those presented here.

application. GP does one thing: When it has been established that X is right for A and that A and B are relevantly similar, the principle requires that it be wrong for A to act or be treated differently from B. Hence GP is formal or neutral in the sense that, taken alone, it does not prescribe the proper content of any decision. Clearly, then, the crucial problems are those involved in establishing the particular premises in specific cases. Those problems, and the question of GP's utility once those problems have been resolved, will be our primary concern. We will also give attention to an argument that a substantive doctrine of equality is concealed by the ostensibly formal or neutral character of GP.

II

It will facilitate the analysis to use a real controversy concerning equality as a source of examples and illustrations. The recent United States Supreme Court decisions concerning apportionment in state legislatures, the "one-man, one-vote" decisions, are well suited to this purpose.[2] In the Court's view, the question before it in the Reynolds Case was whether the Equal Protection clause[3] permits a state to employ a system of apportionment in which the representatives of some districts represent substantially fewer citizens than the representatives of other districts. Although insisting that "mathematical exactitude" is neither desirable nor practicable as a standard for determining the adequacy of a plan, the majority nevertheless found that Equal Protection ". . . requires that a state make an honest and

[2] The leading case is *Reynolds* v. *Sims,* 377 U.S. 533 (1964). A series of companion cases follows immediately in the same volume. I would like to emphasize that it is not the purpose of this paper to present an evaluation of the Court's decisions in these cases. The arguments of the paper would be relevant to such an evaluation, and in it critical statements will be made concerning Chief Justice Warren's argument in the Reynolds case. But the purpose of these remarks will be to question the logic of the argument by which the conclusion is reached, not the conclusion itself. A competent evaluation of the latter would require a much wider investigation than I have undertaken. Also, because I am using the cases purely for illustrative purposes, I have not scrupled to ignore complexities in the Court's reasoning, which would lead beyond present purposes. It is my belief that the following interpretation of Warren's argument can be defended in terms of the text of his opinion, but I have not attempted to defend it here.

[3] Article 14, Section 1.

good faith effort to construct districts in both houses of its legislature as nearly of equal population as practicable."

In terms of GP, the issue can be restated as follows: "Is it constitutional for A (the citizens of legislative district A′ in state Q′) to have X (number of citizens per representative) if B, C, D, . . . N (the citizens of legislative districts B′, C′, D′, . . . N′ in state Q′) do not have X (have Y, a larger or smaller number of citizens per representative)? Under GP the answer could be, for example, "Yes, if X is right for A and if vis-à-vis B, C, D, . . . N, A is (the class of A consists of) relevantly different persons in relevantly different circumstances. The decision, in other words, would seem to turn on establishing the two premises we have identified. We will try to show, however, that what appears to be two tasks is in fact one—establishing the rightness of X.

Many of the arguments of Warren's opinion, by contrast, suggest that in matters of apportionment "right" and the kind of equality demanded by the equal population rule are equivalent; they collapse the first premise of GP into the second. Much of his opinion is designed to show that various differences among citizens such as class, economic status, place of residence, and other characteristics, although alleged to justify differences in treatment, are irrelevant to apportionment. The assumption seems to be that "equal" in the Constitution establishes a presumption that the equal population rule is right (constitutional), and the task is less to defend that presumption than to show that the conditions required for its application (relevant similarity of all citizens) are satisfied. If challenges to the relevant similarity of all citizens can be met, the question of "right" (constitutionality) will be answered. In common with many egalitarian arguments, in other words, Warren treats equality as a sufficient normative principle. To clarify the logic of this argument we will examine some of the key contentions that Warren offers to support his conclusion.[4]

[4] Inasmuch as Warren is interpreting a Constitution to which he must be faithful, it might be thought misleading or worse to say that *he* "treats equality" as a sufficient principle. The Constitution, it might be alleged, established the principle and Warren is required to accept it in his role as judge. If the content of "equal" in "equal protection" were entirely clear, this argument would be persuasive. We will try to adduce logical considerations to show that it is not, and hence that the language used above is appropriate.

Warren's aim is to support the conclusion that all citizens should be treated according to the rule, "Equal representatives for equal population." To do so he asserts: "With respect to the allocation of legislative representation, all voters as citizens of the State stand in the same relation. . . . "[5] Presumably this means that there are no differences among citizens or their situations that would significantly affect their opportunity to participate effectively in or the manner in which they are affected by the system of representation. But the empirical proposition crucial to Warren's argument is highly doubtful. A citizen's relation to the system of representation, his opportunity to participate in and benefit from that process, can be affected by such factors as the predominant interest pattern in his district, whether his party affiliation is that of a majority or a minority party, difficulties of communication with political and governmental centers, informal political tradition, size (in area) of districts, and perhaps others. By adopting the equal population rule Warren *makes it* true that all citizens "stand in the same relation" to the system in the respect required by that rule. But they may stand in very different relations to the system in other respects.

It is of course possible that Warren could justify rejecting these other differences. Indeed it is certain that some of the above-mentioned factors would have to be ignored, especially in establishing constitutional requirements concerning the conditions of participation. It would be difficult and perhaps undesirable to refine those requirements to take account of all of the differences that have been alleged to be relevant.[6] But this possibility, al-

[5] *Reynolds* v. *Sims,* 565. The sentence ends ". . . regardless of where they live." But in the course of the opinion Warren makes the same contention for all differences alleged to be relevant (there is something of an exception for existing political subdivisions) and hence it is not a distortion to broaden the passage by omitting the restriction.

[6] To assume difficulty or undesirability too readily, however, is to risk assuming away the issues concerning proper apportionment. It is instructive in this connection to consider the British practice, a practice that has gone to great lengths in adapting to some of the differences mentioned above. The British have evidently found this practicable, and they have rejected the conclusion that the numerical equality for which Warren opts will provide a base from which the citizen can overcome all other inequalities. For a general account of British practices see Vincent E. Starzinger, "The British Pattern of Apportionment," *The Virginia Quarterly Review,* Vol. 41 (Summer 1965), pp. 321ff.

though relevant to whether Warren's conclusion could be defended, does not support the logic of the argument by which the conclusion is reached.

More particularly, it does not support the assumption that "right" (constitutionality) and "equal" can be equivalent, that establishing the second premise of GP will establish the first premise as well. If the premise that all citizens "stand in the same relation" with regard to representation is not literally true in all respects that affect representation, then treating them according to a rule based on that premise will mean that there will be respects in which they will be treated differently (unequally). In opting for the equal population rule, Warren treats all citizens equally in the respect required by that rule. But in doing so he accepts inequalities with regard, for example, to size (in area) of electoral districts, distance from governmental centers, and perhaps others. Once again, he might be entirely justified in so doing. But his justification cannot be in terms of equality. For he has, if only tacitly, chosen between equalities. He has preferred the equality of the equal population rule to the equality of, say, size of electoral districts.[7] Inasmuch as he is choosing between competing equalities, he has, again if only tacitly, turned to some principle other than equality to make that choice, that is, to decide what is right. "Right" and "equality" are not equivalent; equality is not an independent or a sufficient normative principle.

These considerations can be generalized. They suggest that "right" and "equality" could be equivalent, that establishing the second premise of GP would also establish the first, only if the decision-maker was not faced with the kind of choice Warren had to make. This choice, however, can be avoided only if all those involved in or affected by a decision are similar persons in similar

[7] Here again (cf. note 4 above) it might be argued that Warren did not "choose" between equalities but read the Constitution as *requiring* the equal population rule. Leaving aside general issues concerning the difficulty of interpreting the Constitution, the foregoing argument is intended to show that it is logically impossible for the words of the document, taken alone, to require this finding. Perhaps the usual materials of constitutional interpretation, previous court decisions, intent of the Framers, etc., support Warren's finding. Inasmuch as our concern is not constitutional law but the logic of "equality," this fact, if it is a fact, is of little relevance here and does not affect the above argument.

situations in *all* respects that might affect the results of the decision. Given the diversity of men and their situations, it is difficult to believe that such decisions are a regular occurrence—especially in politics where we are typically concerned with very large numbers. Hence questions arise about the significance of the principle of equality.

It might be useful to restate the foregoing argument in terms closer to those used in our preliminary discussion of GP. When GP is applied to a concrete issue, it serves to raise a comparative question—whether X is right for A *and* for B, C, D, . . . N. The foregoing argument supports our earlier contention that this comparative question cannot be answered until we have determined whether X is right for A. Warren's argument, and all arguments that treat equality as a sufficient standard of right, suggests that X can sometimes be right for A *because* A and B, C, D, . . . N are treated in a like manner. But this argument either begs the question of the first premise of GP—"Is X right for A?"—or simply restates it for B, C, D, . . . N and A. The question whether X is right for A must be independent of GP. It is not a comparative question (that is, not in the sense of GP or equality; it may be comparative in the sense of comparing the merits of alternative policies or actions) and it cannot be answered by a comparative formula or rule. If we treat A wrongly, it will not help to say that we have treated him the same as B, C, D, . . . N.

The same is true of B, C, D, . . . N. The rightness or wrongness of giving X to B, C, D, . . . N could turn on a relationship of equality or inequality between B, C, D, . . . N and A only in the circumstances specified above. (Even in those circumstances questions of ethical naturalism might arise.) Treating them equally in one respect will usually involve treating them unequally in another, possibly more important, respect. If we treat B, C, D, . . . N badly, it will not help that we have treated B, C, D, . . . N and A alike. To decide the rightness or wrongness of X for B, C, D, . . . N, we must repeat for the latter those steps taken to determine whether X was right or wrong for A.

Our reason for contending that establishing the first premise of GP is the main task should now be clear. For the procedure just described establishes the second as well as the first premise. It tells us whether B, C, D, . . . N are, vis-à-vis A, relevantly similar persons in relevantly similar circumstances. Our purpose is to discover whether X is right or wrong for both A and B, C, D, . . . N; and, by hypothesis, we have now done so. What could be more relevant than that the same policy is right or wrong for both? Indeed, what other proper answer could there be to the question of whether they are relevantly similar? The second premise of GP, in short, is properly established through the same procedures as the first, by determining the rightness or wrongness of applying X to B, C, D, . . . N. If that determination accords with the result obtained in making the same decision with regard to applying X to A, then A and B, C, D, . . . N are relevantly similar persons in relevantly similar circumstances and what is right or wrong for B, C, D, . . . N must (logically) be right or wrong for A. If the results of the two determinations differ, A and B, C, D, . . . N are not relevantly similar—and what is right for A is not right for B, C, D, . . . N.

GP demands a certain relationship between A and B, C, D, . . . N; but that relationship is a logically necessary relationship between the results obtained by the application of tests that are independent of GP. The moment we lose sight of this fact we are in danger of mistreating a person, or class of persons, on the irrelevant grounds that he, or it, is receiving the same treatment as others.

III

It might be contended that the foregoing analysis fails to consider certain dimensions and characteristics peculiar to equality questions. We have dissolved equality questions into a series of right-wrong questions connected in a *post hoc,* formal manner. At the least, it might be alleged, this analysis ignores prominent aspects of moral and political thought and practice.

The objection is justified in that one aspect of equality questions is partially masked by the foregoing discussion. The difficulty in the above account is its apparent suggestion that judgments about the rightness or wrongness of X for A are *entirely* independent of, and cannot be upset by, concern with B, C, D, . . . N. There are cases in which this is obviously not true. If A and B, C, D, . . . N are part of the same moral or legal system, giving X to A might have consequences for B, C, D, . . . N, consequences that must be considered in deciding whether it is right to give X to A. In the cases discussed thus far, we have assumed we had accounted for the impact of X on both (all) parties or classes; but there are other types of cases to be considered. Hence the foregoing discussion is incomplete; it needs to be supplemented by an enumeration and further analysis of the combinations that logically can arise under GP. Although this will involve some repetition of earlier arguments, it should summarize the results of those arguments and show what must be added to them in order to handle the objection before us.

We will continue to assume that X is right for A taken alone (that is, without considering the impact on B, C, D, . . . N, of giving X to A). Holding this constant, there are two basic types of situations (A and B) and six types of cases (A. 1-3 and B. 1-3) that can arise under GP.

Situation A.: If X is not right for B, C, D, . . . N, A and B, C, D, . . . N are not relevantly similar, and what is right for A is, by hypothesis, not right for B, C, D, . . . N.

Hence: A.1. To treat A and B, C, D, . . . N alike would be to treat unlike cases alike and would violate GP.

This is the standard type of case in which GP is violated. The Reynolds decision illustrates the point nicely. The Court holds that the equal population rule is the "controlling criterion" under the Equal Protection Clause. All citizens must be treated equally in the respect indicated by that rule. A.1. shows that this rule can be a requirement only if it would not be wrong to treat B, C, D, . . . N in this way. Critics of the decision hold that this condition is not satisfied in some states, that use of the equal population rule will lead to results of the A.1. type. To insist that all citizens be treated equally in this respect might involve treating them unequally in respects more important to the citizens in question and to the system.

To apply the equal population rule to Colorado, for example, might require that citizens of area D′, an isolated mountain district with unique interests, be placed in a legislative district (D) which is uniquely large in terms of area, which poses very difficult transportation and communication problems, in which the overwhelming majority of the population have interests and political affiliations (for example, party) markedly at variance with those of the citizenry in D′, and in which the citizens of D′ share no mechanisms and channels of informal political activity with the other citizens of D. If these conditions do not obtain in other districts of the state, or, as posited, for other citizens in D, the citizenry of D′, although treated equally in terms of the equal population rule, are treated unequally in other respects. Hence to impose the equal population rule upon them would be to treat unlike cases alike. More important, it might be to treat unlike cases alike in a manner that results in treating some groups well and other groups badly.

To treat A and B, C, D, . . . N differently in situation A, for example to give X to A but not to B, C, D, . . . N, might appear to be acceptable under GP. But this is one of the situations prompting the objection that the present analysis ignores distinctive aspects of equality questions. Before we can decide whether it is right to give X to A but not to B, C, D, . . . N, we must decide whether it would be right not to give X to B, C, D, . . . N when one of the conditions is that we are giving X to A. Although we have posited that it would be wrong to give X to B, C, D, . . . N, we do not yet know whether it would be wrong to treat B, C, D, . . . N in the proposed manner, that is, subjecting them to whatever effects would result from giving X to A.

This case (A.2.) is not covered by A.1.; it is an entirely different case and an entirely different problem. But the principle employed to decide A.1. must be used to decide A.2. (and A.3.) as well. We can decide A.2. only by asking whether it is wrong to treat B, C, D, . . . N in this manner.

Hence: A.2. If the policy of giving X to A but not to B, C, D, . . . N is not wrong for B, C, D, . . . N, a permissible classification has been made and it is right to give X to A.

A.2. is the paradigm case of the permissible classification, and a great deal of governmental action falls under it.

A.3. If giving X to A but not to B, C, D, . . . N wrongs B, C, D, . . . N, the classification is not permissible and the policy must be abandoned. Note that it is not the mere fact that A and B, C, D, . . . N are treated differently (unequally) that renders X obnoxious. Or rather, there is only one sense of "unequal" in which this is true, namely that one is treated well and the other badly. They are also treated unequally in many respects under A.2., which is unobjectionable. Again, we can decide only by looking at the effects of the policy for B, C, D, . . . N.

A.2. and A.3. are common in government. A.2. allows a degree of flexibility that is essential if government is to act widely but with discrimination. And the search for A.2.'s sometimes leads to A.3.'s instead. It would be undesirable to tax those with incomes under five thousand dollars at 90 per cent, but it would also be undesirable if we were thereby prevented from taxing multimillionaires at that rate. It would be wrong to conscript men over sixty years of age, but it might be disastrous if we were thereby prevented from drafting men under twenty-six. For another illustration, consider the problem in Reynolds. Strict adherence to the equal population rule renders it impossible for cases of type A.2. to arise in the area of apportionment. Because the rule is satisfied only if everyone in the system is treated equally in this particular respect, the rule must be right for everyone or for no one.

Situation B.: If X is right for B, C, D, . . . N, A and B, C, D, . . . N are relevantly similar persons in relevantly similar circumstances, and what is right for A must be right for B, C, D, . . . N.

Hence: B.1. If X is applied to A and to B, C, D, . . . N, it does not violate GP. If Warren's argument is tenable, the equal population rule falls under this heading. Notice that this is the one and only case in which that rule is consistent with GP as it has been interpreted here.

B.2. If the policy of giving X to A but not to B, C, D, . . . N does not wrong B, C, D, . . . N, a permissible classification has been made and GP is not violated.

This is another situation that prompts an objection to the analysis. But the fact that B.2. is not fully covered by B.1. does not upset the above analysis or indicate that it cannot take account of the distinctive features of equality questions. B.2. is a different policy from B.1., and it must be evaluated independently. But nothing in the analysis prevents us from noticing and evaluating the fact that giving X to A has consequences for B, C, D, . . . N despite the fact that B, C, D, . . . N are not given X.

B.2. is perhaps the most interesting result with regard to equality. The argument is that the fact that X is right for both A and B, C, D, . . . N does not require that it be wrong to give X to A but not to B, C, D, . . . N. To prove that giving X is right is not to prove that withholding it is wrong. To show that it would be wrong to withhold X from B, C, D, . . . N while giving it to A would require a demonstration that the results of withholding X from B, C, D, . . . N, under these circumstances, would be wrong.

Consider the following hypothetical case. School districts A and B would both benefit from receiving X dollars of state aid. State resources are sufficiently great to give X to A and B without hampering other programs, but A has a large population of educationally deprived children and B does not. It would be right to give X to both A and B and to give X to A and not to B would be to treat them unequally in one important respect. But if we are right about B.2., to do the latter would not be wrong. The inequality would not lead to bad results for B and hence it would be based upon a permissible classification.

This point is at the heart of the present argument. "Equally" means "according to one and the same rule." Whenever we treat according to a rule, we will be treating equally in respect to that rule. Clearly then, the crucial question will be "according to what rule should we treat people in this case?" The principle of equality will rarely answer this question because ordinarily we must choose between equalities. If we choose to treat A and B equally in respect to educational achievement, we will treat them unequally in respect to the size of the grant awarded. This must be defended in terms of the relative importance of equal educational achievement as against equal grants of money. To dramatize the example, let us consider whether we could justify giving a larger grant to B despite the fact that B is already ahead of A in terms of educational achievement. To justify such a policy we would have to find another rule that we regard as more important than either size of grant or educational achievement. Let us say that the national defense would be served by concentrating the bulk of our resources in B. If our rule was "maximize contributions to the national defense," we could properly say that A and B had been treated equally if B received a larger grant, unequally if A and B received an identical grant. Hence one could not object to the rule on the ground that it violated the principle of equality. One could only object that the kind of equality served by the rule was less important than other kinds of equality that might have been served in the same situation. Such an objection would involve an appeal from equality to some other principle.[8]

B.3. If to refuse to give X to B, C, D, . . . N while giving it to A wrongs B, C, D, . . N, GP is violated and the policy must be abandoned.

The problem in Reynolds is instructive concerning B.2. and B.3. If B.2. would be unobjectionable under GP, the equal population rule would not be a requirement in situations of this type. It would be acceptable to follow the rule in such cases, but it would not be wrong to depart from it. If Colorado employed the equal population rule in most of its legislative districts, there would be no *prima facie* bar to departure from that rule in dealing with cases such as our hypothetical D′. Special treatment for D′, regardless of how great the departure from the equal population rule, would be condemned only if it could be shown that it wronged non-D′s. If, say, all non-D′s, by every measure other than the equal population rule, were receiving effective representation, special treatment of D′ would no more be a problem than the special treatment we accord to some people virtually every time government acts. For the rule to be a requirement, all departures from it must fall under B.3. This is the position that defenders of a strict "one-man, one-vote" rule, or any comparable egalitarian rule, must defend.

These cases, allowing for recombination among the situations and types, exhaust the logical possibilities under GP. Although they complicate the earlier analysis, none of them upsets that analysis. If correct, the analysis shows that GP, or equality, is not

[8] It has been suggested that this example strains ordinary usage; that "equally" would be used in connection with the effects of the policy on A and B, not its effects on the country at large. One response to this objection would be that the argument considers the effects of the policy on A and B in what Rousseau would call their corporate capacities. Both are members of the system; the system is affected in a particular manner; and hence both, as members, are affected in the same way. It is my view that much public policy is justified, and can only be justified, in this way. But this response concedes more than is necessary to the objection. "According to one and the same rule" is a standard meaning of "equally" and the above argument is in conformity with it. The objection that the argument strains usage could be sustained only if another interpretation of "equally" could be sustained. My suspicion is that an exchange on this point would show not that the argument strains ordinary usage of "equally" but that it departs from widespread conceptions as to which kinds of equality are most important.

a sufficient criterion of a justifiable decision. We must decide whether policies are right or wrong, good or bad; since we will rarely, if ever, be able to treat equally in one respect without treating unequally in others, to equate "right" with "equal" and "wrong" with "unequal" produces the logically absurd result that our decisions must be both right and wrong. This absurdity is avoided by using a criterion other than equality to choose between competing equalities. GP states a logical relationship between the results obtained through use of such a criterion. Hence the analysis indicates that equality is a significant normative criterion only in a *derivative* sense.

IV

The force of the foregoing analysis is primarily negative. It seeks to demonstrate that equality or equal treatment is a derivative criterion inadequate for determining rightness or wrongness. Attempts to elevate it to the status of a sufficient standard involve logical errors that could readily lead to unsatisfactory decisions. The question of how rightness and wrongness are *properly* determined, of how we decide whether X is right for A, is too large for adequate discussion here; but the previous analysis has implications for that question, which we will explore briefly.

We emphasized the difficulties stemming from the fact that equality, as ordinarily understood, is concerned with comparative questions. Such questions are derivative in status and significance. But the argument that led to this conclusion included a more fundamental point, one that has application beyond equality questions. In arguing that substantive (as opposed to formal or neutral as above) egalitarian rules deal with derivative considerations, we suggested that such rules cover only one of many relationships relevant to or created by policies implementing those rules. Our immediate attack was upon the concern with a certain kind of relationship, which, it was suggested, should be replaced by concern with the consequences of treating a person or class of persons in a certain manner. But relationships with which egalitarian rules are concerned are sometimes a consequence of and are almost always affected by the adoption of a policy implementing such a rule. The argument, then, that the rightness of X for A can be determined only by examination of consequences is, in a more basic sense, an argument that egalitarian rules are concerned with an overly narrow range of the consequences of X for A. The equal population rule, for example, establishes, in advance of the application of the policy, not only the criterion of rightness that districting policy must meet but also the aspects of the policy that will be relevant to determining its rightness. A considerable range of the consequences of the policy are classed as irrelevant for purposes of evaluation.[9]

Against the narrowing effect of concern with equality, we suggest the need for a broader, indeed an—in principle—unrestricted, definition of the range of consequences potentially relevant to the evaluation. In morals and in law our primary purpose is to discover whether those affected by our actions and policies are treated well or badly by them. Because our actions and policies have a great variety of effects, any one or all of which might be morally or legally significant, it is dangerous to rule out the possibility that any one of those effects might be crucial for our evaluation.

These considerations suggest that the most general and perhaps the primary principle of moral and legal evaluation would be along the lines, "consider the consequences." Conveniently, Singer has developed a more sophisticated formulation of such a utilitarian principle. He calls it the Principle of Consequences (hereafter, PC).[10] The principle has both a negative and a positive formulation. The negative version reads, "If the consequences of A's doing X are undesirable, A ought not to do X." The positive formulation reads, "If the consequences of A's not doing X are undesirable, then A ought to do X." These two versions result in moral imperatives. There is also a non-imperative version (which Singer rejects as invalid). "If the consequences of A's doing X are desirable, then it would be good if A did X." Since the fact that an action would be morally desirable or good is not in and of itself sufficient to generate an imperative to do that action (other actions might be as good or better) the last formulation would be invalid if it concluded

[9] Once again, this might be a defensible *conclusion* in the Reynolds case or in related types of moral and political decisions. But it is not defensible as an assumption used in the making of such decisions.

[10] Singer, *op. cit.*, pp. 63–67.

that "A ought to do X." But if PC in general is valid, there are no difficulties with the non-imperative version, and it is of considerable importance to public policy.

The strength of PC in the present context, and the only defense that will be offered for it here,[11] is that it provides a standardized statement of some of the findings of the foregoing analysis. As with GP, it is a formal principle. It has no force in any concrete case until the particular premise of the form, "the consequences of this X for this A are desirable," has been established, and the principle itself does not aid us in establishing that premise. But the principle does direct our attention to the most fundamental considerations in any moral decision (including legal decisions with moral dimensions). Before we adopt the equal population rule, or any substantive moral or legal rule, we must test the results of the use of that rule against the requirements of PC. If they fail that test, we search for a more satisfactory rule. The equal population rule, or any other substantive moral or legal rule, will be viewed as a secondary rule useful as a guide in establishing the particular premise of PC. But if a rule fails to satisfy the requirements of PC, if it leads to undesirable consequences in a specific case, it will be abandoned or modified as a rule for that case. If a rule regularly leads to undesirable consequences, it will be abandoned entirely.[12]

With PC available, we can state earlier arguments concerning equality in a more standardized manner. GP states a logical relationship between the conclusions reached through two or more applications of PC. We determine whether X is right for A under PC, and whether it is right for B, C, D, . . . N under PC. If it is right for each, the particular premises of GP are estab-

[11] Singer discusses a number of the issues that arise in connection with PC. See *ibid.*, per Index. I have examined the principle in detail in my recent work, *The Public Interest*, New York: John Wiley, 1966, Chapter 8.

[12] Cf. H. L. A. Hart's notion of "defeasibility." See his "The Ascription of Responsibility and Rights," in Antony Flew (Ed.), *Logic and Language*, First Series, Oxford: Basil Blackwell, 1963. In minimizing equality and other formalist or deontological rules and considerations, the foregoing argument involves a condensed version of act-utilitarianism. There are a number of standard objections to this position, the most relevant here being that the logic of the position eliminates all moral and legal rules and requires that decisions be made exclusively on their individual merits. Hart's discussion of defeasibility rebuts important aspects of this objection, and I have discussed the objection at length in *The Public Interest, op. cit.*

lished and what is right for A is right for B, C, D, . . . N. If not, the second premise is not established. From a moral or legal standpoint, then, GP reduces to "Don't treat people in a manner that violates PC."

If this argument is correct, the formal principle of GP is not subject to the limitations noted in connection with substantive egalitarian rules such as the equal population rule. But the argument might also suggest that GP is trivial. GP has no moral force until PC has been applied, and hence it might appear that it can teach us nothing that we cannot better learn under PC. This inference is unwarranted. If I say that X is red, I do so in virtue of certain properties of X. When I do so, I commit myself to saying that any other object that has the same properties is red. If Y has all the properties that led me to say that X is red, I must say that Y is red as well (or retract my statement concerning X). To deny that Y is red is to contradict myself. There is a logical relationship between the statements "X is red" and "Y is red."

GP states the same logical relationship between two (or more) moral judgments as obtains between "X is red" and "Y is red." The first particular premise of GP is of the form: "This X is right for this A." We establish this premise by applying PC. We say "X is right (or wrong) for A by virtue of the desirability (undesirability) of the consequences of X for A." The second particular premise is established in the same manner. We determine the relevant similarity of A and B by examining the consequences of X for B. Having applied PC, we say "X is right (or wrong) for B by virtue of the desirability (undesirability) of the consequences of X for B." Between the two particular premises of GP stands "must be." "Must be" has the same status as the logical rule that requires us to say "Y is red" if Y has the same properties that led us to say "X is red." GP states a logical relationship between the conclusions reached through two (or more) applications of PC. A logical relationship between two conclusions is not the same thing as the two conclusions themselves, and GP is not the same as two (or more) applications of PC.

Hence GP does not inform us concerning the moral status of any action. But it does not follow that GP lacks utility in discourse concerning morals and politics. If X has properties (P) such that we call it red, and if Y has P, we do not need a rule of logic to tell us that Y is red. When we see that Y has properties P we say that Y is red simply because it has the properties by virtue of which we apply the word "red." But in morals and politics matters are more complicated. To say that X is right for A by virtue of its having desirable consequences is not merely to describe X. It is also to evaluate or commend X. We commend in order to guide conduct, to convince people, including ourselves, to act in a manner that might be contrary to our inclinations. To say "X is right for A because P" is not simply to describe X; it is to say that A ought to do X. And hence, under the logical rule of GP, it is to say that every A ought to do X. If I say "Jones ought to give $1000 to charity because P" and if my giving $1000 to charity produces P, then I must give $1000 (or retract my statement concerning Jones). But giving $1000 to charity is apt to conflict with my inclinations in a manner that saying "Y is red" is not likely to do. I am apt to want to say that giving $1000 to charity is not right for me. But GP provides a logical principle, neutral to the moral question, that can discipline my thinking in a salutary manner. "Didn't you say Jones ought to give $1000 to charity?" "Yes." "How does his case differ from yours?" If I am unable to show a relevant difference, if I am unable to show that for A to do X produces P, whereas for me to do X produces Q, then it will be inconsistent for me to refuse to do X. Hence although I do not share Hare's lofty estimate of the significance of GP (he calls it the universalizability principle) for moral discourse, I do not see how it can be regarded as trivial.[13] Because GP is a formalized statement of the principle of equality, the same conclusion applies to that principle.

V

Our conclusion concerning equality runs counter to a pervasive and influential strand in Western thought. The primary justifications for the conclusion are those presented in

[13] R. M. Hare has developed these points in great detail in his two books on ethics: *Language of Morals,* Oxford: Oxford University Press, 1952; *Freedom and Reason,* Oxford: Oxford University Press, 1963. But Hare makes no explicit use of PC or any equivalent principle, and this leads to serious difficulties. See especially his discussion of "fanatics" in *Freedom and Reason.*

the course of the analysis, but it will be useful to deal more explicitly with some of the basic contentions with which the present conclusions conflict.[14] Conveniently, some of these have been restated in recent papers by William Frankena and Gregory Vlastos.[15] Consideration of these arguments will lead to the question mentioned earlier: whether, despite the foregoing analysis, a substantive moral principle lurks beneath the logical rule to which we have reduced GP and equality.

Both Frankena and Vlastos contend that an aspect of treating men well is to treat them equally in a non-derivative sense. In Frankena's words, "all men are to be treated as equals, not because they are equal in any respect, but simply because they are human."[16] Equal treatment in this fundamental sense is regarded as an irreducible value that depends in no way upon utilitarian considerations. Both writers concede that very dissimilar treatment is often consistent with, indeed demanded by, recognition of equal worth, and that in most cases relevant equalities are to be identified by reference to good treatment. But they insist that there are actions or policies that would be unjustified solely because they are inconsistent with "recognition of the equality or equal intrinsic value of every human person-

[14] It should be noted that the "egalitarian tradition" is anything but well defined. For recent attempts to sort out some of the important strands in the tradition, see the papers by Richard Wollheim and Sir Isaiah Berlin in Frederick Olafson (Ed.), *Justice and Social Policy,* Englewood Cliffs, N.J.: Prentice-Hall, 1961. See also S. I. Benn and R. S. Peters, *Social Principles and the Democratic State,* Chapter 5, London: George Allen and Unwin, 1959. For a general historical survey, see Sanford A. Lakoff, *Equality in Political Philosophy,* Cambridge: Harvard University Press, 1964. Berlin and Benn and Peters make a number of arguments closely akin to those presented above.

[15] See Richard Brandt (Ed.), *Social Justice,* Englewood Cliffs, N.J.: Prentice-Hall, 1962. These papers are concerned with questions of justice and rights, and there is some suggestion in them that these questions are regarded as distinct from those of rightness and wrongness, goodness and badness, which are discussed here. (This suggestion has been made more strongly by H. L. A. Hart in another symposium related to present concerns. See the papers of Hart, Frankena, and Stuart M. Brown, Jr. in *Philosophical Review,* Vol. 64, No. 2 [April, 1955], pp. 1ff.) The present writer is not convinced that there is a radical distinction between the logics of these different concepts. It is impossible to enter into the question here, but it may be that some of the disagreements between the aforementioned papers and the present one are to be explained in this way.

[16] Brandt, *op. cit.,* p. 19.

ality."[17] One of Vlastos' examples is that cruel treatment, even of a cruel man, is always unjustified because it singles the man out for treatment contrary to a moral rule based not on any merits distinctive of the man but on recognition of human equality, on the man's "birthright as a human being." In such cases, if their analysis is correct, good treatment would be derivative of equal treatment rather than the reverse.[18] If I understand them correctly, it is only in respect to such cases that their position is at odds with that argued here.

As suggested by the phrase "equal intrinsic value of every human personality," Frankena and Vlastos are arguing against meritarian justifications for certain kinds of unequal treatment. Now if "intrinsic value" is to provide a meaningful ground for moral judgments, it must be identified sufficiently to allow us to know what it will support and when it has been disregarded or violated.[19] The task of identifying it has proved to be a difficult one, productive of considerable controversy.[20] However defined, respect for intrinsic value might require, in particular situations,

[17] *Ibid.*, p. 14.

[18] The non-derivative, irreducible status of "all men are to be treated as equals" suggests that this imperative is unanalyzable. In this respect the argument is deontological or formalistic in character and it raises the problems associated with deontological positions generally, particularly the problem of how one could defend or justify the imperative or action taken in accord with it.

[19] Cf. Benn and Peters, *op. cit.*, especially p. 109. See also R. M. Hare, *Freedom and Reason*, especially pp. 211–213.

[20] Frankena identifies it as follows: "I accepted as part of my own view the principle that all men are to be treated as equals, not because they are equal in any respect but simply because they are human. They are human because they have emotions and desires, and are able to think, and hence are capable of enjoying a good life in a sense in which other animals are not. They are human because their lives may be 'significant' in the manner which William James made so graphic . . . : 'Wherever a process of life communicates an eagerness to him who lives it, there the life becomes genuinely significant. Sometimes the eagerness is more knit up with the motor activities, sometimes with the perceptions, sometimes with the imagination, sometimes with reflective thought. But wherever it is found . . . there *is* importance in the only real and positive sense in which importance anywhere can be.' By the good life is meant not so much the morally good life as the happy or satisfactory life. As I see it, it is the fact that all men are similarly capable of enjoying a good life in this sense that justifies the *prima facie* requirement that they be treated as equals." Brandt, *op. cit.*, p. 19. For Vlastos' statement see *ibid.*, pp. 47–52. See also the related arguments of H. L. A. Hart, *loc. cit.*

that one person's intrinsic value be served, another's not served at all or disserved. If (a) the intrinsic value of one man differs in its manifestations from that of others, its manifestation as a specific need, interest, or demand might conflict with the needs, interests, and demands of others. Even if (b) the intrinsic worth of all men is the same in nature and manifestation, scarcity of resources, administrative difficulties, human shortcomings, and the like might make it impossible to serve all men equally. In the case of either (a) or (b), if a decision must be made, the decision-maker will require a principle that will allow him to justify serving the intrinsic value of one person or set of persons over the intrinsic value of others. By hypothesis, equality of intrinsic value cannot provide that principle. This is the main reason that equal rights for which equal worth provides the ground[21] are *prima facie* rights, the claim to which can be defeated in particular circumstances.[22]

The doctrine of equal intrinsic value, then, can provide a basis for moral and political decisions and policy only if the various manifestations of value are self-regarding in significance or if there is a harmony between them such that all can be served equally well. I am not prepared to deny the possibility of such cases. But moral and political questions arise primarily where other-regarding behavior and conflicts of needs, interests, and demands are present. In such cases the doctrine of equal intrinsic value cannot provide a basis for decision. To assert that particular decisions of this kind are justified because they serve equality is to assert that one kind of equality is preferable to other kinds of equality that might have been served in the same situation. If the foregoing analysis is correct, the more fundamental principle is: "Treat people well as demanded by PC."

This conclusion leads to a final problem. Assuming that the present argument is correct and that equal treatment (GP) must be interpreted through good or right treatment (PC), it nevertheless appears that GP commits us to treating everyone equally in the sense of equally well. Hence it might appear that a sub-

[21] See Vlastos, *op. cit.*, pp. 50–52.

[22] See especially Frankena's paper in the symposium cited above, pp. 227–232 and *passim*. See also the very important note 44 on p. 52 of Vlastos' paper.

stantive moral principle underlies GP even as it has been interpreted here.

We want to treat people equally in the sense of equally well. To determine whether we are treating people well, we examine the consequences of policies and actions and evaluate them in terms of moral standards. Satisfaction of the standard or standards selected in any case will often require treating people dissimilarly in respect to standards that have been rejected in the case in question. If we institute selective conscription, we might draft all healthy, unmarried men between the ages of eighteen and twenty-six on the grounds that such a classification best serves the standard, "Maintain national security." If we draft all those who fall into the class, and none who do not, we will be treating the citizenry equally in terms of that standard. But our policy might produce inequalities in terms of other standards. Draftee A is in the throes of a passionate love affair, the interruption of which causes him enormous psychic distress; Draftee B, on the other hand, is bored with his life and finds military service a welcome relief. Drafting them both creates inequality in terms of the standard "minimize psychic distress." We justify this inequality by arguing that the national security requires it and that in the circumstances the rule "maintain national security" is more important than the rule "minimize psychic distress." But if the national security is our standard, we cannot justify treating A differently from B in terms of that standard. If the consequences of drafting A and B were the same with regard to service of the national security, we could not exempt A on grounds of his love affair. We must treat everyone equally in the sense that we must apply our standards *impartially* to all. To defer A but draft B under the posited conditions would be to give preference to A in a manner that could not be justified in terms of the standard.

The last phrase, together with a notion of impartiality, is the key to the significance of the notion that we must treat people equally well. We attempt to decide how to treat people by looking at the consequences of different policies for them and evaluating those policies in the light of standards that we can defend. Because men are extraordinarily diverse in person and circumstance, adhering to our standards often requires that men be treated very differently. But we treat men differently because we have good reasons for doing so, namely that treating well according to the best standards that we can construct requires that we do so. Where no such reason can be offered, it would be impossible to defend or justify unequal treatment. In such cases we describe differences in treatment as based on partiality or bias. The question then becomes: "Why are we opposed to partiality and bias?"

Vlastos and Frankena contend that our opposition stems from a sense of the equal intrinsic value of all men. Granting that we have or should have such a "sense," for the reasons noted this position will rarely if ever help us to reach and defend a decision. A second position is suggested by Hare, who argues that the requirement of generalization (universalization) is a purely logical requirement and that departure from it is a logical mistake.[23] On this view we are against partiality and bias because they lead to violations of the logical rule that it is inconsistent to treat unlike cases alike and like cases unalike. We are against violations of the rules of logic. But in the moral realm violation of these rules coincides with violation of the moral rule PC. And we are against violations of PC because we are in favor of treating people well and against treating them badly. Treating them unequally in the sense identified by violations of GP will be to treat them badly, and hence we are against it for moral as well as logical reasons. It is because partiality and bias are productive of such results that we are right in objecting to them. Hence there is a respect in which unequal treatment is morally wrong, but it is derivative in the sense argued throughout the paper.

[23] See R. M. Hare, *op. cit.*, especially pp. 10–13 and 30–35.

[2]

American Philosophical Quarterly
Volume 7, Number 2, April 1970

EGALITARIANISM AS A DESCRIPTIVE CONCEPT

FELIX E. OPPENHEIM

LIKE "democracy" or "freedom," the term "equality" has a laudatory connotation. Hence the tendency to apply it to those, and only those, institutions or policies which one wishes to commend, and to qualify as inegalitarian those of which one disapproves. The question then arises: Is saying that, say, a sales tax or a graduated income tax is egalitarian like maintaining that one or the other is equitable? If so, persons with different views as to whether a certain policy is equitable are bound to disagree as to whether it is egalitarian, and communication is likely to break down. If, on the other hand, it were possible to set down descriptive criteria of equality, it would become possible to discuss in a meaningful way whether egalitarianism in general is just, or whether egalitarian principles of a particular kind are desirable. It seems, therefore, worthwhile trying to explicate the concept of egalitarianism so that it yields criteria which are not only empirical, but also general. Then we can ascertain whether any given rule—sales tax or graduated income tax, universal military training or student deferment, to each according to his work or need—is egalitarian or inegalitarian, regardless of whether it is equitable or just or desirable on some other grounds.

I. "Equality" in the Expression to be Defined

First, let us determine the expression we want to define. Here we must distinguish. "Equality" can be predicated either of certain characteristics of persons, or of distributions made by one actor to at least two others, or of rules stipulating how such distributions are to be made. "Equality" in the first two meanings presents no problem from the point of view of our topic, and we shall be mainly concerned with equality as a property of rules of distribution.

A. *Equality of Personal Characteristics*

When two or more persons are said to be equal with respect to age or citizenship or race or income or aptitude or need, this simply means that they have the same age or nationality or color or income or ability or need[1]—or that they are substantially similar in such respects. When Hobbes says that "nature has made men so equal in the faculties of the body and mind"[2] that anyone can kill, but not outwit, another, he means that all men have substantially the same physical and mental power, and that differences are insignificant. Persons of different age or race or ability are considered unequal in those respects. Human beings can be said to be equal or unequal only with respect to certain characteristics which must be specified. It is elliptic, and hence meaningless to say that "all men are equal." With respect to any given characteristic, some men may be equal, but all are unequal. The only characteristic which they all share is a common "human nature," but that is a tautological statement.

Equality and inequality of characteristics are no doubt descriptive concepts. Indeed, whether *A* and *B* have the same age or nationality or income can be empirically ascertained. So can assertions that *A* has greater ability or aptitude than *B*. These are characterizing value judgments:[3] such statements are descriptive, not normative.

B. *Equality of Treatment*

Whether two or more persons are being "treated equally" or not is also an empirical question. *A* and *B* are treated equally by *C*, if *C* allots to *A* and *B* the same specified benefit (e.g., one vote) or burden (e.g., one year's military service), or the same amount of some specified benefit or burden (e.g., salary, tax burden). If *A* is let to vote but not *B*, if *A* is drafted but *B* exempted, if *A* receives a

[1] To this point, cf. Hugo A. Bedau, "Egalitarianism and the Idea of Equality" in J. Roland Pennock and John W. Chapman (eds.), *Nomos IX; Equality* (New York, Atherton Press, 1967), pp. 3–27; esp. p. 7.

[2] Thomas Hobbes, *Leviathan*, ch. XIII.

[3] Ernest Nagel, *The Structure of Science* (New York, Harcourt, Brace & World, Inc., 1961), pp. 492–494.

higher salary than *B*, then *A* and *B* are treated unequally in those respects.

Whether *A* and *B* are to receive the same treatment will often depend on some general rule of distribution. *With respect to a given rule*, *A* and *B* are treated equally, not if they receive the same treatment, but if the rule is applied impartially to both. This is the concept of "Equality *before* the law, which lays down that we should treat each case in accordance with an antecedently promulgated rule."[4] Equality before a law limiting suffrage to whites requires that any white and no black citizen be allowed to vote. Equality before the law does not demand that the law itself be egalitarian.

C. *Egalitarian Rules of Distribution*

Our concern is not with egalitarian or inegalitarian treatment relative to a given rule, but with the egalitarian or inegalitarian character of the rules themselves.

Like "just," "egalitarian" can be predicated only of those rules which stipulate how certain benefits or burdens are to be allocated among persons. One may ask whether it is morally right or wrong to legalize or to outlaw abortion or divorce, but not whether such policies are just or unjust,[5] or whether they are egalitarian or inegalitarian. These latter categories can be applied to principles which stipulate how benefits (e.g., voting rights, salaries) or burdens (e.g. the duty to pay taxes, or to serve in the armed forces) are to be allotted.[6]

Rules of distribution have the general form: some specified benefit (e.g., franchise) or burden (e.g., a sales tax) is to be allocated or withheld from any person, depending on whether he has or lacks some specified characteristic (e.g., being a citizen over twenty-one, being white, buying cigarettes). Or: the amount of some specified benefit (e.g., salary) or burden (e.g., income tax) to anyone shall be a function of the amount or degree to which he has a certain characteristic (e.g., his ability, his income). Our question then is: Is there a criterion which permits us to classify any actual or conceivable rule of distribution into egalitarian and inegalitarian ones, independently of any valuational or normative considerations?

II. Traditional Criteria of Egalitarianism

Let us examine some of the criteria which have traditionally been applied, if often only implicitly.

A. *Equal Shares to All*

According to the most extreme view, a moral or legal system is egalitarian if *all* benefits and burdens are to be distributed in equal amounts to *all*. This is Aristotle's principle of numerical equality—"being treated equally or identically in the number and volume of things you get"[7]—applied to all things anyone is to receive—or has to relinquish. It is the utilitarian principle—"everybody to count for one, nobody to count for more than one"[8]—in the distribution of all benefits and burdens. Equal treatment of all in every respect has been advocated by some 19th century anarchists: equality of occupation (intellectuals to participate in manual work), of consumption (all to eat and dress alike), and especially of education would ultimately wipe out existing inequalities of personal characteristics such as those of talent and intelligence and would eventually mold a uniform human species.[9]

However, practically all rules of distribution are concerned with *certain* benefits or burdens, to be allocated to *certain* persons. Even principles as general as those of the American and French Revolutions proclaim that the same basic legal *rights* are to be given to all, and that means to all citizens in any given political system by their respective governments. If egalitarianism meant equal shares of everything to all, practically all existing rules would be inegalitarian.

B. *Equal Shares to Equals*[10]

Aristotle himself enlarged the criterion of egalitarianism to include rules which allot "equal

[4] J. R. Lucas, *The Principles of Politics* (Oxford, Clarendon Press, 1966), p. 246.

[5] Cf. William K. Frankena, "The Concept of Social Justice" in Richard B. Brandt (ed.), *Social Justice* (Englewood Cliffs, N. J., Prentice-Hall, 1962), p. 4.

[6] I use the terms "benefits" and "burdens" to refer to anything which can be distributed to or exacted from several persons and which is generally valued by them either positively or negatively.

[7] *Politics*, 1301 b.

[8] John Stuart Mill, *Utilitarianism*, ch. V.

[9] Cf. Isaiah Berlin, "Equality as an Ideal" reprinted in Frederick A. Olafson, *Justice and Social Policy* (Englewood Cliffs, N.J., Prentice-Hall, 1961), pp. 128–150; see p. 139.

[10] To this point, and some of the following, cf. Felix E. Oppenheim, "The Concept of Equality," *International Encyclopedia of the Social Sciences*, vol. 5 (1968), pp. 102–107.

shares to equals"; i.e., equal shares of some specified kind to all who are equal with respect to some specified characteristic. Conversely, a rule is inegalitarian "when either equals are awarded unequal shares or unequals equal shares."[11]

Here, the opposite criticism applies. Every existing, and even every conceivable rule of distribution turns out to be egalitarian in this sense, since every one allocates the same benefit or burden to all who have the same specified characteristic, but not to those who are unequal in that respect. Universal suffrage means that every adult citizen shall have one vote, but that minors and aliens shall have none. Suffrage to whites means that the right to vote is given to all white adult citizens, but not to colored persons. Conversely, an inegalitarian rule in this sense is a logical impossibility. A rule cannot stipulate that equals—in the sense of: those who have the characteristic specified by the rule—shall be awarded unequal shares, and unequals equal shares.[12] To practice racial discrimination is to give the same treatment to those of the same color, and to give unequal shares to those who are unequal with respect to this characteristic.

C. *Equal Shares to a Relatively Large Group*

Since every rule of distribution designates a certain class of persons who are to be treated equally, it could be argued, as it is by Isaiah Berlin, that one rule is more egalitarian than another if it insures "that a larger number of persons (or classes of persons) shall receive similar treatment in specified circumstances."[13] To be more specific: a distribution of *benefits* is the more egalitarian, the larger the class of persons who receive it, as compared with the number of those excluded. Universal suffrage which excludes only minors and aliens is more egalitarian than a system which excludes Negroes in addition. Disenfranchising women is more inegalitarian than disenfranchising Negroes if the latter constitute less than half the population, but less inegalitarian if the majority is colored. Locke, who advocated equal political rights for property owners, was more egalitarian than his predecessors, but less so than later advocates of universal suffrage. On the other hand, a rule which allots *burdens* is the more egalitarian the larger the class of persons on whom it is imposed. Exempting students from the draft is less egalitarian than drafting them also.

This criterion has the great advantage that egalitarianism and inegalitarianism become comparative concepts. From the point of view of empirical science, this is an advance from merely classificatory concepts, possibly leading to quantification. The disadvantage is that all rules of the type, "to each according to his need" would become highly inegalitarian, unless it so happens that a fairly large proportion of the population had the same, and high, degree of need. A sales tax would be very egalitarian; but a graduated income tax, very inegalitarian, since it divides taxpayers not merely into two classes but into a large number of brackets and imposes the greatest tax burden on the usually small number of those with the highest income. Only in the unlikely case that the great majority falls within the highest bracket would a graduated income tax become more egalitarian. Even the principle of equality of opportunity would, in spite of its name, be inegalitarian, since it provides greater advantages to those who lack certain opportunities than to those who already have them.

D. *Proportional Equality*

Yet, we are inclined to consider more benefits for the needier or a graduated income tax egalitarian. They would be, if egalitarianism were taken in the sense of Aristotle's "proportional equality" or "equality of ratios."[14] A rule of distribution may be said to fulfill this requirement, provided the amount of benefit or burden allotted to anyone is a monotonically increasing function of the personal characteristic specified by the rule; i.e., the more of the characteristic, the greater the share. Any two persons are treated equally in this sense, provided the difference in the amount allotted to each is similarly correlated to the degree to which they differ with respect to the specified characteristic.

However, every conceivable rule would become egalitarian by this criterion, just as it would according to the principle of equal shares to

[11] *Ethics*, 1131 a.

[12] Cf. also John Rawls, "Justice as Fairness" reprinted in Olafson, *op. cit.*, pp. 80–107. "Now, that similar particular cases, as defined by a practice, should be treated similarly as they arise, . . . is involved in the very notion of an activity in accordance with rules" (p. 82).

[13] Berlin, *loc. cit.*, p. 135.

[14] *Politics*, 1301 b.

equals. Indeed, all rules of distribution not only allot "equal shares to equals" and "unequal shares to unequals," but also allots them "in proportion to" the latter's inequalities. Both rules, "to each according to his need" and "to each according to his height," assign different shares to different persons in the proportion in which they differ as to need or height. A flat rate and a graduated income tax both fulfill the requirement of proportional equality. Marx's ideal was the principle, "to each according to his need," rather than "to each according to his work." Yet, he did not deny that the latter rule, too, is egalitarian, since "the right of the producers [to receive means of consumption] is *proportional* to the labor they supply; the equality consists in the fact that measurement is made with an *equal standard*, labor."[15] It is, therefore, an egalitarian principle, even though "it tacitly recognizes unequal individual endowment and thus productive capacity as natural privileges."[16] Rules which establish only two categories are also egalitarian by this standard. Both universal suffrage and suffrage for whites only treat all persons in proportion to their inequality, with respect to the specified characteristic. Numerical equality is then but a special case of proportional equality.[17]

E. *To Each According to His Desert*

Aristotle sometimes contrasts equality, not with proportional equality in general, but with "equality proportionate to desert."[18] Amounts of benefits are to be proportionate to the degree to which beneficiaries have—not whatever characteristic a rule might specify—but one specific characteristic; namely, relative desert. The more deserving a person, the greater his reward, and equal shares to persons of equal desert. Any criterion of distribution which disregards desert is then not truly egalitarian.

This time, it can, of course, not be argued that every rule turns out to be egalitarian. The criticism is rather that egalitarianism is here defined in valuational rather than in descriptive terms. Aristotle himself considers a distribution egalitarian in this sense, if "the relative values of the things given correspond to those of the persons receiving."[19] Now, the relative value of *things* given can usually be objectively ascertained and measured; and so can personal characteristics such as age or income, and even intelligence or aptitude for a certain task. On the other hand, the relative value of a *person* (receiving); i.e., the degree of his desert is clearly a matter of subjective valuation, not of objective assessment. Statements to the effect that *A* is more deserving than (or twice as deserving as) *B*, in the sense that *A* is of greater value or moral worth, are genuine, not characterizing, value judgments.

Implicit here is the Platonic–Aristotelian doctrine that men are essentially of unequal value or desert, in contrast to the later Stoic view of the equal worth or dignity of every human being. On the basis of the criterion under discussion, equality, e.g., of political rights, would be egalitarian to the latter and inegalitarian according to the former view. Again, if whites are considered "superior" to Negroes (in overall desert, not, e.g., in intelligence), then racial discrimination is egalitarian; the same policy would be inegalitarian to those who do not regard a person's worth as depending on his color.

F. *Unequal Distributions Corresponding to Relevant Differences*

At present the most widely held version of proportional equality is the following: a rule of distribution is egalitarian if, and only if, differences in allotments correspond to *relevant* differences in personal characteristics; in other words, provided the specified characteristic is relevant to the kind of benefits or burdens to be distributed. Thus, age and citizenship are said to be relevant to voting rights; it is, therefore, egalitarian to limit the franchise to adult citizens. Wealth is relevant to taxation; hence, a flat rate or a graduated income tax is egalitarian. Conversely, a rule is inegalitarian if it is either based on irrelevant differences of characteristics or disregards relevant ones. Sex or color or wealth are irrelvant to voting; restricting the franchise to men or whites or poll tax payers is

[15] "Critique of the Gotha Program" in Lewis S. Feuer (ed.), *Karl Marx and Friedrich Engels, Basic Writings in Politics and Philosophy* (Garden City, N.Y., Anchor Books, 1959), p. 118. Italics in original.

[16] *Ibid.*, p. 119.

[17] Frankena considers rules satisfying the criterion of proportional equality *inegalitarian*. Cf. *Some Beliefs About Justice* (Department of Philosophy, University of Kansas, 1966), p. 7.

[18] *Politics*, 1301 a.

[19] *Politics*, 1280 a. The passage begins: "A just distribution is one in which. . . ." However, "justice" and "equality" are synonymous to Aristotle. Cf. below.

inegalitarian. Wealth *is* relevant to taxation; hence, a sales tax is inegalitarian, since it taxes poor and wealthy buyers at the same rate.

Like personal desert, relevance of a personal characteristic is an evaluative, not a descriptive, term. While the ascription of characteristics such as a certain age or income to a person is a matter of fact, judgments to the effect that such characteristics are relevant or irrelevant to some kind of distribution are valuational, not factual. That age is relevant to voting, but color not, means nothing more than that it is just to require a minimum age for voting, but unjust to base franchise on color. It is inegalitarian—and that means that it is unjust—to treat persons unequally who share a "relevant" characteristic; but unequal awards to persons who differ in some "relevant" respect are egalitarian, i.e., just. Or, in a recent formulation, "a difference in treatment requires *justification* in terms of *relevant* and sufficient differences between the claimants."[20] Advocates and opponents of racial discrimination are likely to disagree as to whether race is a "relevant" difference and whether discrimination is just. On the basis of the definition under discussion, they also would have to disagree as to whether such a policy is egalitarian.

This valuational interpretation of the concept of relevance has recently been challenged. For example, Bernard Williams holds it "quite certainly false" to claim "that the question whether a certain consideration is *relevant* to a moral issue is an evaluative question." He argues as follows:

> The principle that men should be differentially treated in respect to welfare merely on grounds of their color is not a special sort of moral principle, but (if anything) a purely arbitrary assertion of will, like of some Caligulan ruler who decided to execute everyone whose name contained three 'R's.[21]

A racist's advocacy of racial discrimination in welfare matters need not be arbitrary at all, but may well be rational—in the sense of consistent with his overall evaluations and other normative principles. To deny that such principles are *moral* ones is to apply the term "moral" itself in an emotive sense to only those normative views to which one happens to subscribe.

Perhaps Williams means only to assert that the grounds on which such a normative principle would be defended, or criticized, reduces to purely empirical propositions. Indeed, he argues that, "if any reasons are given at all" for racial discrimination,

> they will be reasons that seek to correlate the fact of blackness with certain other considerations which are at least candidates for relevance to the question of how a man should be treated: such as insensitivity, brute stupidity, ineducable irresponsibility, etc.[22]

I do not deny that a statement such as "color is relevant to intelligence" is descriptive. It means that intelligence is a function of color, and this statement can be empirically tested, and disconfirmed. But here it is intelligence, not color, which is considered relevant; e.g., to voting rights. Unlike "color is relevant to intelligence," "intelligence is relevant to voting rights" is normative, just as "color is relevant to voting rights" is normative.[23] It means that franchise ought to depend on intelligence or on color, and that a rule to that effect is just. To call a rule based on differences judged relevant *egalitarian* (rather than just) does not alter the normative character of the statement.

More recently, W. T. Blackstone has explicated the concept of relevance as follows:[24]

> To say "*x* is relevant," when we are speaking about the treatment of persons, means "*x is* actually or potentially related in an instrumentally helpful or harmful way to the attainment of a given end and *consequently ought* to be taken into consideration in the decision to treat someone in a certain way."

I agree with the author that the first part of this definition is descriptive and the second part prescriptive. But I disagree with "consequently." I deny that a statement of the "is" type of relevance entails one of the "ought" type. Let us take his own example:

> If, for example, race or color were cited as grounds for the differential treatment of persons in regard to

[20] Morris Ginsberg, *On Justice in Society* (Baltimore, Md., Penguin Books, 1965), p. 79. Italics added.

[21] Bernard Williams, "The Idea of Equality" in Peter Laslett and W. G. Runciman (eds.), *Philosophy, Politics and Society* (Oxford, Basil Blackwell, 1962) [2nd series], pp. 110–131; see p. 113.

[22] *Ibid.*

[23] Williams holds that "few can be found who will explain their practice merely by saying, 'But they're black: and it is my moral principle to treat black men differently from others'." (*Ibid.*) Discrimination might be justified precisely in this way by segregationists who concede that Negroes are potentially no less sensitive, intelligent, and responsible than whites.

[24] W. T. Blackstone, "On the Meaning and Justification of the Equality Principle," *Ethics*, vol. 77 (1967), pp. 239–253. The quotations are on p. 241. Italics added.

educational opportunities and it were shown that color or race has nothing to do with educability, then the factual presupposition of those who invoke these criteria would have been shown to be false and those criteria themselves to be irrelevant (in the factual sense of "relevant").

"Color is relevant to educability" is a factual statement, and "educability is relevant to educational opportunity" is normative. But the former statement does not entail the latter. Someone may agree that color "has nothing to do with"; i.e., is not relevant to, educability. Yet, he may hold without inconsistency that greater educational opportunities should be given to the more educable, *or to whites*, or that all should have the same education (i.e., that no group should receive preferential treatment). Blackstone himself concedes:

> It could easily be the case that individuals agree on the factual part of a judgment of relevance (i.e., that certain facts are instrumentally related to certain goals) and yet disagree on the prescriptive part of that judgment (i.e., on what goal is desirable).

This seems to contradict the first-quoted statement ("consequently"). "Relevance" is not a descriptive criterion of egalitarianism as a characteristic of rules of distribution.

G. *Unequal Distributions Which Are Just*

Egalitarianism is sometimes defined directly in terms of justice (rather than indirectly, via relevance). According to a recent article by a political theorist, "the true opposite of equality is arbitrary, i.e., unjustifiable or inequitable inequality of treatment."[25] It would follow that justifiable or equitable inequality of treatment is "truly" egalitarian. Whether racial discrimination is egalitarian or inegalitarian would again depend on whether it is considered just or unjust.

This is an instance of what I should like to call "the definist fallacy in reverse." The definist fallacy itself consists of defining a value word; e.g., "good" or "desirable," by reference to descriptive terms; e.g., "happiness" or "approval." Now, if "good" means the same as "conducive to happiness," or "desirable" the same as "approved by the majority," it would be self-contradictory to say that something which promotes happiness is bad, or that something is undesirable but approved by the majority. Aristotle's statement that "the unjust is unequal, the just is equal"[26] is another instance of this fallacy. Here the normative concept of justice is defined in terms of egalitarianism which Aristotle himself considers a descriptive term, as we have seen ("giving equal shares to equals"). Again, it is not self-contradictory to say that a graduated income tax is inegalitarian yet just.[27]

Here we have the reverse procedure. Egalitarianism, a concept which we want to function descriptively, is defined by the normative concept of justice. If "rule *x* is egalitarian" means the same as "rule *x* is just (or justifiable or equitable)," then it is self-contradictory to consider a graduated income tax just and inegalitarian, or a sales tax inequitable but egalitarian.

Egalitarianism has been identified, even more broadly, with moral rightness. According to J. R. Lucas, a law may

> be said to be unequal, in that the categories are wrongly specified, or the distinctions wrongly drawn, so that the law . . . discriminates between classes of people who ought not to be discriminated between.[28]

Accordingly, people who disagree as to the rightness or wrongness of some discrimination are bound to disagree as to whether a law to that effect is egalitarian or inegalitarian.

H. *Procedural Equality*

Equality is also linked with justice by those who regard egalitarianism as a "procedural" principle: "Treat people equally unless and until there is a justification for treating them unequally."[29] Taken in this sense, "egalitarianism" does not refer to a *characteristic* of rules of distribution at all, but to a rule of distribution itself; namely: "All persons are to be treated alike, unless good reasons can be found for treating them differently."[30] It is true that this "Equality Injunction is not itself a positive

[25] W. Von Leyden, "On Justifying Inequality," *Political Studies*, vol. 11 (1963), pp. 56–70; see p. 67.

[26] *Ethics*, 1131 a.

[27] The same criticism seems to me to apply to Rawls's "Justice as Fairness," *loc. cit.* Rawls, too, defines justice in terms of equality (equal right to liberty, inequalities being justified only under certain conditions).

[28] *Op. cit.*, p. 256.

[29] Frankena, *loc. cit.*, p. 8.

[30] Monroe C. Beardsley, "Equality and Obedience to Law" in Sidney Hook (ed.), *Law and Philosophy* (New York, New York University Press, 1964), pp. 35–42; see p. 36.

rule of ethics, but a rule for adopting rules."[31] It is, nevertheless, a *normative* rule (for adopting substantive rules).

This principle is not only purely normative but also purely procedural, compatible with whatever substantive discriminatory rules of distribution may be held "justified" or based on "good reasons." Such a criterion of egalitarianism does not enable us to classify substantive rules of distribution into egalitarian and inegalitarian ones.[32]

The search for an adequate explication of the concept of equality has been fruitless so far. To repeat briefly: if egalitarianism were defined by "equal shares to all," hardly any rule would be egalitarian; if it meant "equal shares to equals" or "proportional equality," *every* rule would be; and *any* rule could be egalitarian on the basis of definitions referring to desert, or relevant differences, or justice. Procedural equality does not even designate a characteristic of rules of distribution. "Equal shares to a relatively large group" remains the least unsatisfactory definition, but I have indicated that its application leads to results which are often counter-intuitive. Indeed, even advocates of racial discrimination are likely to consider it inegalitarian (yet just) to restrict welfare benefits to whites regardless of need (even if the great majority of the population is white), but egalitarian (though unjust) to make welfare payments to the needy regardless of race (even if the needy are a small minority). I believe that it is possible to find a general descriptive criterion of egalitarianism which captures such distinctions.

III. Proposed Criterion of Egalitarianism

All the definitions we have examined so far consider only how much of some specified benefit or burden is to be allotted to any two persons, *A* and *B*. Rules of distribution may also be considered from the point of view of the end result. How much will *A* and *B* retain after the rule has been applied to them? How are benefits or burdens to be redistributed between *A* and *B*? We must then distinguish between three stages: (1) the original distribution, (2) some rule of redistribution being applied, and (3) the final distribution resulting from 2. *Example* 1: (1) *A* has 8 units; *B* has 2; (2) take 3 from *A*; give 3 to *B*; (3) both *A* and *B* end up with 5.

A. *Simple Criterion*

A rule of redistribution might be said to be egalitarian, if it equalizes, or at least reduces the difference between initial holdings. Example 1 would be an instance of an egalitarian redistribution, since the initial difference between the holdings of *A* and *B*, namely, 6 (8−2), is reduced to 0 (5−5). So would *Example* 2: Take 3 from *A* (who has 8) and nothing from *B* (who has 2)—since the difference between their holdings at the end (5−2 = 3) is smaller than it was at the start (8−2 = 6). Conversely, a redistribution which leaves previous inequalities of benefits or burdens unaffected or increases the difference would be inegalitarian. *Example* 3: Take 1 from *A* and 1 from *B* (the initial difference between their holdings, namely, 6, remains unaffected). *Example* 4: Take 1 from *A* and 2 from *B* (the difference increases from 6 to 7).

These examples show that a rule of redistribution can be said to be egalitarian or inegalitarian only relative to some previous distribution. Egalitarianism becomes an ordering concept, an advantage which it shares with the "least unsatisfactory definition" examined under IIC. With respect to a given distribution, a rule of redistribution is the more egalitarian, the smaller the difference between holdings at the end in comparison with those at the start. The redistribution in example 1 is more egalitarian than in example 2, more inegalitarian in example 4 than in example 3.

The examples also illustrate that equal allotments may lead to inegalitarian redistributions, and vice versa. A sales tax (example 3) is inegalitarian, since it weighs heavier on the poorer buyers and does not reduce differences in wealth. Conversely, a graduated income tax (example 2) tends to equalize previous holdings and is as such egalitarian by this criterion. This definition of an egalitarian rule does then remedy precisely the

[31] *Ibid.* Similarly: "understood in this way, the principle of equality does not prescribe positively that all human beings be treated alike; it is a presumption against treating them differently, in any respect, until good grounds for distinction have been shown. . . . [It is] a rule of procedure for making decisions: Presume equality until there is reason to presume otherwise. But this is a formal, not a substantive rule." S. I. Benn and R. S. Peters, *Social Principles and the Democratic State* (London, George Allen & Unwin, 1959), p. 111. Yet it is a *rule*, not a characteristic of rules.

[32] Furthermore, as we have seen, no rule stipulates that "all persons are to be treated alike." Every rule of distribution treats *some* people equally and others unequally.

defects of the definition examined under IIC.[33] Like the former, it is couched exclusively in descriptive terms, and is therefore valuationally neutral.

B. *More Adequte Criterion*

This rather simple criterion does, however, lead to counter-intuitive results in certain instances. *Example* 5: *A* has 97 units and *B* has 3; the difference between their holdings is 94. Taking 3 from *A* and 2 from *B* reduces their holdings to 94 and 1, respectively, and the difference between their holdings is now smaller than before; namely, 93 (instead of 94). Although more is taken from *A* who starts with more than from *B* who starts with less, we hardly would consider such a redistribution egalitarian.

Now, let us look at percentage differences between holdings. If the total of units at the beginning is 100, taking 3 from *A* (who has 97) and 2 from *B* (who has 3) reduces this total to 95. *A* is then left with about 99 per cent of this total (94/95), and *B* with about 1 per cent (1/95). The percentage difference between the final holdings of *A* and *B* (99−1) is 98; this is *larger* than 94, the percentage difference between their initial holdings. According to this criterion, the redistribution turns out to be inegalitarian.[34]

This result is more in line with our general conception of egalitarianism. Indeed, if the difference between initial holdings is very large, taking more from those who have more does not necessarily make the redistribution egalitarian. I propose, therefore, to consider a rule of redistribution egalitarian if it reduces and inegalitarian if it increases the *percentage* difference between the holdings of those to whom the rule is being applied. With respect to a given initial distribution, a rule of redistribution is then the more egalitarian, the smaller the difference between the percentage holdings at the end in comparison with the difference at the start. A sales tax is more clearly inegalitarian according to the present criterion than according to the previous.[35] Even a graduated income tax may be inegalitarian according to the present criterion (as in example 5). To be egalitarian, those in the highest brackets must pay proportionally very much, and those in the lowest very little (or nothing, as in example 2).[36]

IV. Some Egalitarian and Inegalitarian Principles

Let us examine a few of the more important rules of redistribution in the light of the proposed criterion of egalitarianism.

A. *Equalization of Wealth*

Full equalization of commodities, even when it is held desirable, is generally considered utopian. Even if this goal were realized at one moment, differences would soon reappear, if only because "men are unequal" as to personal endowments; hence, power and influence are bound to remain unequally distributed under every political and economic system. Equalization of wealth usually means merely reducing rather than removing existing inequalities of possessions. According to the proposed definition, this kind of redistribution, although less egalitarian, is egalitarian just the same. In Rousseau's words:

> By equality, we should understand, not that the degree of power and riches be absolutely identical for everybody, but that . . . no citizen be wealthy enough to buy another, and none poor enough to be forced to sell himself.[37]

On the other hand, "not even the equal distribution of money will lead to equal happiness."[38]

[33] Definition IIC remains applicable when benefits and burdens are not quantifiable. E.g., extending the franchise from white to black citizens, or lowering the voting age, is egalitarian, since there is an increase in the proportion of citizens who receive the benefit relative to those who do not. This also satisfies the present criterion of egalitarianism, since the difference between initial position (having or lacking the franchise) is being reduced.

[34] I acknowledge my gratitude to T. J. Pempel, graduate student at Columbia University, for having suggested to me this improved criterion.

[35] The percentage difference between the initial holdings of *A* and *B* in example 3 is 60 (80–20). The total of their holdings is reduced from 100 per cent to 80 per cent. *A* ends up with 87.5 per cent (70/80), and *B* with 12.5 per cent (10/80). The percentage difference between their holdings has *increased* from 60 to 75 (whereas the absolute difference between their holdings remains the same).

[36] In example 2, the total of holdings is reduced to 70 per cent. *A* ends up with 50/70 = 70 per cent, and *B* with 20/70 = 30 per cent. The percentage difference is *reduced* from 60 to 50. The rule is egalitarian according to both criteria (whereas example 5 is egalitarian on the basis of the first and inegalitarian on the basis of the second criterion).

[37] Jean-Jacques Rousseau, *The Social Contract*, Bk. II, ch. XI.

[38] John Hospers, *Human Conduct* (New York, Harcourt, Brace & World, 1961), p. 424.

Besides, happiness or satisfaction or utility are not tangible benefits which can be distributed or redistributed to *A* and *B* by *C*, either equally or unequally.

B. *Equality of Opportunity*

Like utilities, opportunities cannot, strictly speaking, be given or distributed to *A* and *B* by *C*. "*A* has the opportunity to achieve *x*" means that there are no obstacles in his way of achieving *x*, so that he can do *x* if he wants to. *C* gives *A* the opportunity to reach *x* if he removes such obstacles and thereby enables *A* to achieve *x*, so that, whether *A* reaches *x* depends only on his native and acquired ability and on his effort. *A* and *B* have equal opportunity to win a race if they start from the same line. If *A* is initially behind *B*, he must be moved forward to the common starting line to have the same opportunity as *B*.

The principle of equality, or rather equalization, of opportunities is thus concerned with the redistribution of access to the various positions in society, not with the allocation of the positions themselves. The problem is: how to match individuals with unequal endowments with positions yielding unequal remuneration or power and prestige. The solution is to open them up to all on a competitive basis. The assumption is that, if everyone is given an equal start, the position everyone will occupy at the end will depend exclusively on how fast and how far he can run.

Classical liberalism held that equality of opportunity could be implemented by means of an *equal* allocation of the basic legal rights of "life, liberty, and property." If only legal privileges are abolished and equality of legal rights established, no obstacle will stand in the way of everyone's *pursuit* of happiness; i.e., everyone's ability to accede to the position commensurate with his highest ability.

Later it was realized that equality of rights is not sufficient to open up to the socially disadvantaged the opportunities open to the socially privileged. Unequal distributions are required to bring the former up to the common starting level: legal privileges and material benefits for the economically underprivileged, such as "head start" programs. To the extent to which such policies lead to an equalization of opportunities, they are egalitarian.

C. *Equal Satisfaction of Basic Needs*

The principle of equalization of opportunities is linked to another principle of equalization: the equal satisfaction of basic needs. While personal needs vary in kind and extent, there is a minimum of basic needs which are substantially identical for all in a given society at a given time. However, persons are unequal with respect to their *unsatisfied* basic needs. "Unequal distribution of resources would be required to equalize benefits in cases of unequal need."[39] The greater someone's unsatisfied basic need, the greater the benefits he receives. Those whose basic needs are already more nearly satisfied may not receive anything and may even have to give up some superfluities to provide for the former's necessities. The end result of such unequal distributions is, again, greater equalization of wealth and of opportunities.

D. *To Each According to His Merit*

Contemporary proponents of the democratic welfare state tend to combine the two egalitarian principles of equal satisfaction of basic needs and of equality of opportunity with another rule of redistribution: to each according to his merit. Once everyone's minimum needs have been taken care of, and all have been given an equal chance, the race is on, and the position everyone occupies at the end will depend only on his aptitude or "merit," again in theory at least. Unlike a person's "desert," his "merit" in the sense of proficiency at some specified task can in principle be objectively determined. But like "to each according to his desert," "to each according to his merit" is an inegalitarian rule of redistribution.

Schematically, we may then distinguish between the following stages: (1) an initial unequal distribution of commodities; (2) giving more to the needier, resulting in (3) a more egalitarian redistribution: equal satisfaction of basic needs, equality of opportunity; (4) from there on: an inegalitarian final redistribution: to each according to his merit.

This concept of equality is not only general and descriptive, but also valuationally neutral. For example, the author of *The Rise of Meritocracy* advocates "not an aristocracy of birth, not a plutocracy of wealth, but a true meritocracy of talent."[40] By the proposed criterion, all three of these principles are inegalitarian, the one he pro-

[39] Gregory Vlastos, "Justice and Equality" in Brandt, *op. cit.*, pp. 31–72; see p. 43.
[40] Michael Young, *The Rise of the Meritocracy: 1870–2033* (Baltimore, Md., Penguin Books, 1961), p. 21.

pounds as well as the two he rejects. On the other hand, advocates of "meritocracy" do in general not want to extend this principle to political participation; they remain in favor of equal suffrage, regardless of "merit."

This leads to the conclusion that modern democratic theory as a whole cannot be qualified as either egalitarian or inegalitarian, but is a mixture of both kinds of principles: equalization up to a certain level (by means of unequal distributions); inegalitarian redistributions beyond. It is, therefore, less inegalitarian than ideologies which base inequality of treatment on hereditary status, color, religion, or wealth.

There is, of course, no contradiction in calling meritocracy both inegalitarian and just. It may also be deemed unjust, yet desirable for other reasons—unjust because a person's merit depends in part on factors over which he has no control, such as innate intelligence and education or training (at least in the absence of full equality of educational opportunities)—desirable nevertheless on utilitarian grounds, because incentives to higher productivity will increase the welfare of all.

It has often been argued that men are equal and, therefore, egalitarianism just, or that inegalitarianism is equitable because men are unequal. For example, John Schaar, in a recent article, takes "the large discrepancy between the observed facts of inequality and the policy or value of equality as a serious intellectual embarrassment."[41] As if it were inconsistent to hold that men should be given equal opportunities even though they are of unequal intelligence—or unequal salaries in spite of their equal basic needs. Normative principles cannot be derived from factual generalizations; equality or inequality of some personal characteristic does not entail the desirability of either egalitarianism or inegalitarianism.

Mistaken arguments of this kind are often the result of confused language. There is the tendency to use factual statements for expressing normative views. We have seen that the allegation that "men are equal," if taken in the factual sense, is either meaningless, or tautological, or false. However, this adage serves more often as a rhetorical device to disguise the normative principle that men should be treated equally—in some respect which is often left unspecified. Then there is the temptation to use the factual statement that such and such a principle is egalitarian for the purpose of commending that particular rule. Conversely, valuational terms are being used to refer to some advocated goal; e.g., persons are to be treated according to their *desert*, or treated equally unless there are *relevant* differences, or unless unequal treatment is *justfiied.* When such value words are left unspecified, no substantive normative principle is being propounded.

Value words should be used exclusively to express the *advocacy* of some goal or principle; the *advocated* state of affairs should be characterized exclusively by descriptive terms. Following this practice would make for much-needed clarity in our moral discourse.

University of Massachusetts

Received November 19, 1968

[41] John H. Schaar, "Some Ways of Thinking About Equality," *Journal of Politics*, vol. 26 (1964), pp. 867–895; see p. 868.

[3]

Principles of Equality

J. RAZ

This paper is not about equality. It is about egalitarianism and about principles of equality. I shall not discuss or question the sense in which men are or should be equal. Nor will I query any claims that men are or should be equal in some respects or others. I shall, however, try to explain the sense in which a political morality can be said to be egalitarian and to unravel the presuppositions of egalitarianism.

The starting point is the existence within the western cultural heritage of an egalitarian tradition. Certain moral and political theories have come to be thought of as egalitarian. I shall suggest that one should distinguish between rhetorical and strict egalitarian theories and that the latter are marked by the special role that principles of a certain kind which I shall call principles of equality have within their framework. Principles of equality, it will become evident, form a part of many non-egalitarian theories, and in all of them they form an egalitarian element. It is when they dominate a theory that it is a strictly egalitarian theory.

I. *The Problem*

We assume a pre-analytical—a naive—ability to tell which theories are egalitarian. This is our ability to recognize theories as belonging to a certain historical tradition. We aim to account for the egalitarian character of these theories through the predominance within them of principles of a special kind. In other words we shall explain the egalitarian character of theories through the egalitarian character of some important principles they contain. The first task is to find out which principles can be usefully regarded as principles of equality. I am using this qualified expression since in a sense most principles can be regarded as principles of equality simply in virtue of their generality. We are looking for principles which, first, are related to equality in a way absent in all other principles and, secondly, are capable of accounting for the egalitarian character of egalitarian theories.

The theories we characterize as egalitarian or not are moral theories—complete moral theories omitting only their doctrines of the ascription of responsibility (i.e. of praise and blame). Such theories are complete if they entail an answer to all questions concerning what one ought to do (based solely on moral considerations) and a complete justification of such answers. Naturally, most of the theories actually under discussion by philosophers are merely skeletal theories providing answers to some such questions and imposing constraints on the acceptable answers to the others. Moral theories are sets of principles with their justifications. Principles are commonly described as normative statements specifying a condition of application and a normative consequence. When the condition of application is met the normative consequences (e.g. that someone ought to behave in a certain way or has a certain right) follows, other things being equal. Principles, that is, have only a prima facie force.

Since different statements can stipulate extensionally equivalent conditions, it may be better to regard principles as classes of statements identified by designating extensionally equivalent conditions for the application of the same normative consequence.[1] Each statement in the class states or describes the principle, but not all do so perspicaciously. A statement of a principle is a perspicacious statement if the condition of application it specifies is also the ground for the normative consequence or if it indicates the nature of the ground. The ground for a normative consequence is the reason which justifies that consequence. Only rarely will the stated condition for application suffice to identify the ground completely. Suppose that the reason for treating with equal respect creatures of a certain kind is that they are capable of having an image of themselves as they are and as they want to be and that they are capable of planning and controlling (to a certain extent) the course of their own lives. This then is the ground for the normative consequence. But the principle will rarely be stated in this—its perspicacious—form. Instead it will normally be stated as 'All men are entitled to equal respect'. This is less perspicacious than the above but more than 'All featherless bipeds are entitled to equal respect'. It provides a better indication than this last statement of the nature of the ground. When referring to principles and their form we shall have in mind

1 There are other uses of 'principles' which need not concern us here. See e.g. my 'Legal Principles and the Limits of Law', *Yale Law Journal*, 1972, p. 823 and p. 838.

completely perspicacious statements of the principles, though the examples—for reasons of brevity—will be only relatively perspicacious.

Moral theories include two (overlapping and interrelated) parts: a doctrine of virtue determining how one must act and live to be morally virtuous, and a doctrine of well being or of welfare determining how others should be treated for their own well being and whose responsibility it is so to treat them. The doctrine of virtue concerns the good of the agent, the doctrine of welfare—one's good as an object of the action of others. I shall assume, with all other writers on equality, that principles of equality are welfare principles. Welfare principles are themselves of two kinds: aggregative, governing the production and conservation of benefits and resources, and distributive, determining their proper distribution in the relevant population. I shall assume that the doctrine of distribution dominates the principles of aggregation. The goals in terms of production and conservation of benefits and resources are those necessary for the realization of the ideal distribution.[1] Principles of ideal distribution (i.e. of the best or the optimal distribution of benefits) are the foundation of the doctrine of distribution. The rest are principles assigning responsibilities and devising strategies for the realization of the ideal distribution. Again I shall adopt the common assumption that the egalitarian element of a theory is in its principles of distribution.

A person is entitled to G (1) if it is better, other things being equal, that he will have G than not have it, (2) if the reason for this is at least partly that it is to his benefit to have it and (3) if there is someone who is required to provide him with G. (So that for some at least it is not merely supererogatory to provide him with G). The second condition shows that principles of entitlement are principles of distribution. They may be either principles of ideal distribution, i.e. based on the ground that their satisfaction will tend to make their subjects better off overall and this is a justifiable end in itself, or principles of distribution justified instrumentally, that is, on the ground that making their subjects

1 It may be justified to sacrifice ideal distribution and tolerate greater inequalities if it would enable an increase in some aggregative goal, but only if this will enable a better conformity to distributive ideals in other respects, e.g., more people will have more of what they should have. Remember, too, that 'benefits' is used broadly to include opportunities, the care, concern or respect of others etc. and not merely 'material benefits'.

better off by giving them the benefit they stipulate is justified (whether or not satisfaction of the principle will make the subjects better off overall) as a means for some further goal. Principles of entitlement which are part of the theory of ideal distribution are principles of desert. Consider the principle: to each according to his intelligence. Many urge its acceptance on instrumental grounds: to give more benefits to a person is to promote his well being at least in some respect. It may be that overall the more intelligent will be better off (more harmonious, modest, socially happy, etc.) if they do not have more than those less intelligent. In any case they may not deserve to be better off—their being better off because they are more intelligent is not good in itself. Yet it is said to be justified as a means towards some other goal. Others may regard the principle as a principle of desert. I shall assume, in order to simplify the discussion, that principles of equality are principles of entitlement though my conclusions will not be affected if one regards egalitarian principles as including other kinds of principles of distribution.

Principles of entitlement fall into two types, positive and negative. The general form of the positive ones is:

(1) All Fs are entitled to G.

The general form of the negative ones is:

(2) Being or not being an F is irrelevant to one's entitlement to G.

Such principles are present in all theories, egalitarian and non-egalitarian alike. What kind of principles of entitlement are principles of equality? Consider the following:

(3.1) All those who are equally F are entitled to equal G.
(3.2) All those who are equally F are equally entitled to G.

Corresponding formulae can easily be produced for negative principles. What is the significance of the mention of equality in (3.1) and (3.2)? Compare the following statements:

(1a) Human beings are entitled to education.
(3.1a) Those who are equally human are entitled to equal education.
(3.2a) Those who are equally human are equally entitled to education.

(1b) Intelligent people are entitled to university places.
(3.1b) Equally intelligent people are entitled to equal (comparable) university places.
(3.2b) Equally intelligent people are equally entitled to university places.

Inserting 'equally' in the specification of the ground, whether or not accompanied by a similar insertion in the statement of the consequence, suggests that the ground admits of degrees and that the degree to which one has the property which is the ground determines the degree to which one is entitled to the benefit, i.e. the strength of one's claim to it (3.2a, 3.2b) or, the amount or quality of the benefit to which one is entitled (3.1a, 3.1b).

(3.1) and (3.2) are the general forms of principles couched in comparative terms where the degree to which one possesses the quality which is the ground determines the strength of one's entitlement or its extent. (1), when interpreted narrowly to exclude (3.1) and (3.2), is the general form of principles couched in classificatory terms where the property, possession of which is the ground, cannot be possessed in different degrees or where it does not matter to one's entitlement to what degree it is possessed. Surely egalitarians and non-egalitarians may both wish to endorse principles of both kinds.

Slightly different are statements of the following form:

(4.1) All Fs are entitled to equal G.
(4.2) All Fs are equally entitled to G.

These often amount to a combination of a positive and a negative principle: F is a ground for an entitlement to G and nothing overrides it. For example, 'Every human being is equally entitled to education' may imply that no quality but that of being human is relevant to a claim to education.

In various contexts 'equal' and its cognates contribute variously to the sense of the expression. Some further cases will be examined in sections 4 and 5 below. But for the most part those contributions do not help to uncover the peculiar quality of principles of equality we are looking for. This is not really surprising. All principles are (sets of) statements of general reasons. As such they apply equally to all those who meet their condition of application. Generality implies equality of application to a class. Adding 'equally' to the statement of the conditions or consequences of a principle does not necessarily turn it into one which has more to do with equality.

2. *Equality as Universal Entitlement*

The above argument does not establish much. It certainly does not prove that principles which are traditionally thought of as egalitarian lack a property related to our idea of equality which is not possessed by other principles. All I have argued for is that if there is such a quality it can not be identified through the exclusive study of the use of 'equal' and like terms in the formulation of the principles. Indeed many principles commonly thought of as egalitarian (e.g. free medicine or education for all) are normally stated using no such expressions at all.

Some philosophers have suggested that egalitarian principles are principles of universal entitlement and principles entailed by them. In virtue of their generality all principles apply equally to classes of people. By the same token, however, they distinguish between those who meet their conditions of application and those who do not. Not so universal principles. Those apply to all and thus establish the equality of all with respect to the normative consequences they stipulate. No one is excluded. Who must be the subjects of a principle if it is to be universal? One suggestion may be that 'all' should include everything and the content of the principle be allowed to determine whether it is vacuously fulfilled in some cases. 'All are entitled to have their interests respected' would apply vacuously to stones because they have no interests. This suggestion would however, allow too many principles to count as egalitarian principles. According to it, for example, 'All are entitled to have their property respected' is an egalitarian principle.

Another suggestion is that a principle is universal if it applies to all moral subjects. 'Moral subjects' is not to be equated with 'moral agents.' One has to be a moral agent to be the subject of a principle requiring action, i.e. of the doctrine of virtue. Even creatures who are not moral agents may be subjects of the doctrine of welfare and thus of principles of entitlement. Who are the moral subjects? I don't think there is any independent way to identify them. They are simply the subjects of moral principles. The test of universality means, therefore, that a principle of entitlement is a principle of equality if it applies to all moral subjects, i.e. if there is no valid moral principle whose subjects are not also subjects of the principle under consideration.

The comparison must be with all *valid* moral principles. It will not do to say that a principle is egalitarian relative to a person *A* if *A* accepts it and there is no principle accepted by *A* whose subjects are not also the subjects of that principle. Either a principle is egalitarian or it is not. It cannot be egalitarian in so far as A holds it and not be so with respect to B when B comes to believe in it.

A principle will not be considered egalitarian unless it applies to all normal human beings. We could take it as agreed that universal principles will encompass at least this group. They may also apply to others, perhaps to all persons or to all living creatures. To assume that all persons and animals are moral subjects may create a difficulty for the universality conception of principles of equality. Egalitarianism is not necessarily restricted to humans. A principle stating that equal respect is due to all living creatures is readily conceded to be an egalitarian principle. But so are principles restricted to human beings such as 'All men are entitled to equal opportunities'. Sometimes such principles are derivable from truly universal principles, but suppose that in a particular case this is not so? Let us, however, waive this point for another. 'Everyone is entitled to his property' is egalitarian even by this test. So we must strengthen it by stipulating that to be universal the conditions of application of the principle should be such that it applies non-vacuously to every moral subject (at least during a certain period of his life and if he wants it to apply to him). Not every one has property or can acquire it if he wants to. This definition is meant to guarantee that principles which qualify by it are truly universal, that every moral agent does in fact qualify under them to benefits and is not excluded except with his consent.

Even so not all universal principles can be regarded as egalitarian for though they all guarantee some benefit to all they don't entitle all to the same benefits: To each according to his intelligence, strength or beauty are all universal principles, provided they entitle each person to something, however little. It is clear that universality is not by itself sufficient to make a principle into an egalitarian one. Nor are all universal principles of desert (i.e. principles of *ideal* distribution) egalitarian: Meritocratic principles are held by some to be principles of desert. The universal conception (i.e. that ultimate principles of desert are all universal) has great appeal. Its appeal seems to me to be spurious and derives from a confusion between it and moral humanism. A moral theory is humanistic if its doctrine of well-being is concerned with the well-being of at least all human beings.[1] Humanism is consistent with the view that some people should have more opportunities or resources than others because they have greater need for them or are better able to profit from them (a meritocratic conception of desert).

Many different moral theories are humanistic and only some of those are egalitarian. But some popular so-called egalitarian principles amount to nothing more than an affirmation of humanism. Statements to the effect that all are entitled to equal concern or respect or care or to equal treatment or equal protection, etc., by their common interpretations mean little more than that every person should count and that benefits and advantages should not be distributed on grounds excluding the well being of some people. Admittedly many non-egalitarian positions are incompatible with humanism. Some racial, sexist, etc. views are based on ultimate principles of desert endowing certain groups of people with entitlements denied to others on grounds of race, sex, etc. To that extent humanism excludes certain kinds of inegalitarian positions, but as we saw it is consistent with many others. Consider Bentham's utilitarianism which is definitely humanistic. It applies to all moral agents and prescribes equal respect for them all in the sense of considering each pleasure and the avoidance of each pain as of equal intrinsic value regardless of whose pleasure or pain they are, and depending only on their intensity and duration. The inegalitarian results of this theory are well known. A situation in which few people have many pleasures while others have but few is as good as one in which all have equal pleasure provided the sum total of pleasure in both is equal. Often it is indifferent whether we should save from pain a person who (in utilitarian terms) already has above average benefits or one who is much worse off. Often we should direct resources to help normal competent people and away from the handicapped for often one has to invest more to cause a certain amount of pleasure to the handicapped than to the normal.

1 I am not assuming that all distributive principles of humanistic theories are principles of entitlement, nor should humanism be understood to exclude special duties based on special relationships (parental etc.).

3. *Principles of Equal Distribution in Conflict*

Assuming that egalitarian theories are humanistic theories containing principles of equality in dominating position we are back with our basic question: What kind of principles are egalitarian? Perhaps (5.1) represents the characteristic form of such principles

(5.1) If there are n Fs each is entitled to 1/n of all the G.

Comparing (5.1) with (1), 'All Fs are entitled to G', two questions present themselves: (a) Is (5.1) a distinct kind of principle? (b) Do principles of this form deserve the title of egalitarian principles? It cannot be claimed that whereas (1) determines that each F is entitled to a share of G it lacks any distributive aspect and does not determine the relative or absolute size of his share. Remember that (1) (as well as (5.1) represents the perspicacious form of a kind of principle. Therefore, it is not merely the case that Fs are entitled to G but also that they have this entitlement because they are Fs. Being an F is the ground of the entitlement. The ground of an entitlement determines its nature. It determines what counts as satisfaction of or respect for the entitlement, i.e. it determines what the entitlement is an entitlement for. Since all Fs have an entitlement to G based on the same ground they have the same entitlement. Hence, if their entitlement is completely satisfied or respected they will in fact be receiving an equal amount of G each. This argument assumes that being an F is not a matter of degree (one cannot be more or less an F) or that if it is, the degree to which one is an F does not affect one's entitlement to G. In other cases principles of type (3.1) and (3.2) apply and they too have their distributional implications as we saw in section 1 above.

Naturally, since every principle has merely a prima-facie force, it can happen that because of the operation of some other principles the all things considered entitlement to G of the Fs is not equal. But this is true of principles of type (5.1) as much as of those of type (1).

The situation is transformed in cases of scarcity, i.e., where it is impossible to satisfy all the justified claims to the full. Such situations give rise to conflicts of reasons and principles of type (5.1) provide more determinate guidance as to their resolution than principles of type (1). Imagine that there are 2 Fs and 4 units of G and that each F is entitled to 3 units (his claim will be completely met if he has 3 units). Each F can have one G without denying the other of anything he is entitled to. But they compete for the other two units. Each has a claim to both and none has a better claim than the other. In so far as (1) is concerned, giving both units to one F or one unit each are equally good ways of distributing the two. (5.1) requires giving one each. Like (1), (5.1) is a principle of entitlement but it is also a principle of conflict. Let us separate these two elements in (5.1) and formulate a principle which is just a principle of conflict resolution:

(5.2) In scarcity each who has equal entitlement is entitled to an equal share.

The smallest unit is one which makes a difference. It need not coincide with any natural limit of divisibility. (Normally it will be a shoe—not half a shoe). When one unit is claimed by two one of which has the better claim (e.g. he will suffer from the cold more) then the principle of entitlement settles the conflict in his favour—there is much more reason to give the benefit to him. He has the better reason according to the principle of entitlement itself. Thus principles of entitlement themselves act as principles of conflict resolution. In fact they are sufficient to settle most conflicts. Sometimes they result in holding two claims as equal and in that case they would not dictate a preference for equal distribution to any other distribution. In such cases we could say that the reason on which the principle is grounded has exhausted itself. So far as it is concerned there are several distributions which are equally good. That is where principles of type (5.2) may be invoked to give more determinate guidance.

Having established that principles of the (5.2) variety are of a type distinct from other principles of entitlement we must turn to examine whether they can be regarded as egalitarian principles. Such principles differ from ordinary principles of entitlement of type (1) in that the scope of the entitlement they stipulate depends on the actual number of people who qualify under them to the entitlement. Type (1) principles are not similarly affected. If all are entitled to a house then every person is entitled to a house and he is entitled to a house regardless of the number of people who actually qualify for a house under the principle. Naturally the number of qualifiers affects one's chances of having one's claims fulfilled but it does not affect the claim itself. This difference explains why (5.2) is a form of principles of equality in a sense that does not apply to type (1) principles—be they universal or not. Under type (1) principles each person's entitlement is independent of that of other people. He has it because the reason for the entitlement applies to him. Other people may or may not have the same entitlement. If they do that is because there is in their case too a reason (the same one) to give them G. When several people qualify under a principle the principle generates equality of entitlement but that is entirely fortuitous and accidental. A person's entitlement would be the same were he the only one entitled under the principle.

(5.2) type principles, on the other hand, are designed to achieve equality. Each of their subjects' entitlement is adjusted according to the total number of those who qualify to make sure that each has an equal share of the benefit. This feature can be present in principles which do not confer entitlements.[1] When they do confer entitlements one can say that equality is not only their result but also their purpose—they are designed to achieve equality between their subjects with respect to the benefit with which they are concerned.

4. *Principles of Non-Discrimination*

We identified one kind of egalitarian principle. Such principles can be called principles of equal distribution in conflict but it should be understood that this term is used narrowly to designate only principles of type (5.2). These principles can be regarded as egalitarian because it is their purpose to ensure equality within their sphere of application. It is (part of) the reason for each of their subjects' entitlement to his allotted share that giving him that share will make him equal in his entitlement to the others. Equality is (a part of) the ground on which such principles are based.

In the case of principles of equal distribution the dependence of the scope of the entitlement on the number of persons entitled is an indication that equality is their ground. (As well as the fact that they stipulate *equal* distribution between the qualifiers. A principle may stipulate distribution in different proportions between qualifiers). This fact is typical of egalitarian principles of

1 E.g. if such principles are justified on grounds of envy or diminishing marginal utilities *and if* these are held not to establish entitlements.

conflict. Egalitarian principles in the sense of principles whose ground is equality need not be principles of conflict and there may be other indications of their nature. One important kind of egalitarian principles are principles of non-discrimination, but here again I am using the term narrowly to designate principles of type (6) only:

> (6) All Fs who do not have G are entitled to G if some Fs have G.[1]

Principles of non-discrimination, unlike principles of equal distribution, are not sensitive to the number of qualifiers. Instead, they are sensitive to existing inequalities between members of the relevant group with respect to the relevant benefit. Ordinary principles of entitlement are indifferent to the existing distribution of their benefits. If all are entitled to food, accommodation, education, etc., then their entitlement is the same regardless of whether they have no food, education, etc., or but little or enough or whether some have more than others. If the entitlement is based on need then each is entitled just to his needs. Unless the actual distribution of the benefit affects the nature or extent of the need for it (which it may do) it is irrelevant to the entitlement. Actual distribution determines whose claims have been met and whose have not. Thus they determine only the incidence of unmet claims and their strength (though this—as we saw above—is important in scarcity).

The sensitivity of principles of non-discrimination to existing distributions is the crucial pointer to their character as egalitarian principles. Being an F by itself does not qualify one to G. It is the actually existing inequality of distribution which creates the entitlement. The entitlement is designed to eliminate a specific kind of existing discrimination. Such principles reflect the view that it is wrong or unjust for some Fs to have G while others have not. Such inequalities must be remedied in one of two ways. Either depriving those Fs who have G of it, or giving G to all the other Fs. So long as some Fs have the benefit while others are denied it the principle applies and the rest of the Fs are entitled to G. If their claims are met the inequality is eliminated.

1 There can be other kinds of principles of non-discrimination sharing the essential features of (6). E.g. All Fs who don't have G are entitled to it if some non-Fs have it.

(6)-type principles, however, do not in themselves give those Fs who happen to have the benefit an entitlement to it. The mere accident of having a benefit is rarely thought a sufficient ground of title to it. Even conservative principles (in the sense of 'conservation principles') usually rely on the harm deprivation will cause or on the likelihood that a redistribution will be for the worse. Instead of achieving equality by giving the benefit to those who lack it one can equally (in so far as the (6)-principle itself is concerned) achieve it by denying the benefit to those who have it thus preventing the entitlement under the principle from arising. Therefore, such principles of non-discrimination do often lead to waste. If there isn't enough of the benefit to go around then whatever of it we have should be wasted rather than given to, or be allowed to be retained by, some. It is true that the principles themselves do not require waste but often the only way to avoid violating them is to create or allow waste. Needless to say we are here concerned with the non-discrimination principles themselves. There may be other principles proscribing waste which may have to be balanced against the non-discrimination principles. In any case, it is only the effect of other principles which can explain our preference for giving the benefit to those who lack it to denying it to those who have it. This preference cannot be explained on the basis of the non-discrimination principles themselves.

5. *Rhetorical Egalitarianism*

Some principles of entitlement such as (5.2) and (6) are designed to promote equality as such. I'll occasionally call such principles 'strictly egalitarian'. Theories dominated by them are strictly egalitarian theories. The main claim of this article is that in its core the egalitarian tradition in western thought is strictly egalitarian, i.e. dominated by principles or types like (5.2) and (6). I know of no way in which such a claim can be proved. In the next section several important egalitarian principles will be shown to incorporate a principle of non-discrimination. Yet it cannot be denied that equality is invoked on other grounds as well. Nor is it surprising: all principles of entitlement generate equality (in some respect) as an incidental by-product since all who have equal qualification under them have an equal entitlement. Furthermore, some principles are naturally expressed using 'equality' and related terms without having anything to do with egalitarianism. Such are (3)-type principles encountered above, like 'Equally able people are entitled to equal remuneration', and other principles which allow for degrees of entitlement.

Arguments and claims invoking 'equality' but not relying on strictly egalitarian principles are rhetorical. This is not meant in a derogatory sense. There need be nothing wrong with such invocations of equality. It is simply that they are not claims designed to promote equality but rather to promote the cause of those who qualify under an independently valid principle. They invoke equality sometimes to facilitate exposition (as in claims based on (3)-type principles) and often to gain from the good name 'equality' has in our culture. It was mentioned above that principles of equal respect or concern, etc. often amount to little more than assertion of humanism (and humanism in one form or another is rarely rejected by anyone in our culture). Such principles can be expressed with equal ease without invoking equality. They are not designed to increase equality but to encourage recognition that the well being of all human beings counts. Yet given the current fashion for equality they are often couched in egalitarian terminology. If this makes them more attractive so much the better. The price we pay is in intellectual confusion since their 'egalitarian' formulation is less perspicacious, i.e., less revealing of their true grounds, than some 'non-egalitarian' formulations of the same principles: 'Being human is the only ground for respect' is a more explicit rendering of 'All humans are entitled to equal respect'.

Rhetorical invocation of equality is linguistically proper in a variety of contexts. A parent who gives the medicine to the healthy child and not to the sick one, or who deceives one of his children and not the others is treating them unequally. A person who keeps his promises to one person and breaks his promises to another is, likewise, treating them unequally. But in all these cases the wrong is the same as where a parent has only one child and he deceives him or refrains from giving him the medicine when the child is ill or when a person always breaks his promises to all. Accusing a person of unequal treatment in such and many other contexts is permissible if he behaved wrongly or badly towards some while behaving properly towards others. To accuse him of unequal treatment, however, is not to identify the nature of the wrong: It could be any wrong and it is *definitely not* the wrong of creating or perpetuating inequalities. As my examples show the same wrong can exist in situations involving no inequality.

In these and in many other contexts in which equality is invoked it functions contextually rather than normatively. It indicates features of the situation in which the wrong is perpetrated which have nothing to do with the reasons for it being a wrong, nothing to do with the kind of wrong it is. This is not to say that such invocations of equality do not have useful argumentative functions. They are sometimes used as *ad hominem* arguments: you seem to acknowledge the force of the reason in one case so why do you deny it in the other? They also indicate sometimes that something can be done to improve things. Here I have in mind not so much charges of unequal treatment as of inequality in the way things are: Poverty may be no worse in a society where it afflicts only some than in a society where all are poor. It is bad or regrettable in both to the same degree and for the same reasons. The charge of inequality which can be levelled only against one of these societies is used here rhetorically: the wrong is poverty and its attendant suffering and degradation, not the inequality. But the fact of inequality is an indication that there may be resources in the inegalitarian society which can be used to remedy the situation.

I hope that these comments—and they are not meant to be exhaustive of the uses of 'equality'—vindicate my claim that I am using 'rhetorical' literally and not pejoratively. The important point is that in all those cases the offence is other than inequality and the action to be taken is not designed to achieve equality but some other good.

6. *Strict Egalitarianism*

The previous section illustrated some of the rhetorical uses of 'equality' namely those where despite appearances the wrong to be righted is not inequality, where the ground or reason for action is not the maximization of equality. It is crucial not to confuse the point of these comments with another often voiced in criticism of egalitarianism: that it is empty for all equalities are in some respect or another, and the only question is in what respect should people be equal, and that anyway any equality in some respects means inequality in others.[1] All this is here presupposed. The

1 This fact is fatal to the view that the essence of egalitarianism is that equality needs no justification, only inequalities require justification.

point of the last section isn't that we all promote equality in some respect or other but rather that insofar as we rely on principles which aren't strictly egalitarian in the sense explained we don't promote equality as a goal at all, it is merely a by-product.

The purpose of the present section is to show that in its core the egalitarian tradition in western culture was always based on the dominating position of strictly egalitarian principles. I shall concentrate on principles of non-discrimination.[1] My aim is to show that (6)-type principles, and others which are egalitarian in the same sense and can be regarded as variations on (6), are omnipresent in the main line of egalitarian theories. I am assuming throughout that only humanistic theories are egalitarian.

Consider first the following principle:

(A) All are entitled to equal welfare.

Normal assertions of this sentence are best interpreted as implicit endorsement of two principles combined:

(1c) All are entitled to the maximum welfare there can be.
(6a) If some people are better off than others then those who are less well off are entitled to the extra benefits necessary to bring them to the level of welfare enjoyed by those who are better off.

The combined operation of both principles is (a) to favour securing as much welfare all round as possible; (b) when new benefits are created they should be allocated to the worst off (they have the stronger claim being supported both by (1c) and (6a) whereas the better off are supported by (1c) only); (c) when new benefits can't be produced the principles can be satisfied by transferring benefits to the less well off. If (6a) is taken to always override (1c) when they conflict then the principles also require: (d) when not enough benefits can be created or transferred some should be taken from the better off and wasted to prevent (6a) from coming into operation and, (e) production of new benefits should not be undertaken and a lower level of welfare all round should be preferred if this is necessary to prevent creating or preserving inequalities of welfare.

1 (5.2) principles when sensibly applied (i.e. separately for every beneficiary unit) yield in practice the same result as (6)-type principles appropriately framed to apply to such cases. (5.2) can for practical purposes be regarded as a special case of (6).

Needless to say different supporters of (A) assign different weights to its component principles, often allowing some inequality for the sake of a higher level of welfare for some, all or many. Many egalitarian principles conform to the same pattern: they are a combination of an ordinary principle of entitlement (type (1)) and a principle of non-discrimination governing its application and dominating it totally or only relatively. Thus:

(B) All are entitled to equal opportunities

is normally understood as a combination of

(1d) All are entitled to all the opportunities there can be; and

(6b) If some have more opportunities than others then those who have less are entitled to additional opportunities to bring them to the level of those who have more.

Some may query my interpretations of (A) and (B) on the ground that 'All are entitled to maximum welfare (or opportunities)' is an aggregative, not a distributive, principle and is not part of (A) or (B) but separate from them. This is a mistake. It is true that 'All are entitled to maximum welfare', etc., trivially entails that as many benefits as possible should be produced. This, which it is appropriate to name the principle of unlimited growth, is an aggregative principle but it is entirely unintelligible unless one assumes some distributive principle such as 'It is good that each person shall have as many benefits as possible'. Furthermore, (A) and (B) don't merely assert that if opportunities or other benefits are to be had at all they should be had in equal measure. They also assert that people are entitled to have them.

It can't be proved that principles of non-discrimination are embedded in all the core egalitarian views. All one can do is to provide some illustrations of the way common egalitarian principles when analysed are seen to include principles of non-discrimination. Here is a further example:

(7) Inequality in the distribution of G to Fs is justified only if it benefits all Fs (or alternatively: only if it benefits the least advantaged F).

(7) is but a weak version of a principle of non-discrimination of the (6) variety. By (6) it follows that if someone has a certain benefit this fact by itself entitles others to the benefit. Hence it

follows that none should have it unless all can have it. That giving the benefit to one will be instrumental in providing it to all is but one way of satisfying the non-discrimination principle. (7) is a weak non-discrimination principle for it does not insist that the benefit to be given to the person who has produced benefits for others shall not be greater than theirs.

(C) Inequality in the distribution of any benefit is justified only if it benefits all.

Here a (7)-type principle is generalized to range over all benefits and its sponsors usually give it absolute dominance so that it cannot be overridden by any other moral principle. A theory thus dominated by (C) may be only weakly egalitarian—tolerating as it does many inequalities—but it is egalitarian in the strict sense. Supporters of (C) usually interpret it to mean:

(1c) All are entitled to the maximum welfare there can be; and

(6a) If some people are better off than others then those others are entitled to the extra benefits necessary to bring them to the level of the better off; and

(D) When (1c) and (6a) are in conflict (6a) is overridden (i.e. inequalities are tolerated) provided all benefit to a certain degree in consequence.

In other words (C) is usually read as (A), plus a rule for resolving conflicts between the two components of (A).

7. *The Presuppositions of Egalitarianism*

Moral theories are strictly egalitarian if they are dominated by principles of non-discrimination. This domination means that the principles are never or relatively rarely overridden in conflict situations. In a sense this means that egalitarian principles are all important within such theories. In another sense these principles are secondary for they merely regulate the application of primary principles of entitlements—to opportunities, happiness, welfare, etc. Egalitarian views may differ in the details of the egalitarian principles they endorse, and also in their basic principles of entitlement. But not every principle of entitlement can form the foundation of an egalitarian theory for not every principle of entitlement can be sensibly regulated by a principle of non-discrimination, only insatiable principles can.

A satiable principle is one the demands of which in respect of a particular moment in time can be completely met, such that whatever might have happened they could not be satisfied to higher degree. An insatiable principle is one which it is always possible in principle to satisfy more. Compare:

(E) All are entitled to maximum pleasure,

with

(F) All are entitled to the satisfaction of their needs.

It is reasonable to assume that (F) is satiable whereas (E) is not, that is, it is possible that at a certain time all a person's needs are completely satisfied, but he can always have more pleasure. Satiable principles have different implications from insatiable ones for conflict situations. The further one is from the point of satiation the stronger is one's claim to that to which one is entitled. Those whose unmet need to G is greater have the stronger claim to the next G. No similar way of assessing the strength of reasons is available for insatiable principles, for there is no point of satiation one's distance from which can be measured. Nor does it, in many cases, make sense to talk of a zero point distance from which can be measured; 'A life with no pleasure at all' doesn't make much sense. In such cases we judge the strength of competing reasons through comparative judgments: those who have less have the stronger claim, etc. This is precisely what principles of non-discrimination tell us to do and it is to such principles that we often appeal to regulate the application of insatiable principles.

It would be wrong to suggest that principles of non-discrimination have only one use: to regulate the operation of insatiable principles of entitlement. They have miscellaneous other uses as well, especially as educational devices. They also have symbolic or expressive functions in small and intimate groups (as when one refuses an advantage because one's friend cannot share it) or with respect to positions of symbolic value (President, etc.). They may have other legitimate uses, but there is no doubt that their most important political use is as the egalitarian component in egalitarian theories. A strict egalitarian may take a principle of non-discrimination as the only fundamental principle of his theory of distribution—rather than regarding it as regulating the application of independent, insatiable, principles. But such pure egalitarian positions have too many absurd consequences to be taken seriously: They regard a person as entitled not to be harmed only if and because not everybody is harmed, etc. In other words they admit of no independent good or evil for the recipient. Their only grounds of entitlement are relational. The more reasonable egalitarian theories consist of insatiable principles of entitlement coupled with strictly egalitarian principles (which are themselves satiable) regulating their operation.

It is no accident and not a result of a mere logical technicality that strict egalitarianism is bound up with insatiable principles. The strict egalitarian's presupposition of insatiable principles reveals his commitment to the consumer's conception of man (as I shall rhetorically call it). It is this commitment to a consumer's view of man which is the main weakness of strict egalitarian theories. But this is a large topic and it is one affecting not strict egalitarianism only but all moral theories based on insatiable principles, most notably all varieties of utilitarianism. So I shall not discuss the problem here. I shall conclude by trying to show why strict egalitarianism presupposes the conception of man as a consumer, and even this I shall do only by arguing from examples.

Most of the popular egalitarian principles belong to one of four types: (a) All are entitled to equal respect: (b) All are entitled to equal opportunities: (c) All are entitled to equal welfare: (d) To each according to his needs. Principles of equal respect, as we saw, are affirmations of humanism and as such they are second order principles, i.e. statements of what kinds of principles are acceptable. They are not themselves principles of entitlement specifying grounds for specific entitlements. Humanism means that since all people count and since entitlements are for the good of the person concerned they must be such that none is excluded. Supporters of equal respect see in them more than has been suggested here for they proceed to develop a view of the good and regard it as implied by the principle of respect itself. It is better, however, to separate one's substantive doctrine of well being (when is a person well off) from the principle that the entitled should include all humans.

Principles of equal opportunities encounter greater problems of definition than the others since ultimately only genetic identity and identity of every feature of the environment provide equal opportunities. Once the required clarifications are provided these principles can be seen to become in the hands of some an extension of need principles (d) and in the hands of others a welfare principle (c) weakened to give way in conflict in most situations not concerned with securing equal opportunities.

'To each according to his needs'[1] is a satiable principle and as such cannot be the only principle of a complete doctrine of welfare: What is to be done with the surplus resources once the satisfaction of people's needs have been guaranteed? What is to be done with those resources which cannot be directed to the satisfaction of needs even when not all the needs have been satisfied? These are not merely theoretical questions, they face all affluent societies. Even if it is agreed that the satisfaction of needs should take precedence over all other principles it is evident that one needs other principles as well. The same can be said of the 'equal welfare to all' doctrine when welfare is interpreted as a satiable concept. There is, e.g., a conception of happiness by which a happy man cannot be made happier. Though it may be true that it is always possible to have more and more intense pleasures or to have more of one's preferences satisfied nevertheless these extra pleasures or satisfactions will not make a happy person happier or as such contribute to his well being. But there is also the insatiable interpretation of the equal welfare doctrine by which the more net pleasures one has the better off one is, the greater is one's welfare. This is the more common interpretation and it is the one assumed above and below.

The equal welfare doctrine bases entitlement on ability to consume—nothing more than a person's ability to have more pleasure justifies his entitlement to have more pleasure. The same is not true of the needs principle. First, the satisfaction of needs is necessary for survival and ability to function as a person. Secondly, the principle is only part of a complete doctrine of well being which will necessarily include other principles not based only on consumer demand. Being part of one coherent doctrine principles of needs themselves acquire strength and justification from the rest of the doctrine, i.e. from principles not based on consumer demand. The doctrine of equal welfare being a complete doctrine of well being in itself is not justified by anything beyond consumer demand. This is precisely the reason for which it has to include a principle of non-discrimination. Being insatiable

1 It hardly needs pointing out that many avowed supporters of the need principle fail to understand it properly and are in fact believers in the equal welfare doctrine.

and non-discriminating between items of consumer demand on any other ground but strength of demand or consumer satisfaction it is bound to lead to the distorted humanism exemplified by Benthamite utilitarianism. To avoid this it must be supported by strict egalitarian principles. The needs principle on the other hand being both satiable and not based on consumer demand needs no mechanical support from a strictly egalitarian principle and is egalitarian in the rhetorical sense only.

Thus strict egalitarianism inevitably involves embracing insatiable principles embodying the consumer conception of man and like them it presupposes three of the most common yet very doubtful beliefs of contemporary society: that humanism is incompatible with any basis for entitlement other than subjective ability to enjoy and welcome that to which one is supposedly entitled; that toleration (or moral scepticism) leads to a principle of the transparency of well being namely that subject to minor qualifications a person is better off if and only if he believes that he is better off; and that a person has a goal is a good reason for him to pursue it and therefore (because of humanistic principles) a reason for others to help him pursue it. If strict egalitarian theories are open to objections these paradoxically concern not their egalitarian component but principles of entitlement common to them and to many other moral theories like utilitarianism.[1]

1 I am grateful to R. M. Dworkin, P. M. S. Hacker, Ch. McCrudden and D. Parfit for many helpful comments on an earlier draft of this article.

BALLIOL COLLEGE, OXFORD

[4]

Australasian Journal of Philosophy
Vol, 56, No. 1; May 1978

Derek Browne

NONEGALITARIAN JUSTICE

The existence of a fundamental and intimate connection between justice and equality is widely regarded as axiomatic. Aristotle was merely echoing a commonplace when he said, 'If, then, the unjust is unequal, the just is equal, as all men suppose it to be, even apart from argument.'[1] All men—all philosophers at least—have indeed supposed this to be so, not only apart from argument, but in the absence of argument as well. Given the venerability of the claim, we would reasonably expect modern philosophy to have advanced accurate analyses of the relationships between justice and equality, injustice and inequality. There are no such analyses, for the very good reason—as I will argue—that the relationship does not exist in any fundamental way.

'Egalitarianism' can be regarded as a name or as a descriptive label. People may use names how they like: my concern is with those theories of justice which could appropriately be called by the descriptive label. Such theories will be those which employ a morally significant concept or principle of equality in an underivative role: precisely what this amounts to will emerge as I proceed. I will argue that no morally reasonable and theoretically defensible theory of justice is egalitarian in this sense. I will argue, not the normative thesis that justice is inegalitarian, but the conceptual thesis that justice is nonegalitarian.

The connection which obtains between justice and equality could be any of three different kinds: (i) *analytic egalitarianism* defends a logically necessary connection between them; (ii) *normative egalitarianism* defends a substantive or morally necessary connection; (iii) *nonegalitarianism* asserts that, to the extent that there is a connection, it is merely contingent and adventitious, of neither conceptual nor moral significance. I will defend the third of these against the second, having rejected the first as quite confused.

The theory of social justice (as I define it) is concerned with the considerations which govern the proper distribution to persons of particular sorts of benefits and burdens, and specifically of the benefits and burdens which arise from social interaction and co-operation. The major distributable benefits can, roughly but usefully, be seen as wealth, power, and prestige—that is, as economic, political, and social goods respectively.

[1] Aristotle, *Ethica Nichomachea*, trans. W. D. Ross, Bk. 5, Ch. 3.

Particular assignable conditions which are exclusive of these goods, or to which are attached only sparse shares of them, constitute the major distributable burdens. I shall consider only the distribution of goods.

The fundamental axiom, which is definitive of the concept of justice, is usefully captured by the traditional requirement that each person must be rendered his due. The concept of a man's dues, however, is rather more closely related to particular substantive theories of justice, which tie distributions to merit or desert, than is advisable in a principle which seeks (as far as possible) to be substantively neutral. I shall sometimes speak, accordingly, of a person's *claims* as the basic determinant of what he should, in justice, be given. The first task of a theory of justice is to analyse this fundamental concept of dues or claims, and to provide a formal characterisation of the principles which govern the distribution of benefits and burdens in accordance with dues. A morally substantive theory of justice will seek to supply the actual normative principles which assess the dues of men, and which determine the appropriate kinds of distributions. My arguments against the egalitarian theory of justice are basically conceptual, intended to demonstrate that no principles of equality can supply answers to distributive problems, whereas certain kinds of nonegalitarian principles can give these answers. At the same time, some normative background must be presupposed, but it is to be hoped that it is an uncontroversial one. Egalitarian distributive principles are possible, but the final objection to most of them is that the distributions they would warrant are morally unacceptable. In the remaining cases, the argument will be that although there are egalitarian principles which will produce distributions that we can accept to be just, these principles rest on an utterly mysterious appeal to the intrinsic value of equality. The same distributions can be accounted for by nonegalitarian principles which have the considerable advantage of resting on a nonmysterious and morally transparent base.

Developing an analysis which was first advanced by J. Feinberg,[2] distributive principles will be distinguished into comparative and noncomparative kinds.

The simplest case, and (for my theory) the fundamental case, is that in which a *noncomparative* principle is sufficient to determine the requirements of justice. Such a principle is one which can determine what an individual's claims or dues are, independently of any comparisons between him and other people. A person's own attributes and achievements, his own situation and circumstances, give rise to his dues, to the claims for goods that he has. He might, for instance, have a claim to a certain volume of food, based simply on his actual present state of deprivation, his *need* for food. A principle of justice might assert the moral propriety—the propriety in (social) justice—of his having that food or its being made available to him, on the basis of the fact

[2] Joel Feinberg, 'Noncomparative Justice', *The Philosophical Review* 81 (1974), p. 297f.

that his current state of deprivation has that particular moral significance. Such a principle would be a noncomparative one. The impossible demand is not made of such principles that they must have no reference to anything at all beyond the present case. What is forbidden is the determination of a person's dues by a comparison of him and his attributes or situation with some other relevant person. More precisely, a *comparative* principle will be one in which such comparisons between people are logically ineliminable.

Noncomparative principles will normally establish a person's dues by assessing him against some standard or criterion which assigns some measure of goods to him, as appropriate to his possession of some attribute. It is only in the wider context of criteria of human starvation and of deprivation generally that we are able to determine the moral significance of a particular person's condition and the morally appropriate response to that condition. In some cases of noncomparative principles, the criteria appealed to might be formally and officially laid down: a person's right to a certain income, for instance, follows from the fact that an industrial award establishes the minimum wage which is legally payable to anyone in his situation doing his job.

Noncomparative principles do not, in any significant way, involve or presuppose any morally important relationships between people. Usually, when the concepts of equality and inequality are conceived to have moral significance, the terms that they relate are people. Accordingly, noncomparative principles are not egalitarian in that sense; I will later on provide reasons for denying that they are genuinely egalitarian in any other, more subtle (or more attenuated) sense.

If human beings lived under conditions of abundance, then justice would be wholly noncomparative. If there were enough goods to meet entirely all the claims that people made on them, then no problems of distribution would arise that could not be solved purely by noncomparative principles.

The conditions which generally prevail in this world, however, are conditions of scarcity. Because natural resources and socially-produced goods are limited, it is not always possible to assign to each individual everything to which he has a claim. The theory of *social* justice is concerned to formulate the principles that govern the distribution of goods under those conditions where the social existence of individuals renders inadequate the noncomparative principles of individual justice.

The need for *comparative* principles arises because the goods available for ditribution are not sufficient to meet all the claims made on them. Comparisons between different people, between the claims that each makes, are necessary to establish how the goods available should be divided. The assignment of goods to one individual can no longer proceed in isolation from the possibility that an assignment of some of those very same goods to other individuals might also be required. How much one person ought, on balance, to have, will crucially depend on the total volume of the goods available, and the nature and urgency of other people's claims to them.

There are two different ways of formulating the concept of a comparative principle. The first of these ways is suggested by Feinberg's analysis,[3] and is the one which coheres best with egalitarian preoccupations. It regards comparative principles as *parallel* to noncomparative ones: just as the latter measure dues under conditions of abundance,[4] so comparative principles measure dues under conditions of scarcity. Accordingly, if a distributive principle, in measuring what is due to one person, necessarily makes reference to another person, then it will be comparative. The dues of the former person will be established, at least in part, by reference either to the volume of goods which justice requires to be made to the other person or which he stands to obtain anyway, or to the holdings of those goods that he already has. The dues of people, their claims in justice, are necessarily relative to the actual conditions of scarcity which prevail.

I will defend the alternative conception of comparative principles. The application of these principles presupposes the prior application of noncomparative principles, which are seen as yielding *prima facie* claims to goods. The role of comparative principles is a mediatory one, to weigh these competing *prima facie* claims against each other, and to determine the fair division of goods under conditions where it is not possible to make to each the full assignment that is actually due to him. The nature and urgency of the competing claims, and the total volume of goods available, will be the central determinants of what justice requires under these conditions.

In spite of the dust which egalitarians have stirred up on the issue, the concept of equality itself is not at all a problematic one. Equality is a relational attribute which holds between any two individuals in respect to any attribute which they have in common. It is logically necessary that each of the individuals related has that further attribute, the same attribute in each case. In many cases the existence of an equality between two individuals can be established without comparing them, provided only that their individual possession of the property in question can be established by an inspection or consideration of each in turn. If two men both have black beards, then they are the same (are equal) in that respect, but it is not necessary to compare them to establish the fact. Comparisions between the individuals concerned are necessary, however, if that attribute is actually defined, or its degree is determined on the basis of some relational facts about those individuals. If we know that one man is twice as heavy as the other or that they are equally tall, but we do not know the actual weight or height of either, then we know irreducibly comparative facts about them.

Expressions referring just to comparative facts have two characteristic shortcomings. In the first place, they are under-determining descriptions, in

[3] *Ibid.* The account in *Social Justice* (Englewood Cliffs, N.J., 1973) is slightly different: cf. p. 98.

[4] Such a good as fair appraisals of character — the kind of good Feinberg ('Noncomparative Justice', *op. cit.*) regards as the major object of noncomparative justice — is always unlimited in this sense.

the sense that they do not advance a *determinate* description of the attributes of each individual (their actual weight or height). Secondly, purely comparative descriptions (with the possible exception of ones assigning spatial or temporal positions) are given entirely in terms of consequential or derivative characteristics. That is, the possession by each of the two men of a definite height is both necessary and sufficient for the existence of the relational property of their being of equal height. The intrinsic physical extension that each man has is ontologically a more basic property than is the consequential or derivative relational property which holds between them.

A genuinely egalitarian theory of justice must derive its distributive principles from, or seek to implement, some relations of equality, relations the moral significance of which is not consequential or derivative. This reasonable requirement disqualifies all analytically egalitarian theories. Especially in the Aristotelian form which is most familiar, these theories misinterpret the fact that equality is a consequential property. Provided that a situation includes individuals who instantiate the right selection of nonrelational properties, then that situation will, of necessity, be characterised by relations of equality and inequality. Accept (for the sake of argument) the possibility of a distribution which is just wholly in virtue of noncomparative facts, in the sense that only those facts have underivative moral significance. It must be the case that some relations of equality or inequality will hold consequentially, and some of these relations might even appear to be morally significant, for some reason. On these thin grounds, the analytic egalitarian insists that the justice of the distribution is egalitarian.

This theory does not take seriously the triviality of the necessities involved. If the justice of some distribution depends, finally, only on the fact that each individual in it has some particular claims, and possesses the quantity of goods appropriate to those claims, then the justice is entirely noncomparative and nonegalitarian. For justice to be irreducibly egalitarian, the normative claim must be made that the relational property *itself* has some additional moral significance over and above that of the attributes it relates. Moral importance must attach, not (or not only) to the fact that each individual has a certain attribute, but (also) to the fact that those individuals are *equal* (in some respect). If this condition is not met, the property of equality will not be a morally ultimate and ineliminable part of the theory, and there will be no justification for speaking of an egalitarian theory of justice.

A basic but gross confusion has beset the literature on justice at least since the time of Aristotle. Against this confusion it must be insisted that the relational attribute of equality is not identical with, but is just one instance of, the relational attribute of proportionality. It is true that the two relations of equality and inequality exhaust the field between them; but unless it is already presupposed that *equality* has some special significance, the designation of all other proportionalities as 'inequalities' is tendentious and unjustified. The egalitarian cannot rest his case on the argument that any comparative principles of distribution which are necessary depend on relations of proportionality. This confusion appears in its most famous form in the analytic egalitarianism of 'the Aristotelian principle'. This celebrated piece of philosophical doggerel states that equals are to be treated equally, and unequals are to be treated unequally. In spite of the inflated claims made on its behalf—that it is a necessary, purely formal constituent of every theory of justice[5]—the principle is a wholly trivial implication of some more basic facts, whose real significance it distorts and obscures. In Aristotle's developed theory, the principle presupposes a background of noncomparative justice. Each individual person has certain dues—a certain definite merit, as Aristotle says—in accordance with which a certain definite volume of goods should be allocated to him. The Aristotelian principle is an analytic implication of this. For if two individuals each have certain dues to which, in accordance with noncomparative principles, goods are properly apportioned, then if their dues are alike, their shares of the good will be alike; and if their dues are different, their shares of the good will be proportionately different. Justice here consists in the proper noncomparative assignment of goods to dues. Other things being equal, that is sufficient for justice to be done. The various relations of proportionality which obtain have, in this case, no moral significance: to describe the situation in terms of proportionalities would accordingly be inappropriate and misleading. To go on to describe those relations of proportionality as relations of equality and inequality is doubly unjustified. The Aristotelian principle would be acceptable if it had *already* been shown that equalities between people were crucial in justice, and that the other relations of proportionality which obtained were, for some reason, important in respect to the extent of their departure from this basic equality. The Aristotelian principle cannot, however, be used to establish in the first place that equality has this basic significance. The objection applies to analytic egalitarianism generally.

Substantively egalitarian principles might be advanced within either of the two conceptions of comparative principles distinguished earlier. On the orthodox conception favoured (seemingly) be Feinberg, a clear example of an egalitarian principle would be one which regarded a certain volume of goods as due to one person because he was the equal in some crucial respects of some other person. Under my conception of comparative principles, on the other hand, a representative egalitarian principle would be one which required that, although the *prima facie* claims of two individuals could not both be met in full, they should in any case be treated equally, in some respect.

My general strategy will be as follows. I wiil show that no comparative principle of the kind allowed by Feinberg can solve, in a morally acceptable way, the problem of what is due in justice to any person. It follows immediately that no *egalitarian* comparative principle of that kind will be

[5] For one instance among many, see William K. Frankena, 'The Concept of Social Justice', in *Social Justice*, ed. R. B. Brandt (Englewood Cliffs, N.J., 1962).

satisfactory either. Turning to my own conception of the comparative principles of social justice, I will argue that egalitarian principles are not necessary, that there is *at least* one model of nonegalitarian justice which is neater and morally more plausible than any egalitarian model.

If conditions of abundance generally prevailed, the problems of social justice would not be particularly pressing. An adequate set of noncomparative principles would yield clear measures of what was due to each person on the basis solely of his individual attributes, situation and achievements. These principles might turn out to be complex and subtle, attaching goods of various kinds and in various quantities to each person, according to fine classifications of the needs or deserts or entitlements of each. Alternatively, the principles might be far more coarse-grained, expressing far less optimism about the ability of fallible creatures to weigh and assess with any accuracy the unique requirements of each individual. It is critically important to appreciate that principles of the latter kind can still be wholly noncomparative. A liberal, Utilitarian theory of justice, for instance, could take the expressed perferences of individuals (not 'the claims that they *have*' but 'the claims that they *make*') as sufficient to establish their claims: if goods were effectively unlimited, no distributive problems would arise, even though no comparisons between individuals had been made. The case for the primacy of noncomparative justice does not rest on the rather risky supposition that fine-grained, objective assessments of each individual's unique dues are possible. Nor does it rest on the belief—so uncongenial to the modern temper—that individual merit or desert is the primary criterion of a person's dues. *Needs*, after all, are individual attributes of individual persons. Under conditions of abundance, the socialist principle of distribution, 'To each according to his needs', is nonegalitarian. It is of course true that, to the exent that each person's needs are roughly the same as everyone else's, the share of goods that each receives will be roughly the same. The socialist principle, then, leads to a distribution which is often in fact equal. It is an example of a theory which is only *contingently* egalitarian, in the sense that whatever equalities prevail are lacking in moral significance. Its historical importance lies in its rejection of *other* grounds (such as property or birth) as criteria of dues. This is a significant moral claim: but it has no special connection with equality.

The proponents of the socialist principle have usually gone on to reject complete equality in *shares* of goods, on the grounds that those who have special needs must be specially treated, and that there is obvious injustice in doing otherwise. This gives the Aristotelian situation where, in respect to *shares* of goods, equals are treated equally and unequals are treated proportionally unequally: but since these two possibilities exhaust the whole range of possibilities when goods are apportioned to dues, that conclusion is a resounding tautology. An egalitarian theory of justice is not one in which just distributions are characterised simply by relations of equality and inequality, because *every* distribution can be characterised in that way. Nor is it a theory in which just distributions *can* be described by reference to relations of equality alone. (The unequal shares which the socialist principle warrants are supposed to be in the interests of promoting equality of *satisfaction.*) In every situation, some relations of equality will obtain, although they will often be trivial relations. An egalitarian theory of justice, to warrant the description, must be one which shows that a distribution is just *because* (or partly because) some relations of equality obtain. The equalities must be part of the very texture of justice itself: they must be ineliminable from, or have an irreducible moral significance in the theory.

It can readily be admitted that comparative principles, and hence egalitarian principles, can provide solutions of a sort—albeit very minimal and very poor solutions—to distributive problems. But it is denied that such principles are sufficient to yield morally acceptable solutions. Their failure to do so is convincing proof that noncomparative principles are at the very least necessary in every case if justice is to be done.

The defect of comparative principles is that, by themselves, they cannot settle what is the *actual amount* that each individual should receive. They can only set a constraint upon the outcome: it is a sufficient condition for a distribution to comply with such a principle that the proper proportional relationship should obtain, whatever the total volume of goods actually involved. If two people are each treated vilely, as vilely in the one case as in the other, they are still being treated equally. If each is denied all privileges, and if respect is shown for the rights of neither, then they are both being treated in perfect conformity with a principle of equal consideration—that is, equal (but minimal) consideration. In general, the requirement that shares of a good are to be equal is satisfied by any of an indefinite number of different situations, including that of equal null shares.

Principles for the distribution of goods can take either of two forms. Some will define the proper distribution by reference directly to the *shares* of the good that are to be provided; the others will define the proper distribution by reference immediately to the *holdings* of a good that people would have if some good (not necessarily the same good) were justly shared out, and thus refer only indirectly to the sizes of those shares themselves—they are the shares necessary to institute those holdings.

The theories of distribution which have the strongest title to be described as egalitarian are those which require equality in either the provision or the holdings of goods. Thus, any theory which asserts that an equal division of a good is (*prima facie*) just will be egalitarian. Similarly, any theory which asserts that the shares of a good which is to be distributed must be adjusted so that, while those shares might themselves be unequal, the resulting holdings of that (or some other) good will be equal, will also be egalitarian. No theory which defines the just distribution *exclusively in these ways can, however, be a morally adequate theory. There is no situation in which justice is done simply by ensuring that the shares or the holdings are equal, but disregarding the amount* of the good which people receive or hold. Equality is never *sufficient* for justice.

A situation only *raises* questions of justice if certain general conditions are met. If they are not met, then questions about the justice or injustice of that situation (as distinct from its moral goodness or badness in general) simply do not arise. It is one of these necessary conditions that people have claims (or dues) of the right sort. But once this is admitted, it follows that no theory of justice can distribute in total disregard of those claims: in particular, any purely egalitarian theory will be consistent with equal null shares or equal null holdings of a good, even though there is a good available for distribution and some people have significant claims to it. A purely egalitarian theory is a theory of distribution alright (though not one that gives a *single* solution to any problem), but it is totally unacceptable on moral grounds.

A theory of justice might incorporate morally significant relations of equality in some more sophisticated and subtle manner than I have so far considered, although it does run the risk of losing its title to be described as an egalitarian theory. Since an enumerative analysis of all the curious inventions of egalitarians is out of the question here, I will suppose that there are only two other general forms that such a theory might take. Both accept that justice cannot ignore individual dues, and so accept that principles specifically for measuring these dues must be at least a necessary element in any theory. The first of these theories holds that the principles which measure dues *themselves* incorporate or presuppose morally significant relations of equality. The other argues that the comparative principles which are necessary to mediate between conflicting *prima facie* claims must include morally significant relations of equality.

Whether or not the principles which measure dues themselves presuppose principles of equality, is a question most readily answered within the context of particular theories of dues. But certain general arguments can be adduced. The initial point, to be strongly emphasised, is that of the compatibility of the relation of equality with a wide range of different circumstances: the relation holds provided only that the same attribute is possessed by two individuals. Accordingly, if people are to have more or less definite claims in justice, there must be some content to those claims which arises from the individual attributes and circumstances of particular persons, and not simply from the relation in which they stand to others.

It is easy to misrepresent the significance of the fact that, against an assumed background of relevant likenesses, the existence of inequalities between people will often be dramatic evidence that the dues of a less privileged group are not being satisfactorily met. The infrequency with which women occupy the higher positions of authority in government, commerce and the professions, emphasises the way in which assignments to these positions often disregard the rightful claims of women: or alternatively, that the attribute of being female, which is not usually relevant to a person's dues in these areas, has been allowed to carry negative weight. The recognition that there is an inequality between men and women here does not itself identify the injustice, but it does throw into prominence the fact that noncomparative injustices are being done.

Noncomparative justice rests on a base of particular evaluations. For the recognition that a person has certain dues involves an evaluation of the moral significance of his natural condition. However, it is not usually possible in practice to repeat the basic, noncomparative appraisal in every case, with any assurance of fairness, while totally disregarding what evaluations have been made in other cases. What this means is that there is a background of *standards*, of paradigm cases, which is presupposed in the evaluation of people and the ascription to them of dues. The practice *simply* of comparing one person's natural condition and circumstances with the like condition and circumstances of another person cannot give rise to balanced and fair appraisals. Somewhere along the line, a noncomparative evaulation of the moral significance of this natural condition must be made. If we are sensible, we will regard this initial evaluation as conjectural: we will test its typical consequences in other cases. Over time, our evaluative standards are refined—for it is *standards* to which we appeal. Many 'like cases' should be ignored, because what we should be seeking is the *right* evaulation in each case: our guide to this will be those standards which encapsulate the best and fairest evaluations we have so far achieved. If we are in a position to refer *confidently* to standards, then we must be able confidently to identify them as such. If the standards of evaluations we use are embodied in particular cases which function as precedents, then our comparison of other cases with the precedent is justified because we realise that, in *that* case at least, our noncomparative evaluations were as fair as any of which we were capable.

Another opening for the egalitarian suggests itself. Granted that the evaluation of the condition and circumstances of people against standards of needs, deserts, and so on, is not an activity which necessitates reference to any fundamental equalities; but is it not the case that in using the same standards for all people, a crucial moral equality of all persons is presupposed? Egalitarianism seems to be necessary to combat elitism, the view that persons differ in moral worth in such a way that the needs and interests of some are of greater moral importance than the needs and interests of some others. Alternatively an elitist might argue that the good life of some is of more value than that of others, and that this reflects differences in moral significance back onto the like needs and interests of different people.

I will argue that the problems raised by elitism cannot be solved in an egalitarian way. The supposition that they can rests on a characterisation of elitism which renders it a logically incoherent thesis.

It is not logically possible that the fact of being a particular individual could be a bearer of value. Values attach to attributes, or to states or conditions: these are the logically proper subjects of all evaluation. If persons are individuals who have moral value or moral worth while pebbles and space-time points are individuals which do not have such value, this must be the case in virtue of some of the *properties* which persons but not pebbles have.

Values attach to properties, and properties are universals. It follows that attributions of value are necessarily 'universalisable'. The evaluation of a particular individual's condition can only be the evaluation of a particular cluster of properties: to deny to that same cluster of properties the same value when it is instantiated by another individual (and for no reason other than its being differently instantiated) is self-contradictory.

The elitist, of course, is not defeated by this argument, for he has probably never maintained that, although the elect should be treated differntly from the rest, there are no *reasons* why this is so. On the contrary, the elect have certain properties which distinguish them from the rest. Furthermore, these are not just *any* properties, but morally significant ones. Accordingly, people fall into different moral classes, such that the different characteristics of each warrant different modes of treatment. The thesis of universalisability is wholly consistent with this.

Elitism raises serious moral problems: what are the attributes which have value, and what is the precise moral significance of these different attributes? How precisely should we combat the view that men have certain qualities of character which lend to their lives a greater value than the outwardly similar lives of women? Do human beings have properties which distinguish them *in general* from (other) animals, and which justify general discrimination in their favour? These are profoundly difficult problems; their solution, however, has nothing to do with *equality*. Elitism about classes of people is almost certainly false: if it is false, it is because of various facts about the moral significance of the properties which people have. If an 'elitist' preference for human beings over animals *is* justified, then that is also because of the particular moral significance of the differentiating properties which people have. The egalitarian (about all people) denies the truth of elitism. But the position he is arguing for and which the elitist denies is not one in which the concept of equality has any moral significance of its own.

The second and final plausible option for the person who insists that justice is egalitarian lies in the sphere of the principles which are necessary, where conditions of scarcity prevail, to mediate between conflicting *prima facie* claims. Whatever might be the initial plausibility of an appeal to principles of equality, it vanishes on closer analysis.

We are to suppose that each of several persons has a claim of a more or less definite urgency to some good, but that there is not sufficient of the good to meet the claims of all. In assigning some of the goods to each, it is necessary to consider both the total volume of goods available for distribution, and the claims of other persons to some of those goods. What principle of *equality* could conceivably be of help here? None in which that concept has an ineliminable role suggest themselves, if we accept as a constraint on our solution that it should be justifiable by reference to the different claims advanced. Of course we are required to give 'equal consideration to the like claims of all', but that is utterly trivial: what it means is that the sole determinant of the moral urgency of each claim must be, precisely, the moral urgency of that claim. If we simply seek to *equalise* shares or holdings, we might just as well institute equally minimal shares or holdings. If we seek equal-maximal shares or holdings, we have still ignored the very claims which make it a problem of justice in the first place. Sometimes, equal-maximal holdings will be required by justice: but this is to provide what is only a *description* of the just distribution, not an explanation why it is just.

A negative argument (which I believe to be conclusive) is possible here. The egalitarian is challenged actually to produce principles which are genuinely egalitarian, which lead to morally acceptable solutions to distribution problems, and which permit an *explanation* of the justice of those solutions. Rather than rest my case on the failure of egalitarians hitherto to satisfy these requirements, I will outline my own theory of justice under conditions of scarcity.

This theory accepts that the claims which people have to the goods which are socially distributed can be assessed in terms of their moral urgency, relative to the requirements of the minimal good life for people. A person has the goods which are necessary for such a life if no important parts of his rational life-plans are closed to him because of an inadequate holding of some social good. The urgency of a claim is measured by the extent to which an individual's enjoyment of the minimally-defined satisfactory life is *impaired* by his being deprived of some good.

The fundamental principle for dealing with conflicting claims is the simple and intuitively plausible one that claims are to be dealt with in decreasing order of urgency: we must attend to the more urgent before we attend to the less urgent. The only further requirement is that claims be regarded as indefinitely subdivisible, as *classes* of claims. What we usually regard as a single claim to a definite quantity of goods is really a class, a series of claims of decreasing urgency, each of which is a claim to some fairly small quantity of goods. The number of members recognised in each claim-series will depend on the circumstances of particular cases.

Where claims differ in urgency, goods must first be allocated to the more urgent. This might lead to a situation where all the remaining claims are of equal urgency. In this case, the procedure is to allocate some of the goods to one of the persons (to a few members of his claim-series), as a result of which the claims of the others will now be the more urgent ones, and the next portion of the goods will be allocated to them. And so on.

This solution coheres closely with our firmest intuitions about justice. It explains why it is unjust to permit utilitarian trading-off between people, allowing one person to be seriously deprived in order that the total benefit summed over all persons is maximised. The theory also explains why a certain apparently egalitarian picture of justice attracts us. It is plausible to suggest that what justice requires is that goods be allocated initially to the people whose claims are most urgent: none, as socialists say, should have luxuries while others lack necessities. Once everyone has been lifted to the same level of holdings (perhaps of 'holdings of wellbeing'), it is also plausible—at least in some cases, and subject to the requirements of values such as economic efficiency—to distribute the remaining goods in equal-maximal shares. Why all this should be so is explained, directly and elegantly, by my theory. Equality is not *necessary* for justice.

Not only has the egalitarian failed to show that equality is a part of justice, but an alternative, satisfactory and nonegalitarian theory of justice is actually to hand.

University of Canterbury Received November 1977

Part II
Equality of Resources versus Equality of Welfare

[5]

Two Contradictory Ideas of (Political) Equality

Douglas Rae

There are two very different ways of thinking about equality in general and thus two very different ways of thinking about political equality (and equal representation) in particular.[1] The tension between these two views explains some important problems in the jurisprudence of political equality, and the eventual irreconcilability of these two views points toward an ultimate limit on the attainment of political equality and equal representation. One of the two egalitarianisms is attainable, and the other is not; one is sensitive to persons and their differences while the other is not; the attainable equality is, alas, the insensitive one. These points all depend upon there being two views of equality, so I will begin with this distinction.

I. TWO NOTIONS OF EQUALITY

Let us distinguish two fundamentally different ways of conceiving equality, best set apart by way of military example. An army quartermaster receives a memo requiring that every soldier be given "equal" footwear. In one view, he will issue boots with similar leather and construction but cut to fit each man's feet, an allocation based on a comparison of individual needs—for example, the lengths and widths of feet. The aim will be to provide boots of equal value to each (relevantly different) soldier. A second view will lead to a simpler solution: identical boots for all, perhaps size 8½D. In this view, boots are so equal as to rule out all envy and all incentive to switch boots. Surely, even a solider with size 12E feet can gain nothing by trading in his 8½Ds for another's 8½Ds. These two ways of providing equal footwear rest on profoundly different conceptions of equality. Here are some additional "equalities" that further instantiate the two conceptions:

1A) An equal right for every American to speak English	1B) An equal right for every American to speak his native language

1. The following analysis continues and applies a main part of Douglas Rae, "The Egalitarian State: Notes on a System of Contradictory Ideals," *Daedalus* 108 (1979): 37–54, esp. 47 ff. A wider analysis is forthcoming in Douglas Rae; and Douglas Yates, Jennifer Hochschild, Joseph Morone, and Carol Fessler, *Equalities* (Cambridge, Mass.: Harvard University Press, 1981).

Ethics 91 (April 1981): 451–456

2A) An equal scientific education for every child

2B) An education equally suited to the talents and interests of every child

3A) A tax of *N* dollars for everyone, rich and poor alike

3B) A tax amounting to an equal sacrifice at each income level

The "A" solutions follow the conception that leads our quartermaster to find identical shoes for each soldier; the "B," the conception that leads him to issue everyone boots that fit. Each conception turns on its own test of equality, namely:

Test A: Is X as well off with what X gets as X would be with what Y gets? (Likewise, is Y as well off with what Y gets as he would be with what X gets?)[2]

Test B: Is X as well off with what X gets as Y is with what Y gets?[3]

Test B leads us to find language rights or an education or tax rates or boots that will be of the same value for each subject. We will find ourselves treating different people *differently in order to treat them equally.* Test A leads us to insist on the same language rights, the same education, the same taxes, the same boots for all. Here, we will be making sure that no subject has a reason to switch places with another. And if all are treated identically, nobody can rationally entertain a sense of envy. We will, in this view, treat people, who may or may not be different, *identically in order to treat them equally.* There is, as anyone knows even from inarticulate experience with these two conceptions, every difference between them.

Equality arises not as an isolated notion but in contradiction to inequality, and initial inequalities are usually of the vulgar sort best assaulted by what I will call lot-regarding equality (test A). Lot-regarding equality is therefore apt to arise first and to seem sufficient until it in turn conspicuously violates person-regarding equality (test B). Only then does the distinction between these two views of equality come into view. The "dialectic" runs as follows:

A) A vulgar inequality exists and provokes naive envy
A′) A lot-regarding critique of A results in a reform instituting a lot-regarding equality
B) This reformed system gives rise to some person-regarding inequality, provoking sophisticated envy
B′) The disjuncture between the two forms of egalitarianism becomes an issue

2. Suppose that "V_{ji}" reads "the value Mr. J would obtain from having Mr. I's lot," and "V_{ij}" reads "the value Mr. I would get from having Mr. J's lot." Test B, then, requires that $V_{jj} = V_{ji}$ and $V_{ii} = V_{ij}$. This is equivalent to a conception that Hal Varian calls "equity" (see "Equity, Envy and Efficiency," *Journal of Economic Theory* 9: 63–69).

3. More precisely, suppose "V_{ii}" reads "the value Mr. I obtains from his own lot," and "V_{jj}" reads "the value Mr. J obtains from his own lot." Test A is then met if, and only if, it is the case that $V_{ii} = V_{jj}$, for all persons i or j.

I do not say that this is a universal or irresistible pattern, but merely that it is a centrally important one. Its use here is to point up two different forms of envy, each announcing a form of inequality. The first form is naive envy in which the envious person wants to have someone else's lot in society without also taking that person's wants, needs, and ends. I (with my wants, needs, and ends) envy your lot—your political rights, your wealth, your spouse, your income, your place in society. This naive envy is cured by lot-regarding equality (A′ cures A), which is indeed defined by the absence of such envy (test A above). The simplest formula for such a cure is uniformity in whatever it was which provoked naive envy to begin with. Here are the three basic formulae of this move:

A) *Inequalities provocative of naive envy*	A′) *Lot-regarding equalities answering to such envy*
1) X has α and I don't	1′) Both you and X get α (in equal amounts)
2) X has more α than I have	2′) Both you and X get equal amounts of α
3) X has α and I have just β, which is inferior to α	3′) Both you and X get α, or an equal α/β mix

Notice that each of the three formulae can be executed without any specific knowledge about X or the envious man. No matter what traits or tastes or needs or ends the persons possess, we can attain equality by altering the lots of α and β.

Now let us suppose that equality in the lot-regarding sense is attained: "One person one α," "Equal α for all," "Equal α for equal work," "Equal α as a human right." The old unequal-α system is gone, but now there arises a new difficulty. Suppose the formula "Equal α for all" is applied to everyone: Does it follow that this equal share is equally valuable to each of those who receive it equally? Of course not, for some may have greater needs for α and some, lesser needs for α; some may want or need γ instead of α, some may actually find α worthless or offensive. If people are thus relevantly different in their tastes, needs, vulnerabilities, ends, or values, it follows that lot-regarding equality may provoke a more sophisticated form of envy. It is not that the envious person envies another person's lot, but that this person envies the relationship or fit between the other's lot and that same (other) person's needs, tastes, or ends. If you are a round peg in a round hole and I am a square peg in a (lot-equal) round hole, it is not the round hole I envy but the matching relationship between you and your place which I covet. I may now be moved to issue a new series of complaints on the basis of this subtler envy:

B) *(In)equalities provocative of sophisticated envy*	B′) *Person-regarding equalities answering to sophisticated envy*
1) X and I both have α, but he likes α and I instead like γ	1′) X will have α and you will have β so that you and X are equally pleased

2) X and I have equal amounts of α, but I need more of it than he if we are to derive equal welfare from our shares	2′) You will have amount Y and X will have amount X so that your needs are equally served

Thus naive envy gives rise to equal lots; these give rise to sophisticated envy; and this leads to a form of equality based on relevantly different treatment for relevantly different persons. This person-regarding equality is the evident end of Tawney's prescription that

> equality of provision is not identity of provision. It is to be achieved, not by treating different needs in the same way, but by devoting equal care to ensuring that they are met in the different ways most appropriate to them, as is done by a doctor who prescribes different regimens for different constitutions, or a teacher who develops different types of intelligence by different curricula. The more anxiously, indeed, a society endeavours to secure equality of consideration for all its members, the greater will be the differentiation of treatment which, when once their common human needs have been met, it accords to the special needs of different groups and individuals among them.[4]

II. LOT-REGARDING POLITICAL EQUALITY

The reformist history of liberal representation is very largely a struggle toward lot-regarding equality. The institutions of feudalism, right down to the ruralism of American politics before *Reynolds* v. *Sims*, were perfect provocations for lot-equal reform. In every main case, there existed a privileged class or category of citizens whose political entitlements were grounds for the (naive) envy of others. The English rotten borough, the American white primary, the racist grandfather clause, the unequally administered literacy test, the property franchise, the disenfranchisement of Jews, Catholics, and freethinkers, the asymmetries of the U.S. electoral college—all of these and many more such things have provoked the demand for lot-regarding political equality. The solution is always to efface differences. Restricted franchises yield to open (though not wholly universal) ones; rotten Parliamentary boroughs are absorbed by more wholesome ones; grandfather clauses are ruled unconstitutional; unequally populous districts are redrawn on the basis of one person one vote; tilted electoral formulae are displaced by proportional representation; proportional represention by name is even sometimes displaced by proportional representation in fact. The fundamental criterion is one person one vote, a criterion which is not a very fresh breeze at this late date. Mr. Wilson, after all, put it clearly enough in June of 1787:

> . . . equal numbers of people ought to have an equal number of representatives, and different numbers of people different numbers of representatives. . . . Are not the citizens of Pennsylvania equal to

4. R. H. Tawney, *Equality*, 4th ed. (London: George Allen & Unwin, 1952), pp. 49–50.

> those of New Jersey? [Or] does it require 150 of the former to balance 50 of the latter?[5]

This hardly differs from the contemporary doctrine that every person must cast one vote and every vote must be counted equally with every other vote. The dominance of this position is revealed by the fact that all six of Jonathan Still's ingeniously parsed criteria of political equality are special cases of lot-regarding equality with respect to voting. The test is not that two citizens be equally well served by an electoral arrangement, but that neither should be better served by the other's role in that system. If Mr. X is allowed to vote, so should Ms. Y be allowed to vote; if Mr. X shares his electoral arena with 9,312 other persons, so should Y share her arena with 9,312 others; if X's Banzhaf power index is 1/9,312, Y's should also be 1/9,312; if X and Y reverse preferences, yet cast equally weighted votes, this reversal should never affect the result; if any majority favors X's preferred outcome, that should be decisive, no matter which persons constitute the majority; if X's group is represented at a ratio of 1:1,000, so should Y's group be represented at that ratio. The foregoing are indeed Still's criteria, and each expresses a form of lot-regarding equality.

III. INEQUALITY AMONG PERSONS WHO COMMAND EQUAL LOTS

Notice that a great many unpleasant inequalities of a person-regarding sort may be implemented by a lot-regarding system of equality. The poll tax, for example, is lot-equal, since a poor person would gain nothing by trading his dollar obligation for that of a rich person (indeed the poll tax was once defended on equal protection grounds, see *Breedlove* v. *Guttle*, 1937). Likewise, a lot-equal criterion may rule out unequal criteria for the appearance of large and small parties on the ballot. Thus a law requiring 10 percent of the last election's vote for Democrats and Republicans, but 15 percent for other parties, was ruled unconstitutional in *Williams* v. *Rhodes* (1968), and such a law does clearly violate lot-regarding equality. But suppose a new law demanded 30 percent for all parties, large and small alike. That would satisfy lot-regarding equality, but would radically discriminate among parties and thus among their voters. Suppose indeed that the law required a plurality in the last election, thus ruling out all further competition: Would that not satisfy lot-regarding equality? It would, since every party would face the same harsh criterion. In the two examples just considered, the underlying rub is the same, since lot-regarding equality effaces or ignores differences among its subjects. It applies the same poll tax to rich and poor alike, without noting the relevance of their wealth and poverty to its meaning; it applies the same ballot-access criterion to established giants and upstart pygmy parties alike. Indeed, the point of such lot-regarding equality is in both cases to

5. Madison's Journal in *The Records of the Federal Convention of 1787*, ed. M. Farnard (New Haven, Conn.: Yale University Press, 1911), 1:179–80.

generate an unequal impact—to repel poor voters, and to clear away nuisance parties.

Now the foregoing examples may seem peripheral and remediable, but there is another more central and irremediable case to consider. Suppose we have a no-nonsense, Still-Rogowski certified system of lot-regarding political equality, and we have two alternatives, "bla" and "not-bla." Bla wins, but you preferred not-bla. Your complaint is not that your suffrage was disallowed, or your district underrepresented, or anything else of that lot-regarding sort. Your beef is that pro-bla voter X got his way, or that his vote counted as part of the reason for the victory of bla, while your (lot-equal) vote did you no good at all. You harbor no naive envy, since X's role was no different from yours. You instead harbor a sophisticated envy based on the fit between his conformist or common pro-bla ends and your sadly rare anti-bla ends. You agree that the system is lot-equal, but insist that it generates inequality between persons. Just as Anatole France's law forbids rich and poor alike to sleep under bridges at night, political equality forbids the majority and the minority alike to inflict their will upon members of a larger group.

IV. NECESSARY INEQUALITY BETWEEN PERSONS WHO DISAGREE

This difficulty should not, however, be attributed to majority rule any more than to any other "constitution" which (*a*) reaches binary decisions even in the absence of unanimous agreement, and (*b*) lets *any* preferences affect outcomes. Majority rule is one such constitution, but there are innumerable others. The difficulty is that no such system can respond equally to all in making binary decisions (unless it responds to none of them). Given that one view prevails over another, and that both views have supporters, there simply must be person-regarding inequality somewhere in the system. The source of this inequality is ultimately the fact that choice implies inequality between alternatives ("this, not that") and that people cleave to different alternatives, with the result that their preferences must be treated unequally. If we must discriminate between bla and not-bla, we must at least implicitly discriminate between bla-sayers and not-bla-sayers.

If the value of a citizen's place in a polity is a (direct or indirect) function of having his voice prevail, then person-regarding political equality requires either (*a*) that *all* voices be ignored, *b*) that no choices be made, or (*c*) that all voices agree.[6] In all other cases, personal inequality is inevitable. This elemental point may be mystified or complicated by analyses which attend to detail and nuance; it will still remain true after a thousand pages and a thousand years of such discourse.[7]

6. Or, that over time, voices agree and disagree in a crisscrossing braid so that all have their turns on winning and losing sides.

7. It will also be true after all the bizarre attempts to institutionalize person-regarding political equality, such as the ones attacked in United Jewish Organizations v. Carey (1977).

[6]

RONALD DWORKIN

What Is Equality? Part 1: Equality of Welfare

I. Two Theories of Equality

Equality is a popular but mysterious political ideal. People can become equal (or at least more equal) in one way with the consequence that they become unequal (or more unequal) in others. If people have equal income, for example, they will almost certainly differ in the amount of satisfaction they find in their lives, and vice versa. It does not follow, of course, that equality is worthless as an ideal. But it is necessary to state, more exactly than is commonly done, what form of equality is finally important.

This is not a linguistic or even conceptual question. It does not call for a definition of the word "equal" or an analysis of how that word is used in ordinary language. It requires that we distinguish various conceptions of equality, in order to decide which of these conceptions (or which combination) states an attractive political ideal, if any does. That exercise may be described, somewhat differently, using a distinction I have drawn in other contexts. There is a dfference between treating people equally, with respect to one or another commodity or opportunity, and treating them as equals. Someone who argues that people should be more equal in income claims that a community that achieves equality of income is one that really treats people as equals. Someone who urges that people should instead be equally happy offers a different and competing theory about what society deserves that title. The question is then: which of the many different theories of that sort is the best theory?

In this two-part essay I discuss one aspect of that question, which might be called the problem of distributional equality. Suppose some

Philosophy & Public Affairs 10, no. 3
0048-3915/81/030185-62$03.10/1

community must choose between alternative schemes for distributing money and other resources to individuals. Which of the possible schemes treats people as equals? This is only one aspect of the more general problem of equality, because it sets aside a variety of issues that might be called, by way of contrast, issues about political equality. Distributional equality, as I describe it, is not concerned with the distribution of political power, for example, or with individual rights other than rights to some amount or share of resources. It is obvious, I think, that these questions I throw together under the label of political equality are not so independent from issues of distributional equality as the distinction might suggest. Someone who can play no role in determining, for example, whether an environment he cherishes should be preserved from pollution is poorer than someone who can play an important role in that decision. But it nevertheless seems likely that a full theory of equality, embracing a range of issues including political and distributional equality, is best approached by accepting initial, even though somewhat arbitrary, distinctions among these issues.

I shall consider two general theories of distributional equality. The first (which I shall call equality of welfare) holds that a distributional scheme treats people as equals when it distributes or transfers resources among them until no further transfer would leave them more equal in welfare. The second (equality of resources) holds that it treats them as equals when it distributes or transfers so that no further transfer would leave their shares of the total resources more equal. Each of these two theories, as I have just stated them, is very abstract because, as we shall see, there are many different interpretations of what welfare is, and also different theories about what would count as equality of resources. Nevertheless, even in this abstract form, it should be plain that the two theories will offer different advice in many concrete cases.

Suppose, for example, that a man of some wealth has several children, one of whom is blind, another a playboy with expensive tastes, a third a prospective politician with expensive ambitions, another a poet with humble needs, another a sculptor who works in expensive material, and so forth. How shall he draw his will? If he takes equality of welfare as his goal, then he will take these differences among his children into account, so that he will not leave them equal shares. Of course he will have to decide on some interpretation of welfare and whether, for example, expensive tastes should figure in his calculations in the same way as handicaps or expensive ambitions. But if, on the contrary, he takes equality of resources as his goal then, assuming his children have roughly equal wealth already, he may well decide that his goal requires an equal division of his wealth. In any case the questions he will put into himself will then be very different.

It is true that the distinction between the two abstract theories will be less clear-cut in an ordinary political context, particularly when officials have very little information about the actual tastes and ambitions of particular citizens. If a welfare-egalitarian knows nothing of this sort about a large group of citizens, he may sensibly decide that his best strategy for securing equality of welfare would be to establish equality of income. But the theoretical difference between the two abstract theories of equality nevertheless remains important in politics, for a variety of reasons. Officials often do have sufficient general information about the distribution of tastes and handicaps to justify general adjustments to equality of resource (for example by special tax allowances) if their goal is equality of welfare. Even when they do not, some economic structures they might devise would be antecedently better calculated to reduce inequality of welfare, under conditions of uncertainty, and others to reduce inequality of resources. But the main importance of the issue I now raise is theoretical. Egalitarians must decide whether the equality they seek is equality of resource or welfare, or some combination or something very different, in order plausibly to argue that equality is worth having at all.

I do not mean, however, that only pure egalitarians need take any interest in this question. For even those who do not think that equality is the whole story in political morality usually concede that it is part of the story, so that it is at least a point in favor of some political arrangement, even if not decisive or even central, that it reduces inequality. People who assign equality even this modest weight must nevertheless identify what counts as equality. I must emphasize, however, that the two abstract conceptions of equality I shall consider do not exhaust the possible theories of equality, even in combination.

There are other important theories that can be captured only artificially by either of these. Several philosophers, for example, hold meritocratic theories of distributional equality, some of which appeal to what is often called equality of opportunity. This claim takes different forms; but one prominent form holds that people are denied equality when their superior position in either welfare or resources is counted against them in the competition for university places or jobs, for example.

Nevertheless the claims of both equality of welfare and equality of resources are both familiar and apparent, and it is these that I shall consider. In Part 1 of this essay I examine, and on the whole reject, various versions of the former claim. In Part 2, which will be published in a future issue of this journal, I shall develop and endorse a particular version of the latter. I might perhaps add two more caveats. It is widely believed that certain people (for example criminals) do not deserve distributional equality. I do not consider that question, though I do raise some questions about merit or desert in considering what distributional equality is. John Rawls (among others) has questioned whether distributional equality might not require deviations from an equal base when this is in the interests of the then worst-off group, so that, for example, equality of welfare is best served when the worst-off have less welfare than others but more than they would otherwise have. I discuss this claim in the next part, with respect to equality of resources, but not in this one, where I propose that equality in welfare is not a desirable political goal even when inequality in welfare would not improve the position of the worst-off.

II. A First Look

There is an immediate appeal in the idea that insofar as equality is important, it must ultimately be equality of welfare that counts. For the concept of welfare was invented or at least adopted by economists precisely to describe what is fundamental in life rather than what is merely instrumental. It was adopted, in fact, to provide a metric for assigning a proper value to resources: resources are valuable so far as they produce welfare. If we decide on equality, but then define equality in terms of resources unconnected with the welfare they bring, then we seem to be mistaking means for ends, and indulging a fetishistic fascination for what we ought to treat only as instrumental. If we want genuinely to treat people as equals (or so it may seem) then we must contrive to make their lives equally desirable to them, or give them the means to do so, not simply to make the figures in their bank accounts the same.

This immediate attraction of equality of welfare is supported by one aspect of the domestic example I described. When the question arises how wealth should be distributed among children, for example, those who are seriously physically or mentally handicapped do seem to have, in all fairness, a claim to more than others. The ideal of equality of welfare may seem a plausible explanation of why this is so. Because they are handicapped, the blind need more resources to achieve equal welfare. But the same domestic example also provides at least an initially troublesome problem for that ideal. For most people would resist the conclusion that those who have expensive tastes are, for that reason, entitled to a larger share than others. Someone with champagne tastes (as we might describe his condition) also needs more resources to achieve welfare equal to those who prefer beer. But it does not seem fair that he should have more resources on that account. The case of the prospective politician, who needs a great deal of money to achieve his ambitions to do good, or the ambitious sculptor, who needs more expensive materials than the poet, perhaps falls in between. Their case for a larger share of their parent's resources seems stronger than the case of the child with expensive tastes, but weaker than the case of the child who is blind.

The question therefore arises whether the ideal of equality in welfare can be accepted in part, as an ideal that has a place, but not the only place, in a general theory of equality. The theory as a whole might then provide that the handicapped must have more resources, because their welfare will otherwise be lower than it could be, but not the man of champagne tastes. There are a number of ways in which that compromise within the idea of equality might be constructed. We might, for example, accept that in principle social resources should be distributed so that people are as equal in welfare as possible, but provide, by way of exception, that no account should be taken of differences in welfare traceable to certain sources, such as differences in tastes for drink. That gives equality of welfare the dominant place, but it prunes the ideal of certain distinct and unappealing consequences. Or we might, at the other extreme, accept only that differences in welfare from certain specified sources, such as handicaps, should be minimized. On this account equality of welfare would play only a part—perhaps a very minor part—in any general theory of equality, whose main political force must then come from a very different direction.

I shall postpone, until later in this part of the essay, the question of how far such compromises or combinations or qualifications are in fact available and attractive, and also postpone, until then, consideration of the particular problems I mentioned, the problems of expensive tastes and handicaps. But I want to single out and set aside, in advance, one form of objection to the feasibility of compromises of equality of welfare. It might be objected, against any such compromise, that the concept of welfare is insufficiently clear to permit the necessary distinctions. We cannot tell (it might be said) how much any welfare differences between two people who have equal wealth are in fact traceable to differences in the cost of their tastes or in the adequacy of their physical or mental powers, for example. So any theory that embraces equality of welfare must pay attention to people's welfare as a whole rather than welfare derived or lost through any particular source. Obviously, there is much in this sort of objection, though how strong the objection is must depend on the form of compromise proposed. I want, however, to set aside for this essay all objections about the feasibility of distinguishing welfare sources.

I also want to set aside the more general objection, that the concept of welfare is itself, even apart from distinctions as to source, too vague or impractical to provide the basis for any theory of equality. I said earlier that there are many different interpretations or conceptions of welfare, and that a theory of equality of welfare that uses one of these will have very different consequences, and require a very different theoretical support, from a theory that uses another. Some philosophers think of welfare as a matter of pleasure or enjoyment or some other conscious state, for example, while others think of it as success in achieving one's plans. We shall later have to identify the leading conceptions of welfare, and look at the different conceptions of equality of welfare they supply. But we may notice, in advance, that each of the familiar conceptions of welfare raises obvious conceptual and practical problems about testing and comparing the welfare levels of different people. Each of them has the consequence that comparisons of welfare will often be indeterminate: it will often be the case that of two people neither will have less welfare, but their welfare will not be equal. It does not follow, however, that the ideal of equality of welfare on any interpretation, is either incoherent or useless. For that ideal states the political principle that, so far as is possible, no one should have less welfare than anyone else. If that principle is sound, then the ideal of equality of welfare may sensibly leave open the practical problem of how decisions should be made when the comparison of welfare makes sense but its result is unclear. It may also sensibly concede that there will be several cases in which the comparison is even theoretically pointless. Provided these cases are not too numerous, the ideal remains both practically and theoretically important.

III. Conceptions of Equality of Welfare

There are several theories in the field, as I said, about what welfare is, and therefore several conceptions of equality of welfare. I shall divide what I consider the most prominent and plausible such theories into two main groups, without, however, supposing that all the theories in the literature can fit comfortably into one or the other. The first group I shall call success theories of welfare. These suppose that a person's welfare is a matter of his success in fulfilling his preferences, goals, and ambitions, and so equality of success, as a conception of equality of welfare, recommends distribution and transfer of resources until no further transfer can decrease the extent to which people differ in such success. But since people have different sorts of preferences, different versions of equality of success are in principle available.

People have, first, what I shall call political preferences, though

I use that term in a way that is both narrower and more extended than the way it is often used. I mean preferences about how the goods, resources and opportunities of the community should be distributed to others. These preferences may be either formal political theories of the familiar sort, such as the theory that goods should be distributed in accordance with merit or desert, or more informal preferences that are not theories at all, such as the preference many people have that those they like or feel special sympathy for should have more than others. Second, people have what I shall call impersonal preferences, which are preferences about things other than their own or other people's lives or situations. Some people care very much about the advance of scientific knowledge, for example, even though it will not be they (or any person they know) who will make the advance, while others care equally deeply about the conservation of certain kinds of beauty they will never see. Third, people have what I shall call personal preferences, by which I mean their preferences about their own experiences or situation. (I do not deny that these types of preferences might overlap, or that some preferences will resist classification into any of the three categories. Fortunately my arguments will not require the contrary assumption.)

The most unrestricted form of equality of success that I shall consider holds that redistribution should continue until, so far as this is possible, people are equal in the degree to which all their various preferences are fulfilled. I shall then consider the more restricted version that only nonpolitical preferences should be counted in this calculation, and then the still more restricted version that only personal preferences should count. More complex versions of equality of success, which combine the satisfaction of some but not all preferences from the different groups, are of course available, though I hope that the arguments I make will not require me to identify and consider such combinations.

The second class of theories of welfare I shall call conscious state theories. Equality of welfare linked to that sort of theory holds that distribution should attempt to leave people as equal as possible in some aspect or quality of their conscious life. Different conceptions of that ideal are constructed by choosing different accounts or descriptions of the state in question. Bentham and other early utilitarians took welfare to consist in pleasure and the avoidance of pain; equality of welfare, so conceived, would require distribution that tended to make people equal in their balance of pleasure over pain. But most utilitarians and other partisans of the conscious state conception of welfare believe that "pleasure" and "pain" are much too narrow to represent the full range of conscious states that should be included. For example, "pleasure," which suggests a specific kind of sensuous glow, poorly describes the experience produced by a harrowing piece of drama or poetry, an experience people nevertheless sometimes aim to have, and "pain" does not easily capture boredom or unease or depression.

I do not wish to discuss the issues this dispute raises. Instead I shall use the words "enjoyment" and "dissatisfaction" indiscriminately to name the full range of desirable and undesirable conscious states or emotions that any version of a conscious state conception of equality of welfare might suppose to matter. This usage gives those words, of course, a broader sense than they have in ordinary language, but I intend that broad sense, provided only that they must nevertheless name conscious states people might aim to have or avoid for their own sakes, and states that are introspectively identifiable.

People often gain enjoyment or suffer dissatisfaction directly, from sensuous stimulation through sex or food or sun or cold or steel. But they also gain enjoyment or suffer dissatisfaction through the fulfillment or defeat of their preferences of different sorts. So there are unrestricted and restricted versions of the conscious state conception of equality of welfare parallel to the versions I distinguished of conceptions of equality of success. One version aims to make people more equal in enjoyment without restriction as to source, another only in the enjoyment they take directly and from nonpolitical preferences, and another in the enjoyment they take directly and from personal preferences only. As in the case of equality of success, more discriminating versions that combine enjoyment from subdivisions of these different sorts of preferences are also available.

I shall also consider, though only very briefly, a third class of conceptions of equality of welfare, which I shall call objective conceptions. Many subdivisions and further classifications among these three classes of conceptions, beyond those I have just noticed, would have to be considered in any full account of possible theories of welfare, and there are theories of welfare not represented, as I said, in this list at all. But these seem the most plausible candidates for constructing theories of distribution. I shall just mention, however, two sorts of complexities that we should at least bear in mind. First, many (though not all) of the conceptions and versions I have distinguished raise the question of whether equality in that conception is reached when people are in fact equal in welfare so conceived, or rather when they would be equal if they were fully informed of the relevant facts. Does someone attain a given level of success, for purposes of equality of success, when he believes that his preferences have been fulfilled to a given degree, or rather when he would believe that if he knew the facts? I shall try, when questions of that sort might affect the argument, either to discuss both possibilities, or to assume the version that seems to me in context more plausible. Second, many of the conceptions I shall discuss raise problems about time. People's preferences change, for example, so that the question of how far someone's preferences for his life have been fulfilled overall will depend on which set of his preferences is chosen as relevant, or which function of the different preferences he has at different times. I do not believe that any of these temporal problems affect the various points that I shall make, but readers who do should consider whether my arguments hold against alternate versions.

There is, however, a further preliminary question that must detain us longer. We can distinguish two different questions. (1) Is someone's overall welfare–his essential well-being–really just a matter of the amount of his success in fulfilling his preferences (or just a matter of his enjoyment)? (2) Does distributional equality really require aiming to leave people equal in that success (or enjoyment)? The first of these questions takes a certain view of the connection between theories of welfare, such as those I described, and the concept of welfare itself. It supposes that this connection is rather like the connection between theories or conceptions of justice and the concept of justice itself. We agree that justice is an important moral and political ideal, and we ask ourselves which of the different theories about what justice actually consists in is the best such theory.

So we might suppose that (for one or another purpose) the welfare of persons, conceived as their essential well-being, is an important moral and political concept, and then ask ourselves which of the traditional theories (or new theories we might deploy) is the best theory of what welfare, so conceived, actually is.

But the second question does not, in itself, require that we confront—or even acknowledge the sense of—that last question. We may believe that genuine equality requires that people be made equal in their success (or enjoyment) without believing that essential well-being, properly understood, is just a matter of success (or enjoyment). We may, indeed, believe that equality requires equality in success even if we are skeptical about the whole idea of essential well-being, considered to be a deep or further fact about people conceptually independent from their success or enjoyment. That is, we may accept equality of success as an attractive political ideal, even if we reject the very sense of the question whether two people who are equal in success are equal in essential well-being. And we may do so even if we deny that this question is analogous to the question whether producing the highest possible average utility makes an institution just.

I make these remarks because it is important to distinguish between two strategies that someone anxious to defend a particular conception of equality of welfare might use. He might begin, first, by accepting the idea of welfare as essential well-being, and then take, as at least the tentative premise of his argument, the proposition that genuine equality requires people to be equal in essential well-being. He might then argue for a particular theory of welfare (success, for example) as the best theory of what essential well-being consists in, and so conclude that equality requires that people be made equal in success. Or, second, he might argue for some conception of equality of welfare, such as equality of success, in a more direct way. He might take no position on the question whether essential well-being consists in success, or even on the prior question whether that question makes sense. He might argue that, in any case, equality of success is required for reasons of fairness, or for some other reasons having to do with the analysis of equality, that are independent of any theory about the sense or content of essential well-being.

Is it therefore necessary to consider both of these strategies in assessing the case for any particular conception of equality of welfare? I think not, because the defeat of the second strategy (at least in a certain way) must count as a defeat of the first as well. I do not mean myself to claim that the idea of essential well-being, as a concept admitting of different conceptions, is nonsense, so that the first strategy, shorn of nonsense, is just the second. On the contrary, I think that idea, at least as defined by certain contexts, is an important one, and the question of where a person's essential well-being lies, when properly conceived, is sometimes, in those contexts, a question of profound importance. Nor do I think it follows from the conclusion that people should not be made equal in some particular conception of welfare, that this is a poor conception of welfare (conceived as essential well-being). I mean rather to deny something like the opposite claim: that if some conception is a good conception of welfare it follows that people should be made equal in welfare so conceived. This does not follow. I might accept, for example, that people are equal in essential well-being when each is roughly equally successful in achieving a certain set of his preferences, without thereby conceding that an advance towards that situation is even pro tanto an advance towards genuine distributional equality. Even if I initially accept both propositions, I should abandon the latter if I am then persuaded that there are good reasons of political morality for not making people equal in that sort of success, and that these reasons hold whether or not the former proposition is sound. So any arguments capable of defeating the second strategy, by showing that there are strong reasons of political morality why distribution should not aim to make people equal in success, must also count as strong arguments against the first strategy, though not, of course, as arguments defeating the interim conclusion of that strategy: that essential well-being consists in success. In what follows I shall try to oppose the second strategy in this manner.

IV. Success Theories

I want now to examine equality of welfare conceived in the various ways I have described, beginning with the group of theories I called success theories. I should perhaps say once again that I do not intend to make much of the practical difficulties (as such) of applying these or any other conceptions of equality of welfare. If any society dedicated itself to achieving any version of equality of success (or of enjoyment) it could do at best only a rough job, and could have only a rough idea of how well it was doing. Some differences in success would be beyond the reach of political action, and some could be eliminated only by procedures too expensive of other values. Equality of welfare so conceived could be taken only as the ideal of equality, to be used as a standard for deciding which of different practical political arrangements seemed most or least likely to advance that ideal on the whole as a matter of antecedent tendency. But precisely for that reason it is important to test the different conceptions of equality of welfare as ideals. Our question is: If (impossibly) we could achieve equality of welfare in some one of these conceptions, would it be desirable, in the name of equality, to do so?

Political Preferences

I shall begin by considering equality of success in the widest and most unrestricted sense I distinguished, that is, equality in the fulfillment of people's preferences when these include political as well as other forms of preference. We should notice a threshold difficulty in applying this conception of equality in a community in which some people themselves hold, as a matter of their own political preferences, exactly the same theory. Officials could not know whether such a person's political preferences were fulfilled until they knew whether their distribution fulfilled everyone's preferences equally, including his political preferences, and there is danger of a circle here. But I shall assume that equality of welfare, so conceived, might be reached in such a society by trial and error. Resources might be distributed and redistributed until everyone pronounced himself satisfied that equality of success on the widest conception had been achieved.

We should also notice, however, a further threshold difficulty: that it would probably prove impossible to reach a reasonable degree of equality in this conception even by trial-and-error methods in a community whose members held very different and very deeply felt political theories about justice in distribution. For any distribution of goods we might arrange, some group, passionately committed to a different distribution for reasons of political theory, might be profoundly dissatisfied no matter how well they fared personally, while others might be very pleased because they held political theories that approved the result. But because I propose to ignore practical or contingent difficulties, I shall assume a society in which it is possible to achieve rough equality in the amount by which people's unrestricted preferences were fulfilled, that is roughly equal success on this wide conception, either because people all hold roughly the same political theories, or because, though they disagree, anyone's dissatisfaction with a solution on political grounds could be made up by favoritism in his personal situation, without arousing so much antagonism in others as to defeat equality so conceived for that reason.

This latter possibility—that people who lose out because their political theories are rejected could be given more goods for themselves by way of compensation—makes this conception of equality of welfare immediately unattractive, however. Even people otherwise attracted to the idea of equality of welfare, on any conception, would presumably not wish to count gains or losses in welfare traceable to, for example, racial prejudice. So I assume that almost everyone would wish to qualify equality of success at least by stipulating that a bigot should not have more goods than others in virtue of the fact that he would disapprove a situation in which blacks have as much as whites unless his own position were sufficiently favored to make up the difference.

But it is unclear why this stipulation should not apply to all political theories that are in conflict with the general ideal of equality of success, at least, and not just to racial bigotry. It should apply equally to people who think that aristocrats should have more than plebs, or to meritocrats who think that, as a matter of political morality, those who are more talented should have more. Indeed, it should apply even to egalitarians who think that people should be equal in resources or enjoyment or in the success each has in his personal life rather than in the fulfillment of all his preferences including his political preferences. These "wrong" egalitarian theories will of course seem more respectable to officials who have accepted the latter conception of equality than will bigoted or meritocratic theories. But it still seems odd that even wrong egalitarians should have extra resources credited to their personal account just to make up for the fact that their overall approval of the situation would otherwise be lower than those who hold the political theory assumed to be correct, and on which any claims the former might make to extra resources must in some sense rely. It seems odd (among other reasons) because a good society is one which treats the conception of equality that society endorses, not simply as a preference some people might have, and therefore as a source of fulfillment others might be denied who should then be compensated in other ways, but as a matter of justice that should be accepted by everyone because it is right. Such a society will not compensate people for having preferences that its fundamental political institutions declare it is wrong for them to have.

The reason why racial bigotry should not count, as a justification for giving the bigot more in personal goods, is that this political theory or attitude is condemned by the proper conception of equality, not that the bigot is necessarily insincere or unreflective or personally wicked. But then other forms of nonegalitarian political theory, and even misconceived forms of egalitarian theory, should be discounted in the same way. Suppose, moreover, that no one has a nonegalitarian or wrong-egalitarian political theory of any formal sort, but that some people are merely selfish and have no political convictions even in the extended sense, so that their overall approval of the state of affairs after any distribution is just a matter of their own private situation, while others are benevolent, so that their overall approval is increased by, say, the elimination of poverty in the society. Unless we refuse to take that benevolence into account, as a positive source of success in meeting the preferences overall of those who are benevolent, we shall end once again by giving those who are selfish more for themselves, to compensate for the success others have from that benevolence. But it is surely a mark against any conception of equality that it recommends a distribution in which people have more for themselves the more they disapprove or are unmoved by equality.

Consider, finally, a different situation. Suppose no one holds, in any case very deeply, any formal political theory, but each is generally benevolent. Many people, however, by way of what I called a political theory in the extended sense, sympathize especially with the situation of one group of those less fortunate than themselves–say, orphans– and have special preferences that these be looked after well. If these preferences are allowed to count, this must have one or the other of two results. Either orphans will, just for this reason, receive somewhat better treatment than equality would itself have required in the absence of these special preferences, at the inevitable expense of other groups–including those disadvantaged in other ways, such as, say, cripples; or, if this is ruled out on egalitarian grounds, those who care more about orphans than about cripples will be given extra resources to make up for the failure to fulfill this discrete preference (which extra resources they then may or may not contribute to orphans). Neither of these results does credit to an egalitarian theory.

So we have good reason to reject the unrestricted conception of equality of success, by eliminating from the calculation of comparative success both formal and informal political preferences, at least for communities whose members differ in these political preferences, which is to say for almost all actual communities with which we might be concerned. We might just pause to consider, however, whether we must reject that conception for all other communities as well. Suppose a community in which people by and large hold the same political preferences. If these common preferences endorse equality of success, including success in political preferences, then that theory for all practical purposes collapses into the more restricted theory that people should succeed equally in their nonpolitical preferences. For if a distribution is reached that everyone regards with roughly equal overall approval, and the force of individual political convictions, in each person's judgment of how well he regards it, is simply to approve the result because everyone else does regard it equally, then the distribution must be one in which each person regards his own impersonal and personal preferences as equally fulfilled as well. For suppose Arthur is less satisfied with his impersonal and personal situation than Betsy. Arthur can have, by hypothesis, no political theory or attitude that could justify or require a distribution in which he is less satisfied in this way than Betsy is, so Arthur can have no reason to regard the distribution with as much general or overall approval, combining political, impersonal, and personal assessments, as Betsy does.

But suppose the shared political theory is not the ideal of equality of overall approval, but some other, nonegalitarian theory that could provide such a reason. Suppose everyone accepts a caste theory so that, though Amartya is somewhat poorer than others, the distribution leaves his preferences as a whole equally fulfilled because he believes that he, as a member of a lower caste, should have less, so that his preferences as a whole would be worse fulfilled if he had more. Bimal, from a higher caste, would also be less satisfied overall if Amartya had any more. In this situation, unrestricted equality of success does recommend a distribution that no other conception of equality of welfare would. But it is unacceptable for that very reason. An inegalitarian political system does not become just because everyone wrongly believes it to be.

Unrestricted equality of success is acceptable only when the political preferences that people happen to have are sound rather than simply popular, which means, of course, that it is in the end an empty ideal, useful only when it rubber-stamps a distribution already and independently shown to be just through some more restricted conception of equality of success or through some other political ideal altogether. Suppose someone denies this and argues that it is good, in and of itself, when everyone approves of a political system highly and equally no matter what that system is. This seems so arbitrary, and so far removed from ordinary political values, as to call into question whether he understands what a political theory is or is for. In any case he does not state an interpretation of equality, let alone an attractive one.

Impersonal Preferences

We must surely restrict equality of success still further by eliminating, from the calculation it proposes, at least some of what I called people's impersonal preferences. For it is plainly not required by equality that people should be equal, even insofar as distribution can achieve this, in the degree to which all their nonpolitical hopes are realized. Suppose Charles very much and very deeply hoped that life would be discovered on Mars. Or that the Great American novel would be written within his lifetime. Or that the Vineyard coast not be eroded by the ocean as it inevitably continues to be. Equality does not require that funds be taken from others, who have more easily fulfilled hopes about how the world will go, and transferred to Charles so that he can, by satisfying other preferences he has, decrease the overall inequality in the degree to which his and their nonpolitical preferences are fulfilled.

Should any impersonal preferences be salvaged from the further restriction this suggests? It might be said that the various impersonal preferences I just took as examples are all impossible dreams, or, in any case, all dreams that the government can do nothing to fulfill. But I cannot see why that matters. If it is right to aim to decrease inequality in disappointment in all genuine nonpolitical aims or preferences, then the government should do what it can in that direction, and though it cannot bring it about that there is life on Mars, it can, as I said, at least partially compensate Charles for his failed hopes by allowing him to be more successful otherwise. In any case, I might have easily taken as examples hopes people have that are not impossible for government to realize, or even particularly difficult. Suppose Charles hopes that no distinctive species will ever become extinct, not because he enjoys looking at a variety of plants and animals, or even because he thinks others do, but just because he believes that the world goes worse when any such species is lost. He would overwhelmingly prefer that a very useful dam not be built at the cost of losing the snail darter. (He has not set out deliberately to cultivate his views about the importance of species. If he had, then this might be thought to raise the special issues about deliberate cultivation of expensive tastes that I shall consider later. He just finds he has these views.) But after the political process has considered the issue and reached its decision the dam is built. Charles' disappointment is now so great (and he cares so little about everything else) that only the payment of a vast sum of public money, which he could use to lobby against further crimes against species, could bring his welfare, conceived as the fulfillment of all nonpolitical preferences, back to the general level of the community as a whole. I do not think that equality requires that transfer, nor do I believe that many, even of those who find appeal in the general ideal of equality of welfare, will think so either.

Of course equality does require that Charles have a certain place in the political process I described. He must have an equal vote in selecting the officials who will make the decision, and an equal opportunity to express his opinions about the decision these officials should take. It is at least arguable, moreover, that the officials should take his disappointment into account, perhaps even weighted for its intensity, in the general cost-benefit balancing they undertake in deciding whether the dam should be built all things considered, that his dissatisfaction should count in a Benthamite calculation and be weighted against the gains to others that the dam would bring. We might wish to go beyond this, perhaps, and say that if the community faces a continuing series of decisions that pit economic efficiency against species preservation, it should not take these decisions discretely, through separate cost-benefit calculations each of which Charles would lose, but as a series in which the community should defer to his opinion at least once. But none of this comes near arguing that the community treats Charles as an equal only if it recognizes his eccentric position in a different way, by undertaking to insure, so far as it can, that his success in finding all his nonpolitical preferences fulfilled remains as high as everyone else's when the series of decisions is completed, no matter how singular his impersonal preferences are. Indeed this proposition contradicts rather than enforces what conventional ideals of political equality recommend, because if the community acknowledged that responsibility, Charles' opinions would very probably play a role far beyond what these traditional ideas provide for them.

But someone might still protest that my arguments depend on assigning to people impersonal preferences that are in the circumstances unreasonable, or, rather, unreasonable to expect the community to honor by compensating for their failure. My arguments do not, it might be said, suggest that reasonable impersonal preferences should not be honored in that way. But this introduces a very different idea into the discussion. For we now need an independent theory about when an impersonal preference is reasonable, or when it is reasonable to compensate for one. It seems likely (from the present discussion) that such a theory will assume that a certain fair share of social resources should be devoted to the concerns of each individual, so that a claim for compensation might be appropriate when this fair share is not in fact put at his disposal, but not if deciding as he wishes, or compensating him for his disappointment would invade the fair share of others. We shall consider, later in this section, the consequences of using the idea of fair shares in this way within a theory of equality of success. It is enough to notice now that some such major refinement would be necessary before any impersonal preferences qualify for the calculation of equality of success.

Nor does it seem implausible to restrict a conception of equality of welfare to success in achieving personal, as distinct from both all political and all impersonal, ambitions. For that distinction is appealing in other ways. Of course people do care, and often care very deeply, about their political and impersonal preferences. But it does not seem callous to say that, insofar as government has either the right or the duty to make people equal, it has the right or duty to make them equal in their personal situation or circumstances, including their political power, rather than in the degree to which their differing political convictions are accepted by the community, or in the degree to which their differing visions of an ideal world are realized. On the contrary, that more limited aim of equality seems the proper aim for a liberal state, though it remains to see what making people equal in their personal circumstances could mean.

Equality of Personal Success

Relative Success. We should therefore consider the most restricted form of equality of success that I shall discuss, which requires that distribution be arranged so that people are as nearly equal as distribution can make them in the degree to which each person's preferences about his own life and circumstances are fulfilled. This conception of equality of welfare presupposes a particular but plausible theory of philosophical psychology. It supposes that people are active agents who distinguish between success or failure in making the choices and decisions open to them personally, on the one hand, and their overall approval or disapproval of the world in general, on the other, and seek to make their own lives as valuable as possible according to their own conception of what makes a life better or worse, while recognizing, perhaps, moral constraints on the pursuit of that goal and competing goals taken from their impersonal preferences. There is no doubt a measure of idealization in this picture; it may never be a fully accurate description of any person's behavior, and it may require significant qualification in many cases. But it seems a better model against which to describe and interpret what people are than the leading and perhaps more familiar alternatives.

So I shall not quarrel with this psychological theory. But we must notice at once a difficulty in the suggestion that the resources of a community should be distributed, so far as possible, to make people equal in the success they have in making their lives valuable in their own eyes. People make their choices, about what sort of a life to lead, against a background of assumptions about the rough type and quantity of resources they will have available with which to lead different sorts of lives. They take that background into account in deciding how much of what kind of experience or personal relationship or achievement of one sort must be sacrificed for experiences or relationships or achievements of another. They therefore need some sense of what resources will be at their disposal under various alternatives before they can fashion anything like the plan for their lives of the sort that this restricted conception of equality of success assumes that they have, at least roughly, already created. Some of these resources are natural: people need to make assumptions about their expected life span, health, talents and capacities, and how these compare with those of others. But they also need to make assumptions about the resources they will have of just the sort society would allocate under any scheme of equality of welfare: wealth, opportunities, and so on. But if someone needs a sense of what wealth and opportunities will be available to him under a certain life before he chooses it, then a scheme for distribution of wealth cannot simply measure what a person should receive by figuring the expense of the life he has chosen.

There is therefore again danger of a fatal circle here. But I propose to set that problem aside, as another instance of the kind of technical problem that I promised not to labor. So I shall suppose, once again, that the problem can be solved in a trial-and-error way. Suppose a society in which people in fact have equal resources. It is discovered that some are much more satisfied with the way their lives are going than others. So resources are taken from some and given to others, on a trial and error basis, until it is true that, if people were fully informed about all the facts of their situation, and each was asked how successful he believed himself to be at fulfilling the plans he has formed given the level of resources he now has, each person would indicate roughly the same level or degree of success.

But this "solution" of the practical difficulty I describe brings to the surface a theoretical problem to which the practical difficulty points. People put different values on personal success and failure, not only as contrasted with their political and moral convictions, and their impersonal goals, but just as part of their personal circumstances or situation. At least they do in one sense of success and failure. For we must now notice an important distinction I have so far neglected. People (at least as conceived in the way just described) choose plans or schemes for their lives, against a background of natural and physical resources they have available, in virtue of which they have discrete goals and make discrete choices. They choose one occupation or job over another, live in one community rather than another, seek out one sort of lover or friend, identify with one group or set of groups, develop one set of skills, take up one set of hobbies or interests, and so forth. Of course, even those people who come closest to the ideal of that model do not make all these choices deliberately, in the light of some overall scheme, and perhaps make none of them entirely deliberately. Luck and occasion and habit will play important roles. But once the choices they do make have been made, these choices define a set of preferences, and we can ask how far someone has succeeded or failed at fulfilling whatever preferences he has fixed in that way. That (I shall say) is the question of his relative success—his success at meeting the discrete goals he has set for himself.

But people make these choices, form these preferences, in the light of a different and more comprehensive ambition, the ambition to make something valuable of the only life they have to lead. It is, I think, misleading to describe this comprehensive ambition as itself only another preference people have. It is too fundamental to fit comfortably under that name; and it is also too lacking in content. Preferences are choices of something preferred to something else; they represent the result of a decision, of a process of making what one wants more concrete. But the ambition to find value in life is not chosen as against alternatives, for there is no alternative in the ordinary sense. Ambition does not make plans more concrete, it is simply the condition of having any plans at all. Once someone has settled on even a tentative or partial scheme for his own life, once his discrete preferences have been fixed in that way, then he can measure his own relative success in a fairly mechanical way, by matching his situation to that scheme. But he cannot tell whether his life has succeeded or failed in finding value simply by matching his achievements to any set target in that way. He must evaluate his life as a whole to discover the value that it has, and this is a judgment that must bring to bear convictions that, however inarticulate these are, and however reluctant he might be to call them this, are best described as philosophical convictions about what can give meaning or value to any particular human life. I shall call the value that someone in this way attributes to his life his judgment of that life's overall success.

People disagree about how important relative success is in achieving overall success. One person might think that the fact that he is likely to be very successful at a particular career (or love affair or sport or other activity) counts strongly in favor of his choosing or pursuing it. If he is uncertain whether to be an artist or lawyer, but believes he would be a brilliant lawyer and only a good artist, he might regard that consideration as decisive for the law. Someone else might weigh relative success much less. He might, in the same circumstances, prefer to be a good artist to being a brilliant lawyer, because he thinks art so much more important than anything lawyers do.

This fact—that people value relative success differently in this way—is relevant here for the following reason. The basic, immediate appeal of equality of welfare, in the abstract form in which I first set it out, lies in the idea that welfare is what really matters to people, as distinct from money and goods, which matter to them only instrumentally, so far as these are useful in producing welfare. Equality of welfare proposes, that is, to make people equal in what is really and fundamentally important to them all. Our earlier conclusion, that in any event the fulfillment of political and impersonal preferences should not figure in any calculation aimed at making people equal in welfare through distribution, might well be thought to damage that appeal. For it restricts the preferences that people are meant to fulfill in equal degree to what I have called personal preferences, and people do not care equally about the fulfillment of their personal preferences as opposed to their political convictions and impersonal goals. Some care more about their personal preferences, as opposed to their other preferences, than others do. But a substantial part of the immediate appeal I describe remains, though the point would now be put slightly differently. Equality of welfare (it might now be said) makes people equal in what they all value equally and fundamentally so far as their own personal situation or circumstances are concerned.

But even that remaining claim is forfeit if equality of welfare is construed as making people equal, so far as distribution can achieve this, in their relative success, that is, in the degree to which they achieve the goals they fix for themselves. On this conception, money is given to one rather than another, or taken from one for another, in order to achieve equality in a respect some value more than others and some value very litle indeed, at the cost of inequality in what some value more. A person of very limited talents might choose a very limited life in which his prospects of success are high because it is so important to be successful at something. Another person will choose almost impossible goals because for him the meaning is the challenge. Equality of relative success proposes to distribute resources–presumably much fewer to the first of these two and much more to the second –so that each has an equal chance of success in meeting these very different kinds of goals.

Suppose someone now replies that the appeal of equality of welfare does not lie where I located it. Its purpose is not to make people equal in what they do value fundamentally, even for their own lives, but rather in what they should value fundamentally. But this change in the claims for equality of welfare achieves nothing. For it is absurd to suppose that people should find value only in relative success without regard to the intrinsic value or importance of the life at which they are relatively successful. Perhaps some people–those with grave handicaps–are so restricted in what they can do that they must choose just so as to be able to be minimally successful at something. But most people should aim to do more than what they would be, relatively, most successful in doing.

Overall Success. This discussion might be thought to suggest a better interpretation of equality of welfare, namely equality of overall rather than relative success. But if we are to explore equality of welfare in that conception we must make a distinction not necessary (or in any case not so plainly necessary) in comparing relative success. We must distinguish a person's own judgment of his overall success (or, if we prefer, the judgment he would make if fully informed of the pertinent ordinary sorts of facts) from the objective judgment of how much overall success in fact he has. A person's own judgment (even if fully informed of the facts) will reflect, as I said, his own philosophical convictions about what gives value to life, and these might be, from the standpoint of the objective judgment, confused or inaccurate or just wrong. I shall suppose, here, that equality of overall success means equality in people's overall success as judged by themselves, from the standpoint of their own perhaps differing philosophical beliefs. I shall later consider, under the title of objective theories of welfare, the different conception that requires equality in the success of their lives judged in some more objective way.

So let us now alter the exercise we have imagined. Now we rearrange resources, as far as we are able, so that when we have finished, each person would, at least if fully informed, offer the same assessment, not of his relative success in achieving the goals he selected for himself, but of his overall success in leading a valuable life. But we must take care in describing just what we take that latter opinion to be. For there are many different beliefs each of which might possibly be thought to count as an assessment of one's own overall success, and it is of crucial importance to decide which of these, if any, should play a role in rearranging resources in the name of equality. Nor can we find much guidance in the literature either of welfare economics or of utilitarianism, which are the natural places to look. For most of those writers who argue or assume that welfare consists in the fulfillment of preferences seem to have had relative rather than overall success in mind, and in any case have not discussed the problems raised by the latter idea when the two ideas are separated. The language they use–

the language of preferences (or wants or desires)—seems too crude to express the special, comprehensive judgment of the value of a life as a whole.

We might begin by distinguishing the question of how valuable someone believes his life has been, taken as a whole, from the question of how much he wants his life to continue. These are certainly different matters. Some people, to be sure, wish their life to end, or are in any case almost indifferent whether it continues, because they regard it as a failure. But others wish to die just because they think their life has been too brilliant to tarnish with a slow decline. And others think that a successful life can be made more successful by the timely use of suicide as a creative act. People can want to end, that is, a life they are proud to have led. Can we say at least that if someone wishes to die he must regard the future life he would otherwise lead as having no or little value? This will certainly be so in most cases, but the connection is nevertheless, I think, contingent. He may only think that though his future life would be quite successful, his life as a whole would be a more successful life if it ended now. Nor does it follow from the fact that someone very much wishes to continue living, for as long as he can imagine, that he thinks that his life is a successful life, or even that his future life will be especially successful. He may, on the contrary, want to live longer because he thinks his life has been unsuccessful, because he needs more time to do anything worth doing, though it is more likely that he simply fears death. The distinction I want to make can be summarized, perhaps, this way. Someone's preferences about the length of his life are just that, preferences that are like his choices of jobs and lovers, fixed as part of the dominating exercise of deciding what life, given background assumptions about resources, would be the most valuable life all things considered. They are not in themselves judgments of overall success or failure.

Can we make a further distinction between the value someone finds in his own life and the value be believes it has *for* him? I am not sure what that latter phrase would mean as part of this contrast. We sometimes say that a person puts a low value on his own life when we mean, not that he is not proud of the life he has or will lead, but rather that he counts the value of that life low compared with the value he puts on his duty or the lives of others. But we are now considering something different, not the value someone puts on his own life as compared with his moral or impersonal values, but as part of the assessment of his own situation.

Perhaps "the value of someone's life for him" means only the intensity of his preference that his life continue. If so, the distinction between the value someone finds in his life and its value for him is the distinction we have already discussed. But someone who uses that phrase may have in mind something more complex and more elusive than that. He may mean to distinguish someone's judgment about the value of any single human life (or indeed human life in general) to the universe as a whole from that person's judgment from the inside, from the standpoint of someone charged with making something valuable of his own life. If so, then it is the latter judgment with which we are now concerned. Or he may mean to distinguish someone's judgment of his own success in that assignment, given his talents and opportunities, from his judgment of whether it was good for him to have had the talents and opportunities and convictions that made him the person whose life would have most value lived that way. It is not hard to imagine lives that illustrate the distinction so understood. It is in fact a cliché that great artists often work, not out of enjoyment (even in the widest sense of enjoyment), but rather in constant misery simply because, in a familiar phrase, it is not possible for them not to write poetry or music or paint. A poet who says this may well think that a life he spent in any other way would be, in the most fundamental sense, a failure. But he might well think that the conspiracy of talents and beliefs that made this true was bad for him, meaning only that his life would be more enjoyable if he lacked these talents or did not have the belief, which he could not however shake, that a life of creating poetry in misery and despair was all things considered the most valuable life for him to lead. Suppose we then ask him the dark question only philosophers and sentimental novelists ask: Would it have been better for you if you had never been born? If he says yes, as he might in some moods, we would know what he meant, and it would not be that he has done nothing valuable with his life. If the distinction between someone's judgment of the value of his life and his judgment of the value it has for him is taken in this way, then it is the former judgment I mean by his judgment of his overall success. But if the distinction cannot be taken in this way, or in any of the other ways I have considered, then I do not understand it, and suspect that it is no distinction at all.

These scrappy remarks are intended to clarify the comparison we must intend when we propose that people should be equal in their overall success. We cannot carry out this comparison simply by discovering two people's own fixed preferences and then matching their situation to these preferences. That is only a comparison of their relative success. We must invite them to make (or ourselves make from their point of view) an overall rather than a relative judgment that takes fixed preferences as part of what is judged rather than the standard of assessment. If we ask them to make that assessment themselves, however, and then try to compare the assessments each makes, we may discover the following difficulty. Suppose we ask Jack and Jill each to evaluate the overall success of his or her own life, and we make plain, by a variety of distinctions, what we mean by overall success as distinct from relative success, enjoyment, how much they wish their lives to continue, and so forth. And we provide them with a set of labels, from "total failure" to "very great success" with several stops in between, from which to choose. We have no guarantee that each will use any one of these labels in the same way as the other, that is, to report what we might independently consider the same judgment. Jack may use one or more of the labels with a different meaning from the meaning Jill uses, and they may be using different scales in judging the intervals between these labels. Jack might suppose, for example, that there is a vast difference between "great success" and "very great success" while Jill understands these terms to enforce only a marginal difference; so that both might use the latter label to report judgments that we, on the basis of further conversation with them, would come to believe were in fact very different judgments. This difficulty, so described, is a difficulty in translation, and I shall suppose that we could in principle conquer it, at least for speakers of our own language, by the further conversations just mentioned.

But of course all this assumes that there is indeed a single kind of overall judgment that we are asking Jack and Jill each to make (or that we propose to make from their standpoint, on their behalf) and that this judgment is in fact a judgment about the inherent value of their life and therefore different from a judgment of relative success or a judgment about how much a person wants his life to continue or how much enjoyment he finds in it. Many people are, of course, skeptical about such judgments so interpreted. If they are right, then the judgments we ask Jack and Jill to make are meaningless judgments. But then equality of overall success is itself meaningless for that reason. (Though someone might still propose, for reasons we need not explore, that people should nevertheless be equal in the character of each's illusion.) If we assume that the skeptics are wrong, however (or even that equality of illusion is the true aim) then equality of overall success must suddenly seem a peculiar goal indeed, at least in the following circumstances.

Suppose that Jack and Jill have equal resources and that they are otherwise roughly similar in every way except in respect of the beliefs I am about to mention. They are both healthy, neither handicapped, both reasonably successful in their chosen occupations, neither outstandingly accomplished or creative. They take roughly the same enjoyment from their day-to-day life. But Jack (who has been much influenced by genre painting) thinks that any ordinary life fully engaged in projects is a life of value, while Jill (perhaps because she has taken Nietzsche to heart) is much more demanding. Jack thinks, for example, that the life of a busy peasant who achieves very little and leaves nothing behind is full of value, while Jill thinks that such a life is only full of failure. If each is asked to assess the overall value of his or her own life, Jack would rate his high and Jill hers low. But there is surely no reason in that fact for transferring resources from Jack to Jill provided only that Jill would then rate her life, while still of little overall success, a bit higher.

It might seem that the difficulty this example exposes arises only from the fact that our procedures attempt to compare judgments of value reached on the basis of very different theories about what gives value to life, which is like comparing apples and oranges. Someone might object that we would do better if we asked Jack and Jill each to make comparative judgments using their own standards for each comparison, and then compared these comparative judgments in some way that would neutralize the difference in their philosophical convictions. This is, I think, a mistake, but we should explore the suggestion nevertheless. We might ask Jack, for example, to compare the value, in his eyes, of his present life with the value of the life he would have under whatever conditions of physical and mental power and whatever collection of material resources and opportunities at his disposal he would take to be ideal. Or we might ask him to compare in this way his present life with the life he would have under what he would take to be the worst conditions. We might ask him how far his life is better than the life he would have if he had no or very few resources or opportunities. We would then put the same questions to Jill. Or we might ask each a rather different sort of question, not asking them to imagine different material circumstances, but rather to compare their present lives with lives in which they would each find no value at all.[1] We might ask each how far his or her life exceeded, in its value, that life. And so on. Once some one of these questions (or perhaps some weighted group of them) had been selected as for some reason especially appropriate for the purpose, equality of overall success, as a political ideal, would recommend redistribution until either the proportion or the flat amount hypothetically reported by way of answer was as close to the same in all cases as could be achieved in that way.

I should say at once that there is room for doubt, at least, whether all or possibly even any of these various questions could actually be answered, or be answered by any but the most philosophically inclined respondents. I shall set the doubts aside, however, and assume that people generally have a sufficient grasp of theories of value to be able intelligently to answer them. But of course the different comparisons the different questions prompted might, if each was harnessed to the ideal of equality of overall success, yield different recommendations for redistribution. Suppose, for example, that Jack thought his present life much better than the worst life he could imagine, but also much worse than the best life, while Jill thought her life not much better than the worst and not much worse than the best. Then the direction of redistribution would depend on which

1. I owe this last suggestion to Derek Parfit, and through him, I understand, to J. P. Griffin and J. McMahon.

of these two comparisons was thought more important for comparing levels of overall success. Even if all the answers to all the questions we could invent pointed in the same direction for redistribution, we should still have to show that at least one of these questions was the right question to ask.

When we look more closely at the questions I listed, however, they turn out to be very much the wrong questions. Suppose Jack and Jill (who, as I imagined, are now roughly equal in resources and enjoyment and relative success in their chosen lives) do disagree radically in judgments about how much more valuable their lives would be if they had everything they could have, for example. Jack believes that with all these resources he could solve the riddle of the origin of the universe, which would be the greatest imaginable achievement for human beings, while Jill believes that riddle unsolvable, and has no comparable dream in hand. So Jack believes his present life is only a small fraction as good as what it could ideally be, but Jill believes her life is not that much worse than what it could possibly be. Surely we have no reason of equality here for transferring resources from Jill to Jack (destroying their assumed equality of resources and enjoyment and relative success) even if such a transfer would cause Jack to rank his new life at a somewhat higher fraction of his ideal solving-the-riddle life.

Suppose that Jack considers his present life much more valuable than any life he would consider to have no value at all, while Jill thinks her life just barely better, on any flat scale of value in life, than a life she would think had absolutely no value. But that this is for the reason already suggested. Jack considers any life fully engaged and active, with as much day-to-day enjoyment as his has, of enormous value, something to be treasured and protected and pursued. He can imagine a life about which he would be indifferent, but it is a life so impoverished that he has no trouble reporting that his life is better than that by a very long chalk. Jill has roughly as much day-to-day pleasure or enjoyment. She is not depressive, but rather, as I said, very demanding in her idea of what life could be deemed a really successful life. She cannot say, when asked seriously to consider this grave question in a philosophical mood, that she thinks her life, for all its apparent richness, is in fact a life of much real value; she can easily imagine a life which she would believe had absolutely no value and cannot say that she honestly thinks her life is really, all things considered, much more valuable than that one. Once again it seems implausible that equality demands that resources be transferred from Jack to Jill.

Why are all these comparative questions so plainly the wrong questions to ask? Because we have not in fact escaped, in switching from flat questions to comparative ones, the difficulty we found in the former. Because the differences between Jack and Jill we have noticed are still differences in their beliefs but not differences in their lives. They are differences in their speculative fantasies about how good or bad their lives would be under very different and bizarre circumstances, or differences in their philosophical convictions about what could give great value to any life; but not, for that reason, a difference in what their lives now are. Each of the judgments Jack or Jill makes, in responding to the different questions we put to them, can be considered a judgment about the value or overall success of their lives. But they are not all the *same* judgment, and none of the judgments we have so far described seems appropriate for a theory of equality of overall success.

I want now to suggest a comparison of the overall success in people's lives, very different from the comparisons suggested in all these questions, that does seem to be connected, at least, to problems of distributional equality. Differences in people's judgments about how well their lives are going overall are differences in their lives, rather than simply differences in their beliefs, only when they are differences, not in fantasy or conviction, but in fulfillment, which is, I take it, a matter of measuring personal success or failure against some standard of what *should* have been, not merely of what conceivably *might* have been. The important, and presently pertinent, comparison seems to me this. People have lives of less overall success if they have more reasonably to regret that they do not have or have not done.

"Reasonably," of course, carries much weight here. But it is all necessary. No one can reasonably regret that he has not had the life that someone with supernatural physical or mental powers, or the life span of Methuselah, would have had. So no one has a less successful life, all things considered, just because he thinks that such a life would be infinitely more valuable, in the philosophical way, than the life he has. But people can reasonably regret that they have not had the normal powers or the normal span of life that most people have. No one can reasonably regret that he has not had the life that someone with an unfairly large share of the world's resources would have led, so no one person's life is less successful than another's because the first thinks his life would be much more valuable in those circumstances while the other does not. But people can reasonably regret not having whatever share of material resources they are entitled to have.

Perhaps the point is now clear. Any proposed account of equality of overall success that does not make the idea of reasonable regret (or some similar idea) pivotal in this way is irrelevant to a sensible theory of equality in distribution. It may develop a concept of overall success useful for some purpose, but not for this purpose. But any proposed account that does make this idea pivotal must include, within its description of equality of overall success, assumptions about what a fair distribution would be, and that means that equality of overall success cannot be used to justify or constitute a theory of fair distribution. I do not mean simply that equality of success could not be applied in some cases without having an independent theory of fair distribution as a supplement for such cases. If the point were only that, then it would show only that equality of overall success could not be the whole story in a theory of distribution. The point is more striking. Equality of overall success cannot be stated as an attractive ideal at all without making the idea of reasonable regret central. But that idea requires an independent theory of fair shares of social resources (this might, for example, be the theory that everyone is entitled to an equal share of resources) which would contradict equality of overall success not in some cases only, but altogether.

Suppose someone contests this important conclusion in the following way. He concedes that the aim of equality of success, properly conceived, is to make people equal in what they have reasonably to regret. But he believes that the idea of reasonable regret can be elucidated in some way that does not require any theory of fair share of resources *other than* some version or refinement of the equal success theory itself. He might propose the following. People cannot reasonably regret that they are not leading the life of someone with supernatural powers. Or the life of a successful sadist. Or a life in which they have resources such that, with those resources, they can achieve a life with less reasonably to regret than others can have with the resources then left for them. This will not do, however. We aim to make people equal in what they have reasonably to regret. Suppose (as before) that Jack and Jill have equal resources. Jack has (as we saw) grand ambitions and, though he does not believe himself entitled to anything in particular, will always regret not having more than he does. We want to know whether Jack and Jill are nevertheless equal in what they have *reasonably* to regret. On the proposed test, we must ask Jack (or ask ourselves from his point of view) how far the life he can now lead falls short of the life he would lead if he had (among other things) the amount of resources such that if he had those resources he would have the same amount reasonably to regret as others would then have. Jack cannot answer that question (nor can we). He can pick some different distribution at random—say a distribution in which he has a million dollars more and others in the aggregate a million less. But he cannot tell whether his new-distribution life is the proper baseline against which to measure his present life without knowing whether the reasonable regret he would feel with one more million is no more than others would reasonably regret with what they could then have, and he cannot tell this without picking some *further* new distribution at random (in which, perhaps, he has two million more) against which to compute his regret at the one-million-more life. And so on into infinite regress. We cannot, of course, repair this failure (as we tried to repair other failures) by some trial-and-error device. For the problem is not that we can offer no noncircular algorithm for reaching an initial distribution to test, but rather that we can offer no method for testing any distribution however reached.[2]

2. In the case of some people, but not Jack, we might be able to find an amount of resources such that their actual regret at not having more is so weak that we do not need to compute their reasonable regret at that amount to say that the latter must be less than the actual regret that others would have if the former had that amount. But many questions arise even then. Is it not necessary still to compute the reasonable regret, as distinguished from the

I conclude that reasonable regret cannot itself figure in the distributional assumptions against which the decision whether some regret is reasonable is to be made. Nor can I think of any other conception or refinement of equality of overall success that can fill that role. If so, then the goal of making people equal in what they have reasonably to regret is self-contradictory in the way I described. I do not mean that comparisons of fulfillment—of how far different people have been able to make a success of their lives in their own eyes—have no place in discussions of equality. Many differences in overall regret—many occasions that people have properly to regret what they have not done—flow from handicaps or bad luck or weakness of will or sudden changes, too late for anything but regret, in people's perceptions of what they really take to be valuable. But it is perhaps the final evil of a genuinely unequal distribution of resources that some people have reason for regret just in the fact that they have been cheated of the chance others have had to make something valuable of their lives. The ideas of fufillment and of reason for regret are competent to express this final argument against inequality only because they are ideas that reflect, in their assumptions, what inequality independently is.

I cannot, of course, prove that no one will invent a test or metric for overall success that will be both pertinent to equality and independent of prior assumptions about equality in distribution. For that reason I considered a fairly wide variety of suggested tests of this sort, hoping to show why I think it unlikely that one can be found. Certainly nothing that I am aware of in the present literature will do. But now suppose someone defending equality of overall success concedes that no such distribution-independent test can be found. I have been assuming that he must then concede that equality of overall success is useless as a distinct political goal because, insofar as it recommends changes from the independent distribution it assumes to be fair, it must recommend distributions it condemns as unfair. But is this too quick a conclusion?

actual regret, of the latter group, before deciding that the former group may not have more than that amount? Can this be done without regress? In any case, the total of amounts fixed as maxima for particular people, in this way, would undoubtedly exceed the total available for distribution. I shall not pursue these complexities further.

Suppose he argues that we must distinguish between the measure of and the means of achieving equality of overall success. He might suggest, for example, that a fair distribution for purposes of computing some person's present overall success is an equal distribution of resources. We compute Jack's and Jill's overall success by asking how far each regards his or her life as less successful than the best life he or she could have if resources were shared equally in society. If Jack's overall success, so measured, is greater than Jill's, we transfer resources from Jack to Jill so far as we can thus reduce the difference. It is true that Jack and Jill will not then have equal resources. Jill will have more resources than Jack. But they will be (more) equal in overall success as measured in the proper, reasonable-regret-oriented, way. There is no contradiction in using the idea of equal resources internally, within the metric for determining overall success, and then actually distributing so as to achieve equality of overall success rather than equality of resources.

But this reply misses the point. The reasonable regret metric for determining overall success makes assumptions about what distribution is fair, about the distribution to which people are *entitled*. If that metric assumes that a fair distribution is an equal distribution of resources, and Jill is then given more than an equal share, she is given more than the theoretical argument supposedly justifying the transfer says is her fair share. Of course Jill might not complain about having more than the share of resources to which she is, by hypothesis, entitled. But Jack will complain about having less than the justification assumes he is entitled to have, and the only way to give Jill more than that share is to give Jack (or someone else) less.

V. Equality of Enjoyment

I now propose to discuss the second group of conceptions of equality of welfare that I distinguished at the outset, which take equality of welfare to consist in equal amounts or degrees of a conscious state. I shall simplify this discussion, as I said there, by taking the concept of enjoyment to stand for a particularly broad version of the conscious state or states in question, and I mean to use that concept in the wide sense I described. Fortunately, this discussion can be simplified in another way as well, for much of the argument I used in considering unrestricted versions of equality of success apply to unrestricted versions of equality of enjoyment as well.

People gain enjoyment, as I said, from the satisfaction of their political and impersonal preferences as well as directly and from their personal preferences, and they suffer dissatisfaction when these political and impersonal preferences are defeated. But the same considerations that argue for a restricted form of equality of success, which does not count success or failure in achieving these preferences in the calculations that theory recommends, argue for similar restrictions in equality of enjoyment. So I shall assume that equality of enjoyment, as a theory of equality in distribution, holds that resources should be distributed, so far as possible, so that people are equal in the enjoyment they take directly and from their beliefs that their personal preferences are achieved.

My first argument against this restricted version of equality of enjoyment is also modeled on the argument I used against equality of relative success. The main appeal of a restricted form of equality of enjoyment lies in the claim that it makes people equal in what they all value equally and fundamentally so far as their personal position is concerned. But that appeal cannot be sustained, because in fact people differ in the importance each attaches to enjoyment even in the widest sense that leaves that term a description of conscious states. When they are made equal in that one respect, they become unequal in other respects many value much more.

For almost everyone, pain or dissatisfaction is an evil and makes life less desirable and valuable. For almost everyone pleasure or enjoyment of some other form is of value, and contributes to the desirability of life. Conscious states of some such form, positive and negative, figure as components of everyone's conception of the good life. But *only* as components, because almost no one pursues only enjoyment or will make any large sacrifice of something else he values to avoid a small amount of pain. And different people give even these conscious states very different weight. Two scholars, for example, may both value creative work, but one may be willing to give up more, by way of social pleasure or the enjoyments of reputation or the satisfaction that comes from completing a piece of research well done, to do work that is in fact more original.

Someone might now object that the first scholar does not value enjoyment less but rather finds it in a different source–not in the delights of society or the glow of fame but in the deeper satisfactions of the pursuit of genuine discovery. But this is plainly not necessarily or even usually so. Some of the most ambitious scholars (and artists and statesmen and athletes) set off in a direction that they predict will bring them only failure, and they know that they will find no delight or satisfaction just in the fact that they are aiming high, but only misery in how far they have fallen short. They may truly say (in the spirit of the poet whose views I described earlier) that they wish that some goal or project had not occurred to them or fallen in their path, or that they did not have the talents that made it necessary for them to pursue it, because then their lives would have been more satisfactory, more enjoyable all things considered. It perverts their report, misunderstands their complex situation, to say that they have actually found more enjoyment in the life they have led. For it is exactly their point that they have led that life in spite of, rather than for, the quality of the conscious life it has brought them.

Now of course not many people are dedicated to some ambition in that particularly strenuous way. But most of us, I think, are dedicated to something whose value to us is not exhausted or captured in the enjoyment its realization will bring, and some are dedicated to more things in that way, or more strongly dedicated, than others. Even when we do enjoy what we have or have done, we often enjoy it because we think it valuable, not vice versa. And we sometimes choose, in the same manner, though not to the same dramatic degree as the most ambitious scholar, a life that we believe will bring less enjoyment because it is in other ways a better life to lead. This is evident, I think, in a psychological fact that in some ways illustrates a different point, but is nevertheless relevant. Suppose you had a genuine choice (which, once made, you would forget) between a life in which you in fact achieved some goal important to you, though you did not realize that you had, and a different life in which you falsely believed that you had achieved that goal and therefore had the enjoyment or satisfaction flowing from that belief.[3] If you make the former choice, as many would, then you rank enjoyment, however described, as less important than something else.

Suppose someone now says, however, that equality of enjoyment is an attractive political goal, not because people all do value that state equally and fundamentally, when they decide what is important for themselves, but rather because they ought to do so. He says, not that the ambitious scholar I described really values enjoyment, but rather that he is mistaken, perhaps even irrational, because he does not. Someone making this objection, that is, abandons what I said is the immediate appeal of any conception of equality of welfare, which is that it claims to make people equal in what they value equally and fundamentally. He argues that the appeal of that political ideal is rather that it makes people equal in what they ought to value equally and fundamentally.

We have, I think, two answers. First, he is wrong in his view about what people ought to value. He is wrong in supposing either that the most valuable life is a life of maximum enjoyment, no matter how generously that conscious state is described, or that everyone ought to hold that view of what is the best life. Second, even if that theory about what people ought to value is more plausible than I think, even if it is in fact true, a political theory of equality based on that conception of the good life is an unattractive theory for a society in which many if not most people reject that conception, and some reject it as alien to their most profound beliefs about the goodness of their own lives.

We may, moreover, find a second argument against the restricted form of equality of enjoyment in the arguments we considered against equality of overall success. Though I have emphasized the error in supposing that ambitious people all take enjoyment in their strenuous lives, or pursue those lives for the sake of the enjoyment they will bring, it is nevertheless true that people of ambition often find dissatisfaction in the failure of their grand aims, and in their regret that they do not have the additional resources of talent and means that

3. See Bernard Williams, "Egoism and Altruism," in *Problems of the Self* (Cambridge: Cambridge University Press, 1973), p. 262.

would make success more likely, whether or not they hold political theories that suppose that they are entitled to more than they have. Though this was not part of my story of Jack and Jill, we might vary that story to suppose, for example, that Jack found keen disappointment and day-to-day dissatisfaction in the fact that he did not have solving-the-riddle talents and means. But it would seem equally wrong to transfer resources to him on account of that greater dissatisfaction as on account of his lower success ratio measured in that way. No one, I think, would want to aim at counting more than the dissatisfaction he found in reasonable regret. But if the arguments I offered earlier are sound, introducing the idea of reasonable regret for the sake of that limitation would introduce a different and inconsistent theory of distribution into the very statement of and justification for equality of enjoyment.

VI. Objective Theories of Welfare

The conceptions of equality of welfare thus far considered are all subjective in the following sense. They may each be enforced without asking whether a person's own consistent and informed evaluation of how far he meets the deployed standard of welfare is correct. Of course the arguments in favor of choosing one or another conception of equality of welfare may assume that people are wrong in what they take to be important, or even in what they would take to be important if fully informed of the pertinent facts. We considered, for example, the argument in favor of equality of enjoyment that people ought to value enjoyment as fundamentally important to their lives, in spite of the fact that many do not. But even if the conscious state conception is defended in that way, it may be applied without any evaluation of the enjoyment in question. It directs officials to produce the distribution such that each person takes equal enjoyment in the life he leads, without asking whether people are right to take enjoyment in what they then would.

Equality of overall success in the form we considered it is also subjective in that way. It aims to make people equal in (as we should now say) the amount or degree by which each person could reasonably regret that he was not leading a life he would deem to be a life of greater value. That judgment is in certain ways, it is true, nonsubjective. It imposes constraints on reasonable regret that the person in question might himself reject, for example. If that person's assessment of what gives value to life changes over the course of his life, the judgment requires some amalgamation or selection among his different judgments. But the judgment does not allow the computation of someone's reasonable regret to be based on assessments of value in life that are wholly foreign to him, that he would reject even if fully informed of the ordinary facts.

I should now mention a version of equality of overall success that is more objective in just that way, however. Someone might propose that people be made equal in the amount of regret they should have about their present lives. On this revised test officials would have to ask whether someone who in fact does not value friendship, for example, and believes his life good though it is solitary and without love, and believes this in spite of the fact that he is aware of the comforts and joy that others find in friendship, is wrong. If so, then resources might be transferred to him, either directly or through special education for him about the values of friendship, on the ground that his overall success is low even though he would count it high, at least before the special education takes hold.

Now we may well object that officials have no business relying on their own judgments about what gives value to life in redistributing wealth. We might believe that such a scheme for redistribution invades autonomy, or is in some other way foreign to the correct liberal principles. But we need not consider these objections, because this more objective version of equality of overall success meets the same argument we used against the more subjective version. Any pertinent test of what someone should regret about the life he is in fact leading, even on the best rather than his own theory about what gives value to life, must rely on assumptions about what resources an individual is entitled to have at his disposal in leading any life at all. So the objective version, like the subjective version, must assume an independent theory of fair distribution, and has no more power to justify giving some people more and others less than what they are entitled to have under that theory. Both versions are self-defeating insofar as they recommend any changes in a distribution independently, under some other theory of distribution, shown to be fair.

I should just mention, though very briefly, another putative conception of equality of welfare that might also be considered an objective conception. This supposes that a person's welfare consists in the resources available to him, broadly conceived, so as to include physical and mental competence, education and opportunities as well as material resources. Or, on some versions, more narrowly conceived so as to include only those that are in fact, whatever people think, most important. It holds that two people occupy the same welfare level if they are both healthy, mentally sound, well-educated, and equally wealthy even though one is for some reason malcontent and even though one makes much less of these resources than the other. This is an objective theory in the sense that it refuses to accept a person's own judgment about his welfare, but rather insists that his welfare is established by at least certain kinds of basic resources at his command.

Equality of welfare, so interpreted, requires only that people be equal in the designated resources. This version of equality of welfare is therefore not different from equality of resources or at least equality in some resources. It is rather a statement of equality of resources in the (misleading) language of welfare. The abstract statement of equality of resources, of course, as I said, leaves open the question of what counts as a resource and how equality of resources is to be measured. These are the complex questions left for Part 2 of this essay. But there is no reason to think that these questions will be easier to answer if we tack on to the ideal of equality of resources the rider that if people are equal in resources, on the correct conception of that ideal they will also be equal in some objective concept of welfare as well.

VII. An Ecumenical Suggestion

I must now consider what might seem a wise and ecumenical suggestion. Perhaps an attractive conception of equality of welfare can be found, not exclusively in one or another of the different conceptions we have now inspected and dismissed, but in some judicious and complex mix of these. In that case the strategy I followed in the last three sections might be the misleading and fallacious strategy of divide and conquer, rejecting each conception of equality of welfare by supposing that unless that conception tells the whole story it may be wholly ignored. Perhaps the ideal of equality of welfare may be considered fairly only by treating the different unrestricted and restricted versions of equality of success and equality of enjoyment each as strands to be considered in a complex package rather than as isolated theories.

It would be foolish to say, in advance, that no new conception of equality of welfare could be described that would make that ideal attractive. We must wait to see what new conceptions are presented. But it is perhaps not foolish to suppose that no successful conception could be formed using the conceptions we considered as components in some larger package. In any case my arguments were intended to reduce confidence in that project. I did not argue simply that no one of the versions I discussed is satisfactory on its own, or that each leads to unappealing consequences if unchecked by some other. If my arguments had been of that character, they would indeed invite the suggestion that these conceptions might be combined so as to supplement or check the shortcomings of each alone. But I meant to support a more radical criticism: that we have no reason to accept any of these versions of equality of welfare as a theory of distributional equality, even pro tanto.

It is of course desirable, in some sense at least, that people's overall success be improved, though philosophers and politicians might disagree whether the subjective or objective versions of that goal should be controlling when the two conflict. But, for reasons explained, neither version can provide other than an idle or self-defeating principle of equality in distribution. Nor, for the same reasons, can either figure as useful components in some complex package of conceptions of equality of welfare. Insofar as equality of overall success figured, even as one component among many, it would figure because some independent test of fair distribution was assumed, and it could not recommend, nor could the package overall, any deviation from that independent test.

The other conceptions of equality of welfare we considered we rejected for different reasons. We found no reason to support the idea that a community should accept the goal of making people more equal in any one of these different ways even when it could do so without damage to any of the others. If that is so then it is unlikely that it should accept the goal of making people more equal in some way that is a composite or compromise among these different ways. Combinations and trade-offs are appropriate when a set of competing goals or principles, each of which has independent appeal, cannot all be satisfied at once. They are not appropriate when no goal or principle has been shown to have independent appeal, at least as a theory of equality, at all.

VIII. Expensive Tastes

I said at the outset that equality of welfare, even as stated simply in the abstract, without specifying any of the conceptions we later distinguished, seems to produce initially troubling counter-examples. The most prominent of these is the problem of expensive tastes (a phrase I shall use, most often, to include expensive ambitions as well). Equality of welfare seems to recommend that those with champagne tastes, who need more income simply to achieve the same level of welfare as those with less expensive tastes, should have more income on that account. But this seems counter-intuitive, and I said that someone generally attracted to the ideal would nevertheless wish to limit or qualify it so that his theory did not have that consequence. I want to return to that suggestion now, not because the problem of expensive tastes is of practical importance in politics, but for two different reasons. First, many readers initially attracted to some conception of equality of welfare may suspect that the arguments I directed against their favorite conception, in the last several sections, would have less force if a limitation or qualification suitable to exclude the expensive tastes consequence had been built into my description of that conception. I think that this suspicion, if indeed it exists, is mistaken, but it is nevertheless worthwhile to consider, for this reason alone, whether such a limitation is in fact possible. Second, there will be readers who are left unpersuaded by my earlier arguments, but would nevertheless abandon their favorite conception of equality of welfare if they believed that it could not in fact be qualified so as to avoid that consequence.

We must be careful to distinguish, when we consider possible qualifications of any such conception, the compromise of a principle from its contradiction. A compromise reflects the weight of some independent and competing principle; a contradiction is a qualification that reflects instead the denial of the original principle itself. The question I want to press is this: Can the principle of equality of welfare be compromised (under any interpretation of what equality of welfare is) in such a way as to block the initially counter-intuitive results of that principle, like the proposition that people with champagne tastes should have more resources? Or is any qualification capable of barring those results rather a contradiction that concedes the final irrelevance of the principle?

Imagine that a particular society has managed to achieve equality of welfare in some chosen conception of that ideal. Suppose also that it has achieved this through a distribution that in fact (perhaps just by coincidence) gives everyone equal wealth. Now suppose that someone (Louis) sets out deliberately to cultivate some taste or ambition he does not now have, but which will be expensive in the sense that once it has been cultivated he will not have as much welfare on the chosen conception as he had before unless he acquires more wealth. These new tastes may be tastes in food and drink: Arrow's well-known example of tastes for plovers' eggs and pre-phylloxera claret.[4] Or they may (more plausibly) be tastes for sports, such as skiing, from which one derives pleasure only after acquiring some skill. Or, in the same vein, for opera. Or for a life dedicated to creative art or exploring or politics. Can Louis be denied extra wealth, taken from those who acquire less expensive tastes (or simply keep those they already have), without contradicting the ideal of equality of welfare that his community has embraced?

Let us first consider how we might explain what Louis has done. No doubt people often put themselves in the way of new tastes carelessly, or on whim, without considering whether they will really be

4. Kenneth J. Arrow, "Some Ordinalist-Utilitarian Notes on Rawls's Theory of Justice," *The Journal of Philosophy* 70, no. 9 (10 May 1973): 254.

better off if they acquire these tastes, or even perversely, knowing that they will be worse off. Even when they think they would be better off, they might be mistaken. But I want to suppose that Louis is not only acting deliberately rather than inadvertently, but is also acting on the basis of the kind of judgment I said people often make when they form and change their preferences. He is trying to make his life a better life in some way. This does not make his claim for extra resources any more appealing or less counter-intuitive, I think. On the contrary, the fact that he is acting so deliberately in his own interests seems to make his claim, if anything, less appealing than the claim of someone who tries an expensive experience on a whim, for the pleasure of the moment, and then finds that he is hooked.

Louis will, of course, have his own ideas of what makes a life better, of where his own essential well-being lies. If his society has chosen one of the discrete conceptions of welfare, such as enjoyment or relative success, however, as the welfare in which people should be equal, then Louis cannot think that his own well-being consists in the maximum amount of welfare in that conception. If he did, his behavior would make no sense. This means that one possibly appealing description of what he has done must be wrong. Many people, first hearing this story, might assume that Louis cultivates expensive tastes in order to steal a march on others, so that it would "reward" improper efforts if he were to receive more income. But if stealing a march means acquiring more welfare than others in the chosen conception, then this is impossible. Of course someone might pretend to like plovers' eggs, though he hates them in fact, in order to gain more income, and then spend that income secretly buying more hens' eggs and thus more enjoyment than others can afford. But the problem of expensive tastes is not the problem of fraud—that problem must be handled separately in any society based on equality of welfare because someone could, after all, pretend to be crippled. If Louis sets out to acquire a taste for plovers' eggs so that, if successful, he will in fact have less welfare on the chosen conception if he does not have them, then he cannot purpose to gain some advantage in that form of welfare over others by this decision. He may of course think that he will in the end get more welfare in that conception from a dollar's worth of plovers' eggs than hens' eggs, costly though the former are. In that case he knows that his income will be reduced if he is successful. Or he may think that he will not gain more welfare per dollar by cultivating a taste for plovers' eggs, but rather less. In that case he knows that his welfare (as always, on the chosen conception) will decline overall (though not by much in a very large community) because the total welfare that can then be produced (of which in the end he can expect only *1/nth*) will decrease. It would be absurd to think that he sets out to reduce his own welfare in order to have a larger income, either absolutely or relative to others. After all, though he may have a larger income than others, they are, by hypothesis, no worse off in the chosen conception of welfare than he is, and he is at least by some degree worse off than he would otherwise have been.

Louis does, as I said, suppose that if he cultivates his new taste his life will be better. But this is because he does not accept that the value of his life is measured just by the welfare in which his society has, for some reason, undertaken to make people equal. It is hard to see how this can justify either the suggestion that he has acted improperly or the decision not to give him more resources but rather to leave him unequal to others in the chosen conception. The choice of that conception was society's choice, not his, and society chose that people be equal in it not the other conception that Louis values more. After all there is no reason to think that people were equal in welfare on Louis's conception even before he developed his new taste, and he may still have less than others have of that even if he is brought back to equality in the chosen conception.

Louis thinks, as I said, that his life would be a more successful life overall—would provide less reason for regret—if he had the expensive taste or ambition even at the small cost in welfare in the chosen conception he would lose if society reestablished equality in that conception for him. Indeed he might think that his life would be more successful overall even if society did not reestablish equality for him. (People develop expensive tastes even in our own society, when they must bear the increased costs themselves.) Suppose the chosen conception is enjoyment. If Louis develops a taste for plovers' eggs, he must believe that a life of satisfying expensive tastes is a better life overall in spite of the fact that it will provide less enjoyment, and might believe it better even if it would provide much less enjoyment.

These may, in fact, be plausible beliefs. Or at least they may be plausible if we substitute, for the contrived examples of plovers' eggs, the sorts of expensive tastes that people do seem to cultivate deliberately and in their own interests, such as a taste for sports that follows from developing skill or a desire for practical power that follows taking an interest in the public weal. It is plausible to suppose that beliefs of that sort figure even in the best accounts of why people in our own economy develop the less admirable expensive tastes—champagne tastes—that figure in the usual examples. For if someone like Louis wishes to lead the life of people in *New York* magazine ads, this must be because he supposes that a life in which rare and costly goods are savored is a life better because it knows a greater variety of pleasures, or more sophisticated pleasures, or, indeed, simply pleasures that others do not know, in spite of containing less pleasure overall.

This explanation of Louis' behavior challenges the importance of the distinction we have thus far been assuming between expensive tastes that are deliberately cultivated and other aspects of personality or person, such as native desires or socially imposed tastes, that affect people's welfare. For the explanation suggests that such tastes are often cultivated in response to beliefs—beliefs about what sort of life is overall more successful—and such beliefs are not themselves cultivated or chosen. Not, that is, in any sense that provides a reason for ignoring differences in welfare caused by these beliefs in a community otherwise committed to evening out differences in welfare. I do not mean that beliefs are afflictions, like blindness, that people find that they have and are stuck with. People reason about their theories of what gives value to life in something of the same way in which they reason about other sorts of beliefs. But they do not choose that a life of service to others, for example, or a life of creative art or scholarship, or a life of exquisite flavors, be the most valuable sort of life for them to lead, and therefore do not choose that they shall believe that it is. We may still distinguish between the voluntary decision someone makes to become a person with certain tastes, or to lead the sort of life likely to have that consequence, and his discovery of tastes and ambitions that he just has. But the distinction is less important than is sometimes thought, because that decision is rarely if ever voluntary all the way down.

If Louis' society aimed to make people equal, not in one of the discrete conceptions of welfare we have thus far been assuming in his story, such as enjoyment or relative success, but in subjective overall success, then we would need a somewhat different account of why he would develop expensive tastes, and of whether it would be fair to deny him extra resources. I argued earlier that any attractive version of equality of overall success must provide a place for the idea of reasonable regret, and that this idea in turn presupposes some independent non-welfare theory defining a fair distribution of resources. If this is right, then no one could claim extra resources for expensive tastes in a community ostensibly governed by equality of overall success. If his share of resources is fair before he cultivates his new taste, his share remains fair after he has done so. But since I want to offer independent arguments in this section, I shall assume that my earlier arguments are unsound, and that an attractive subjective version of equality of overall success can be developed that is not self-defeating in that way.

But then, since the chosen conception is now overall success, we can no longer say that Louis acts as he does because he believes his life would be more successful overall though less successful on the chosen conception. Suppose that before Louis conceived his expensive taste he was satisfied that his life was roughly as successful overall as everyone else's. He then came to believe that his life would be more valuable if he cultivated some expensive hobby, for example. We must ask what he now thinks about the value of the life he had before he formed that belief. He may think that, though his earlier life was just as good as he thought it was, and would remain so if he could not pursue his new hobby, it would be much better if he could. In that case the problem of expensive tastes does not arise. For Louis is claiming additional resources in order to have more welfare than others on the chosen conception, and he does not have even a prima facie claim to that. But he may instead have changed his beliefs about how valuable his life was. He may have read more widely, or reflected more deeply, and come to the conclusion that his former life, for all its former appeal to him, was in fact a worthless and insipid life. He wants to cultivate new and more challenging tastes to repair the defects in his life, as he now understands them. He asks only the resources necessary to make his life as valuable, in his eyes after they have been opened, as other people find their lives. How can a society committed to equality in this respect deny him these resources? It cannot say that he was wrong to continue to reflect on how best to live. An unexamined life is for that very reason a poor life. If Louis had reached his present opinions about value in life before the initial distribution, he would have received then the resources he now seeks. Why should he be refused them now, and be condemned to a life he finds less valuable than everyone else finds theirs?[5]

We might summarize the position we have reached in this way. If the chosen conception is one of the discrete conceptions we considered, other than overall success, then Louis is attempting to improve his welfare on some other conception he values more, while retaining equality in the chosen conception. But if the chosen conception is what really matters for equality, and if in any case others may already have more welfare in the conception Louis prefers, what ground does society have for now refusing him equality in the chosen conception? If the chosen conception is overall success (which is assumed, arguendo, not to be self-defeating) then if a claim for extra resources arises at all, it arises because Louis now believes that the earlier distribution was based on a mistake. He asks no special advantage, but only that society reach the distribution it would have reached if he had been able to see more clearly then. What ground could society have for refusing him that?

One ground perhaps suggests itself, which is the ordinary utilitarian principle that average welfare in society (which we should under-

5. If the chosen conception is some objective version of equality of success, rather than the subjective version discussed in this paragraph, the situation is different still. If the change in Louis' tastes has the consequences that his life is now objectively more successful, then he should have fewer rather than more resources in consequence. If (because Louis' convictions are mistaken) the change makes his life objectively worse, then the claim that someone might make on his behalf, for still further resources for reeducation, seems especially strong. But I am assuming that the objective version of equality of overall success has little appeal for a liberal society, so I shall not pursue this line.

stand to mean welfare in the chosen conception) should be as high as possible. If society "rewards" people who develop expensive tastes by giving them extra resources with which to satisfy these tastes, then people will not be discouraged from doing so. But expensive tastes (by definition) decrease the total welfare that can be produced from a given stock of resources. So the independent principle of utility justifies a compromise with the principle of equality of welfare by recommending that people not be brought to parity of welfare if they develop expensive tastes, in order to discourage them from doing so. If the chosen conception is a discrete conception, this means that people are to be discouraged, for the sake of average utility, from bringing it about that they will need more resources to achieve the same welfare, even though they may think that their lives would be more successful if they did bring that about. If the chosen conception is overall success, judged subjectively, then people are to be discouraged from reexamining their lives in a way that might leave them dissatisfied with the value of the lives they have.

But in fact the principle of utility does not explain what needs explaining here. It can at best explain why compensating those who develop expensive tastes is inefficient. It cannot explain why the ideal of equality does not recommend doing so. It is, after all, a familiar idea in political theory that a just society will make some compromise between efficiency and distribution. It will sometimes tolerate less than perfect equality in order to improve average utility. But the compromise intuitively demanded by the problem of expensive tastes is not such a compromise between efficiency and equality. It is rather a compromise within the idea of equality. Our difficulty is not that, though we believe that equality requires us to pay Louis more because he has forced himself to like champagne, we must deny him equality in order to protect the overall stock of utility. Expensive tastes are embarrassing for the theory that equality means equality of welfare precisely because we believe that equality, considered in itself and apart from questions of efficiency, condemns rather than recommends compensating for deliberately cultivated expensive tastes.

I should also point out, parenthetically, that it is far from plain that the utilitarian principle, by itself, can even provide an explanation of what it does purport to explain, which is why a society that does wish to compromise equality for efficiency would select expensive tastes as the point of sacrifice for equality. Refusing to compensate people who develop expensive tastes will protest average utility only if it succeeds in discouraging at least some people from developing such tastes who would otherwise do so. It is impossible to predict how much of such experimentation would take place in a society dedicated to equality of welfare even without this kind of deterrence, or how effective the deterrence would be. (After all, people develop expensive tastes even in our own society when they do not receive extra resources when they do.) It is also impossible to predict the long-term consequences for utility under any particular assumptions about the success of the deterrence. Any society bent on using noncompensation as a deterrent must set a fairly articulate policy that stipulates reasonably clearly when people whose tastes and ambitions change will be compensated and when they will not be. How would the policy distinguish, for example, between tastes that are deliberately cultivated and those that simply steal up on people? What level of expense—what level of efficiency in producing enjoyment, for example, per dollar cost—would be stipulated as making a taste expensive rather than inexpensive? Beer may very well be less expensive, in this sense, than champagne, but it is also more expensive than water. Suppose the community responds to these difficulties by refusing to compensate for new tastes if people take any positive steps to acquire them or even act in a way that they should know makes their acquisition more likely, whenever these tastes are any more expensive than the tastes, if any, that they replace. If this policy succeeds in discouraging experimentation in tastes to any marked degree, then it might well end, for all we know, in a dull, conformist, unimaginative, and otherwise unattractive community, and a community with less long-term utility as well. There are many reasons for predicting that latter consequence, but I shall mention only the two most obvious. First, some tastes that are expensive when taken up only by a few people become inexpensive—produce more utility per dollar than present tastes—when they become very popular through the example of those few. Second, a society that does become dull and conformist is a society in which no one takes much pleasure in anything, or cares very deeply about achieving the goals that have been taken mechanically from others rather than developed for himself. It is of course not plain that this policy of non-compensation tacked to a general principle of compensation for tastes acquired in a less voluntary way would have these consequences. But that is because no hypothesis about what levels of utility would be achieved by such a society, so different from our own, is worth much, which hardly recommends this explanation of why an equality of welfare society that is also utilitarian would refuse compensation.

So the supposed utilitarian justification of our intuitive conviction, that equality does not require that those who deliberately cultivate expensive tastes have equal welfare after they have done so, fails on two grounds. We still lack a justification for that conviction. But suppose someone now argues in the following way. It is true that people do not choose their beliefs about what would make their lives overall more successful. But they do choose whether and how far to act on these beliefs. Louis knows, or at least ought to know, that if he cultivates some expensive taste in a society dedicated to equality of enjoyment, for example, and is compensated, then that will decrease the enjoyment available for others. If, knowing this, he chooses the more expensive life then he does not *deserve* compensation. He is no longer a member of the company of those who deserve equal enjoyment in their lives.

Louis has a choice. He may choose to keep the presently equal resources I said he had, and settle for a life with the enjoyment he now has but without the tastes or ambitions he proposes to cultivate. Or he may keep his present resources and settle for a life that *he* deems more successful overall than his present life, but one that contains less enjoyment. It is quite unfair that he should have a third choice, that he should be able, at the expense of others, to lead a life that is more expensive than theirs at no sacrifice of enjoyment to himself just because he would, quite naturally, consider *that* life a more successful life overall than either of the other two. The reason why Louis does not deserve compensation is not that the more expensive life he might choose is necessarily a worse life. He might be right in thinking that enjoyment is not all that matters, and that a life poorer in enjoyment may be, just from the personal standpoint, a more successful life overall. We say only that the first two choices are rightly his, but that the third is not.

I myself find this argument both powerful and appealing. It is also an important argument for the following reason. The objection to allowing Louis the third choice described is most naturally put this way. Louis should be free (at least within the limits allowed by a defensible form of paternalism) to make the best sort of life he can with his fair share of social resources. But he should not be free to trespass on the fair shares of others, because that would be unfair to them. But of course once the point is put that way it cannot stand simply as an argument for a compromise to equality of welfare tailored to the problem of expensive tastes. For the idea of fair shares cannot then mean simply shares that give people equal welfare on the chosen conception, because that is exactly the conception to which Louis appeals in asking for extra resources. If fair shares are shares fixed independently of that conception, however, then any compromise using the idea of fair shares becomes a contradiction.

Can the idea of fair shares be defined for this purpose in some way that does not make the shares that produce equal welfare in the chosen conception automatically fair shares, but nevertheless uses that conception in some way that avoids contradicting it? Suppose someone's fair share is taken to be the share that produces equal welfare in that conception, or would produce it if the person in question had not deliberately cultivated an expensive taste. This will not help, as we saw, if the chosen conception is overall success, and Louis believes that the life he would lead if he did not cultivate new tastes would be a life of less overall success than others believe their lives to be. Even if the chosen conception is one of the discrete conceptions, such as enjoyment, defining fair shares in this way will not help. The argument I said I found powerful uses the idea of fair shares not simply to describe the limitation on equality of welfare it recommends, but also to justify that limitation. It proposes to explain why, in spite of the various objections I made earlier in this section, independent and noncontradictory considerations of fairness justify a compromise of equality of welfare. But if the definition of fair shares just assumes that the compromise in question is for some unspecified reason fair, then the appeal to fair shares can itself provide no justification that is not immediately circular. If the idea of fair shares is to do any work, then it must appeal to some independent account of fairness in distribution, and any independent account contradicts the conception to which it is attached, as I said, because it occupies all the space that conception claims for itself. I might add that I think that the most plausible independent account, which I myself had in mind when I said that the argument against Louis' third choice was powerful, is some conception of equality of resources (though of course there are others available such as, for example, some principle that argues that resources are fairly distributed when those with more merit have more).

Perhaps ingenuity could produce some explanation or interpretation of the argument in question—that Louis does not deserve more resources just because he has chosen a more expensive life—which does not use this idea of fair shares or any similar ideas. But any such account would, I suspect, fall before the following further example. Imagine now a society newly dedicated to equality of enjoyment in which, when resources are redistributed in order to achieve equality of enjoyment, Jude has far less money than anyone else because his wants are so simple and so inexpensively satisfied. But one day (perhaps after reading Hemingway) he decides that his life, for all its richness in enjoyment, is a life of less overall success than it might be, and proposes to cultivate a new taste for some challenging sport, such as bullfighting. Suppose that after he does so he finds himself seriously frustrated by his lack of funds with which, for example, to travel to Spain, and asks for more funds through a further redistribution after which he would still, as things fall out, have less than anyone else. Do we now have any grounds for saying that he is undeserving of the increase, when we know that if it is denied he will have both less funds and less enjoyment than anyone else? I doubt anyone will want to say this. But if so then we cannot say that the reason Louis is undeserving of an increase is simply that the taste he has cultivated is expensive. Jude's new taste may be just as expensive. The difference is that Louis asks that more than an equal share of social resources be put at the disposal of his life while Jude asks only that something closer to an equal share be put at the disposal of his. We need the idea of fair shares (in this particular case the idea of an equal share of resources) in order to express the force of this difference.

Suppose that if Jude is given more funds and can travel to Spain he will have not merely as much but *more* welfare than anyone else, in whatever conception has been chosen, including overall success, though he will still have much less money than anyone else as well. Does equality now require that he be denied the additional money? If not, then Jude's case makes an even stronger point. Not only may Jude reestablish equality of welfare in spite of the expensive taste he has deliberately cultivated. He may even succeed, by developing such a taste, in having more welfare than others. In both cases it is the idea of equality of resources that is doing the work.

I hope the moral of this long section is clear. If someone begins anxious to defend some version or conception of equality of welfare, but also wishes to resist the consequence that those who develop expensive tastes should have more, he will come, in the end, to a very different theory of equality. He will find that he must presuppose some other theory that makes his conception of equality of welfare either idle or self-defeating. That is, of course, exactly the conclusion that we reached in studying certain of these conceptions separately. It remains to consider, as I propose to do in the next section, whether there are strong reasons for nevertheless trying to find some small room for equality of welfare within a general and different theory of equality.

IX. Handicaps

I conceded, at the beginning, the immediate appeal of the idea that genuine equality is equality of welfare. One aspect of that immediate appeal may easily have survived the various doubts I have raised, which is the apparent power of equality of welfare to explain why people with physical or mental handicaps (or who otherwise have special needs) should have extra resources. Surely (it might still be said) this is because they are able to achieve less of something that falls within the general ambit of "welfare" than others are on the same share of resources. Perhaps we care about the handicapped because they are able to achieve less enjoyment or relative or overall success, or perhaps it is some discrete combination of these, or all of them. But some tug towards equality of welfare under some interpretation must be part of our intuitions about the handicapped. If so, then this fact might be thought to show that any final theory of equality must provide at least some space for equality of welfare, though perhaps only as a supplement to or qualification of another theory of equality, if only to capture the provisions we insist on making for those who are unfortunate in this particular way.

But it is far from clear that some welfare concept is needed to explain why the handicapped should sometimes have more material resources than the healthy. In the second part of this essay I shall describe a different approach to the problem of handicaps which does not rely on welfare comparisons but which might explain this equally as well. There is no reason to assume, in advance of considering this and other suggestions, that only a welfare-based theory of equality can provide the account that is necessary. In fact (and moreover) a welfare-based theory can provide only a less satisfactory account than might at first appear. The argument we are now considering is that equality of welfare deserves a place, at least, in any general theory of equality, because it so accurately captures our intuitions about how the handicapped should be treated in the name of equality. But is this true? It does seem plausible to say, on any conception of welfare, that people with severe handicaps are likely, as a class, to have less welfare than others. But this is of course true only statistically. In many cases those with handicaps have in consequence less income, and therefore do not have even equal material resources with others. And some people with appalling handicaps need extra income just to survive. But many people with serious handicaps have high levels of welfare on any conception—higher than many others who are not handicapped. That is true, for example, of Tiny Tim and Scrooge. Tim is happier than Scrooge, approves the way the world is going more, is more successful in his own eyes, and so forth.

The intuition I spoke of, however, that those with handicaps should have extra resources, is not limited to those among the handicapped who do in fact have less than average welfare on some conception. If Tim had as much money as Scrooge (when perhaps Tim's welfare would be greater than Scrooge's by an even larger margin) but Tim nevertheless did not have enough money to afford physiotherapy, many of us would think him entitled to extra resources for that purpose. Now of course we might believe this only because our intuitions have been schooled by the statistical fact. On this hypothesis, we feel that the handicapped as a group should have more because their welfare is as a group lower, and we then apply the general intuition to individual cases without checking to see whether the general rule holds. But I do not find that a persuasive account of why we feel as we do. If, when we know that someone handicapped is not particularly low in welfare, we still believe that he is entitled to extra resources in virtue of that handicap, then this is poorly explained by supposing that we have lost the power to discriminate.

So our beliefs about the handicapped are not in fact justified so accurately or powerfully by the idea of equality of welfare as to suggest that any general theory must on that account include some measure of that ideal. The lower-welfare explanation of these beliefs has further shortcomings as well. Suppose that the welfare (on any interpretation) of an entirely paralyzed but conscious person is vastly less than the welfare of anyone else in the community, that putting more and more money at his disposal would steadily increase his welfare but only by very small amounts, and that if he had at his disposal all the resources beyond those needed simply to keep the others alive he would still have vastly less welfare than they. Equality of welfare would recommend this radical transfer, that is, until the latter situation was reached. But it is not plain to me (or I think to others) that equality, considered just on its own, and without regard to the kinds of considerations that sometimes might be thought to override it, really does require or even recommend that radical transfer under these circumstances.

I do not claim (as this last observation recognizes) that any community that embraced equality of welfare in principle would then be committed to the radical transfer. Some other principle the community also accepted (for example the principle of utility) might recommend some compromise with equality here. But where should the line be drawn? It might, perhaps, be left to the practical politics of intuition to draw such a line. But then the victim of total paralysis might well receive nothing at all. The equality principle would in itself offer no reason for the community's accepting an initial utility loss to do him some good that would not also apply to doing him more good, at least in the circumstances I describe in which the marginal utility to him of further transfers does not much decline. The principle would offer almost no guidance to the community here, beyond a call for help equally strident over the whole range of possible transfers to this victim, a call too impractical to honor in full and too unstructured for principled compromise.

Now suppose different facts. There is an expensive piece of equipment that would enable a paraplegic to lead a much more normal life, and the community can afford that equipment at great but not crippling sacrifice to its other needs and projects. The community votes to levy a special tax to provide this machine for him. But he is an excellent and dedicated violinist, and he replies that he would rather have a superb Stradivarius which he could purchase with the same funds. Can the community properly refuse to honor that choice?[6] On any of the conceptions of welfare we might choose, the paraplegic's welfare might in fact be increased more by owning the violin than by having the machine. Even if he knew all the facts he would prefer to have it, it would bring him more enjoyment, make his life both relatively and overall more successful in his own eyes and objectively. The independent principle of utility would recommend the same choice, of course.

But these facts would be embarrassing to a scheme that was not committed generally to equality of welfare, and allowed that ideal only a limited place in order to handle the special problem of handicaps. For consider someone else, not handicapped, who has a low level of welfare on all the same conceptions. He takes little enjoyment from his life, counts it a failure, and so forth, just because, though he has the same amount of wealth as everyone else not handicapped, that is not enough to buy the Stradivarius he covets above all else. If the paraplegic is allowed to use his extra funds to buy the violin, that other person might properly complain. The paraplegic treats the transfer, not as the occasion to remove or mitigate his handicap, but simply as an opportunity to increase his welfare in other ways, and the other violin-lover would seem to have, in his low state of welfare, as much claim to do that as the paraplegic has. But if the community

6. See Scanlon's discussion of this problem in "Preference and Urgency," *The Journal of Philosophy* 72, no. 19 (6 November 1975): 659-661.

denies the handicapped person that use of his extra funds, and requires him to buy the machine instead, its position seems perverse. It grants extra funds to him just on the ground that this will increase his lower than average welfare, and yet denies him the right to increase his welfare, with those funds, as much as he can.

X. Welfarism

If I am right, in the various arguments I have made in this essay, then equality of welfare is not so coherent or attractive an ideal as it is often taken to be. We therefore have reason to consider with some care the alternative ideal of equality of resources. But it is worth stopping now to consider very briefly whether the arguments I have made against equality of welfare might be effective against other forms of welfarism and, in particular, how far they might be effective against utilitarianism. (I am using Amartya Sen's account of welfarism as the general theory that the justice of distributions must be defined exclusively by stipulating some function of individual welfare.)[7]

The different versions of equality of welfare that we have been studying are varieties of welfarism. Utilitarianism, which calls for some maximizing function over some conception of welfare, is another, or rather, another group. Two kinds of justification are in principle available for any form of welfarism. A welfarist theory can be defended on the teleological ground that the stipulated function of the stipulated conception of welfare is something good in itself that ought to be produced for its own sake. Or it can be defended as a particular conception of equality, as a particular theory about when people are being treated as equals. The distinction between these two types of grounds is reasonably clear, I think, in the case of utilitarianism. That theory can be supported in a direct teleological way: not only is pain bad in itself but pleasure (or some other conception of positive welfare) is good in itself, and the more there is of it the better. Or it can be supported as a conception of equality. It is then understood as the theory that people are treated as equals when and only when their pleasures and pains (or components of some other

7. A. K. Sen, "Utilitarianism and Welfarism," *The Journal of Philosophy* 76, no. 9 (September 1979): 463-489.

conception of welfare) are taken into account quantitatively only, each in that sense to count as one and only one. Of course this egalitarian version of utilitarianism cannot, as the teleological version can, purport to supply all of a plausible general political or moral theory. The egalitarian utilitarian would have to explain why it is not as good to aim at maximum average misery as maximum average happiness, for example, or why there is anything to regret in a natural disaster that kills thousands though it improves the situation of a few. But he might find this explanation either in a further political principle, which holds that those who aim at others' misery or failure do not show these others the concern to which human beings, at least, are entitled, or in a distinct morality of outcomes which holds that death or pain or some other kind of suffering is bad in itself, but which uses neither the same conception nor the same metric of welfare as his egalitarian utilitarianism deploys.

The arguments we considered against equality of welfare would seem, at least on a first look, equally effective against utilitarianism when it is understood in that second way, that is, as a conception of equality. Once again we should proceed by stating different interpretations of utilitarianism composed by taking different conceptions of welfare as the maximands for a given community. And once again it will seem implausible only to take gains and losses in enjoyment, for example, or in relative success, as the measure of when people are being treated as equals, because people value welfare in these particular conceptions differently. Nor will it be helpful to take gains and losses in overall success, interpreted either subjectively or objectively, as the measure, because, as we saw, these conceptions of welfare depend on already having accepted a different, independent test of when people are being treated as equals.

These various arguments are plainly beside the point, however, when utilitarianism is supported in the first way, that is, as the teleological theory that welfare on some conception is inherently good in itself. Against that argument my claim, that people cannot be treated as equals by making them equal in some dimension they value unequally, is irrelevant, because what is then in question is only whether welfare on that conception is good in itself. I might add that I think that the teleological ground of utilitarianism, which my arguments do not touch, is much less appealing than the egalitarian ground, which they do. It is the egalitarian ground, I think, rather than the teleological ground, that accounts for whatever appeal utilitarian arguments still have for modern politicians and lawyers.

The distinction between these two types of grounds for welfarist theories might seem less plausible when applied to forms of welfarism other than utilitarianism. But it is available at least in principle, I think, and we can construct a teleological defense of at least some conceptions of equality of welfare. Someone might say that it is simply a good thing when people have the same amount of enjoyment, for example, whether or not everyone agrees that enjoyment is fundamentally important in their own lives, or even, perhaps, whether they ought to agree that it is important in that way. The arguments I have offered do not reach equality of welfare conceived and defended in that way. They are aimed at equality of welfare taken to be a theory about treating people as equals. Equality of welfare, so conceived, is weaker than we might initially have thought. Is equality of resources stronger?

Derek Parfit has been outstandingly generous and helpful in commenting on drafts of this essay, and I have followed his advice at many points. I am also indebted to the Editors of *Philosophy & Public Affairs* for their acute comments.

Part 2 of this article will appear in the next issue.

[7]

RONALD DWORKIN

What is Equality? Part 2: Equality of Resources

I. The Auction

In Part 1 of this essay we considered the claims of equality of welfare as an interpretation of treating people as equals. In Part 2 we shall consider the competing claims of equality of resources. But we shall be occupied, for the most part, simply in defining a suitable conception of equality of resources, and not in defending it except as such definition provides a defense. I shall assume, for this purpose, that equality of resources is a matter of equality in whatever resources are owned privately by individuals. Equality of political power, including equality of power over publicly or commonly owned resources, is therefore treated as a different issue, reserved for discussion on another occasion. This distinction is, of course, arbitrary on any number of grounds. From the standpoint of any sophisticated economic theory, an individual's command over public resources forms part of his private resources. Someone who has power to influence public decisions about the quality of the air he or she breathes, for example, is richer than someone who does not. So an overall theory of equality must find a means of integrating private resources and political power.

Private ownership, moreover, is not a single, unique relationship between a person and a material resource, but an open-textured relationship many aspects of which must be fixed politically. So the question of what division of resources is an equal division must to some degree include the question of what powers someone who is assigned a resource thereby gains, and that in turn must include the further question of his right to veto whatever changes in those powers might be threatened through politics. In the present essay, however,

Philosophy & Public Affairs 10, no. 4
0048-3915/81/040283-63$3.15/1

I shall for the most part assume that the general dimensions of ownership are sufficiently well understood so that the question of what pattern of private ownership constitutes an equal division of private resources can be discussed independently of these complications.

I argue that an equal division of resources presupposes an economic market of some form, mainly as an analytical device but also, to a certain extent, as an actual political institution. That claim may seem sufficiently paradoxical to justify the following preliminary comments. The idea of a market for goods has figured in political and economic theory, since the eighteenth century, in two rather different ways. It has been celebrated, first, as a device for both defining and achieving certain community-wide goals variously described as prosperity, efficiency, and overall utility. It has been hailed, second, as a necessary condition of individual liberty, the condition under which free men and women may exercise individual initiative and choice so that their fates lie in their own hands. The market, that is, has been defended both through arguments of policy, appealing to the overall, community-wide gains it produces, and arguments of principle that appeal instead to some supposed right to liberty.

But the economic market, whether defended in either or both of these ways, has during this same period come to be regarded as the enemy of equality, largely because the forms of economic market systems developed and enforced in industrial countries have permitted and indeed encouraged vast inequality in property. Both political philosophers and ordinary citizens have therefore pictured equality as the antagonist or victim of the values of efficiency and liberty supposedly served by the market, so that wise and moderate politics consists in striking some balance or trade-off between equality and these other values, either by imposing constraints on the market as an economic environment, or by replacing it, in part or altogether, with a different economic system.

I shall try to suggest, on the contrary, that the idea of an economic market, as a device for setting prices for a vast variety of goods and services, must be at the center of any attractive theoretical development of equality of resources. The main point can be shown most quickly by constructing a reasonably simple exercise in equality of resources, deliberately artificial so as to abstract from problems we shall later have to face. Suppose a number of shipwreck survivors are washed up on a desert island which has abundant resources and no native population, and any likely rescue is many years away. These immigrants accept the principle that no one is antecedently entitled to any of these resources, but that they shall instead be divided equally among them. (They do not yet realize, let us say, that it might be wise to keep some resources as owned in common by any state they might create.) They also accept (at least provisionally) the following test of an equal division of resources, which I shall call the envy test. No division of resources is an equal division if, once the division is complete, any immigrant would prefer someone else's bundle of resources to his own bundle.[1]

Now suppose some one immigrant is elected to achieve the division according to that principle. It is unlikely that he can succeed simply by physically dividing the resources of the island into *n* identical bundles of resources. The number of each kind of the nondivisible resources, like milking cows, might not be an exact multiple of *n*, and even in the case of divisible resources, like arable land, some land would be better than others, and some better for one use than another. Suppose, however, that by a great deal of trial and error and care the divider could create *n* bundles of resources, each of which was somewhat different from the others, but was nevertheless such that he could assign one to each immigrant and no one would in fact envy anyone else's bundle.

The distribution might still fail to satisfy the immigrants as an equal distribution, for a reason that is not caught by the envy test. Suppose (to put the point in a dramatic way) the divider achieved his result by transforming all the available resources into a very large stock of plovers' eggs and pre-phylloxera claret (either by magic or trade with a neighboring island that enters the story only for that reason) and divides this glut into identical bundles of baskets and bottles. Many of the immigrants—let us say all but one—are delighted. But if that one hates plovers' eggs and pre-phylloxera claret he will feel that he has not been treated as an equal in the division of resources. The

1. D. Foley, "Resource Allocation and the Public Sector," *Yale Economic Essays* 7 (Spring 1967); H. Varian, "Equity, Energy and Efficiency, *Journal of Economic Theory* (Sept. 1974): 63-91.

envy test is met—he does not prefer any one's bundle to his own—but he prefers what he would have had under some fairer treatment of the initially available resources.

A similar, though less dramatic, piece of unfairness might be produced even without magic or bizarre trades. For the combination of resources that composes each bundle the divider creates will favor some tastes over others, compared with different combinations he might have composed. That is, different sets of *n* bundles might be created by trial and error, each of which would pass the envy test, so that for any such set that the divider chooses, someone will prefer that he had chosen a different set, even though that person would not prefer a different bundle within that set. Trades after the initial distribution may, of course, improve that person's position. But they will be unlikely to bring him to the position he would have had under the set of bundles he would have preferred, because some others will begin with a bundle they prefer to the bundle they would have had in that set, and so will have no reason to trade to that bundle.

So the divider needs a device that will attack two distinct foci of arbitrariness and possible unfairness. The envy test cannot be satisfied by any simple mechanical division of resources. If any more complex division can be found that will satisfy it, many such might be found, so that the choice amongst these would be arbitrary. The same solution will by now have occurred to all readers. The divider needs some form of auction or other market procedure in order to respond to these problems. I shall describe a reasonably straightforward procedure that would seem acceptable if it could be made to work, though as I shall describe it it will be impossibly expensive of time. Suppose the divider hands each of the immigrants an equal and large number of clamshells, which are sufficiently numerous and in themselves valued by no one, to use as counters in a market of the following sort. Each distinct item on the island (not including the immigrants themselves) is listed as a lot to be sold, unless someone notifies the auctioneer (as the divider has now become) of his or her desire to bid for some part of an item, including part, for example, of some piece of land, in which case that part becomes itself a distinct lot. The auctioneer then proposes a set of prices for each lot and discovers whether that set of prices clears all markets, that is, whether there is only one purchaser at that price and all lots are sold. If not, then the auctioneer adjusts his prices until he reaches a set that does clear the markets.[2] But the process does not stop then, because each of the immigrants remains free to change his bids even when an initially market-clearing set of prices is reached, or even to propose different lots. But let us suppose that in time even this leisurely process comes to an end, everyone declares himself satisfied, and goods are distributed accordingly.[3]

Now the envy test will have been met. No one will envy another's set of purchases because, by hypothesis, he could have purchased that bundle with his clamshells instead of his own bundle. Nor is the choice of sets of bundles arbitrary. Many people will be able to imagine a different set of bundles meeting the no-envy test that might have been established, but the actual set of bundles has the merit that each person played, through his purchases against an initially equal stock of counters, an equal role in determining the set of bundles actually chosen. No one is in the position of the person in our earlier example who found himself with nothing but what he hated. Of course, luck plays a certain role in determining how satisfied anyone is with the outcome, against other possibilities he might envision. If plovers' eggs and old claret were the only resources to auction, then the person who hated these would be as badly off as in our earlier example. He would be unlucky that the immigrants had not washed up on an island with more of what he wanted (though lucky, of course, that it did not have even less). But he could not complain that the division of the actual resources they found was unequal.

2. I mean to describe a Walrasian auction in which all productive resources are sold. I do not assume that the immigrants enter into complete forward contingent claims contracts, but only that markets will remain open and will clear in a Walrasian fashion once the auction of productive resources is completed. I make all the assumptions about production and preferences made in G. Debreu, *Theory of Value* (New Haven: Yale University Press, 1959). In fact the auction I describe here will become more complex in virtue of a tax scheme discussed later.

3. The process does not guarantee that the auction will come to an end in this way, because there may be various equilibria. I am supposing that people will come to understand that they cannot do better by further runs of the auction, and will for practical reasons settle on one equilibrium. If I am wrong, then this fact provides one of the aspects of incompleteness I describe in the next section.

He might think himself lucky or unlucky in other ways as well. It would be a matter of luck, for example, how many others shared various of his tastes. If his tastes or ambitions proved relatively popular, this might work in his favor in the auction, if there were economies of scale in the production of what he wanted. Or against him, if what he wanted was scarce. If the immigrants had decided to establish a regime of equality of welfare, instead of equality of resources, then these various pieces of good or bad luck would be shared with others, because distribution would be based, not on any auction of the sort I described, in which luck plays this role, but on a strategy of evening out differences in whatever concept of welfare had been chosen. Equality of resources, however, offers no similar reason for correcting for the contingencies that determine how expensive or frustrating someone's preferences turn out to be.[4]

Under equality of welfare, people are meant to decide what sorts of lives they want independently of information relevant to determining how much their choices will reduce or enhance the ability of others to have what they want.[5] That sort of information becomes relevant only at a second, political level at which administrators then gather all the choices made at the first level to see what distribution will give each of these choices equal success under some concept of welfare taken as the correct dimension of success. Under equality of resources, however, people decide what sorts of lives to pursue against a background of information about the actual cost their choices impose on other people and hence on the total stock of resources that may fairly be used by them. The information left to an independent political level under equality of welfare is therefore brought into the initial level of individual choice under equality of resources. The elements of luck in the auction we have just described are in fact pieces of information of a crucial sort; information that is acquired and used in that process of choice.

4. See, however, the discussion of handicaps below, which recognizes that certain kinds of preferences, which people wish they did not have, may call for compensation as handicaps.

5. See Part I of this essay (*Philosophy & Public Affairs* 10, no. 3 [Summer 1981]) for a discussion of whether equality of welfare can be modified so as to make an exception here for "expensive tastes" deliberately cultivated. I argue that it cannot.

So the contingent facts of raw material and the distribute of tastes are not grounds on which someone might challenge a distribution as unequal. They are rather background facts that determine what equality of resources, in these circumstances, is. Under equality of resources, no test for calculating what equality requires can be abstracted from these background facts and used to test them. The market character of the auction is not simply a convenient or ad hoc device for resolving technical problems that arise for equality of resources in very simple exercises like our desert island case. It is an institutionalized form of the process of discovery and adaptation that is at the center of the ethics of that ideal. Equality of resources supposes that the resources devoted to each person's life should be equal. That goal needs a metric. The auction proposes what the envy test in fact assumes, that the true measure of the social resources devoted to the life of one person is fixed by asking how important, in fact, that resource is for others. It insists that the cost, measured in that way, figure in each person's sense of what is rightly his and in each person's judgment of what life he should lead, given that command of justice. Anyone who insists that equality is violated by any particular profile of initial tastes, therefore, must reject equality of resources, and fall back on equality of welfare.

Of course it is sovereign in this argument, and in this connection between the market and equality of resources, that people enter the market on equal terms. The desert island auction would not have avoided envy, and would have no appeal as a solution to the problem of dividing the resources equally, if the immigrants had struggled ashore with different amounts of money in their pocket, which they were free to use in the auction, or if some had stolen clamshells from others. We must not lose sight of that fact, either in the argument that follows or in any reflections on the application of that argument to contemporary economic systems. But neither should we lose sight, in our dismay over the inequities of those systems, of the important theoretical connection between the market and the concept of equality in resources.

There are, of course, other and very different sorts of objection that might be made to the use of an auction, even an equal auction of the sort I described. It might be said, for example, that the fairness of an auction supposes that the preferences people bring to the auction, or form in its course, are authentic—the true preferences of the agent rather than preferences imposed upon him by the economic system itself. Perhaps an auction of any sort, in which one person bids against another, imposes an illegitimate assumption that what is valuable in life is individual ownership of something rather than more cooperative enterprises of the community or some group within it as a whole. Insofar as this (in part mysterious) objection is pertinent here, however, it is an objection against the idea of private ownership over an extensive domain of resources, which is better considered under the title of political equality, not an objection to the claim that a market of some sort must figure in any satisfactory account of what equality in private ownership is.

II. The Project

Since the device of an equal auction seems promising as a technique for achieving an attractive interpretation of equality of resources in a simple context, like the desert island, the question arises whether it will prove useful in developing a more general account of that ideal. We should ask whether the device could be elaborated to provide a scheme for developing or testing equality of resources in a community that has a dynamic economy, with labor, investment, and trade. What structure must an auction take in such an economy—what adjustments or supplements must be made to the production and trade that would follow such an auction—in order that the results continue to satisfy our initial requirement that an equal share of the resources be available to each citizen?

Our interest in this question is three-fold. First, the project provides an important test of the coherence and completeness of the idea of equality of resources. Suppose no auction or pattern of post-auction trade could be described whose results could be accepted as equality in any society much more complex or less artificial than a simple economy of consumption. Or that no auction could produce equality without constraints and restrictions which violate independent principles of justice. This would tend to suggest, at least, that there is no coherent ideal of equality of resources. Or that the ideal is not politically attractive after all.

We might discover, on the contrary, less comprehensive gaps or defects in the idea. Suppose, for example, that the design for the auction we develop does not uniquely determine a particular distribution, even given a stipulated set of initial resources and a stipulated population with fixed interests and ambitions, but is rather capable of producing significantly different outcomes depending on the order of decisions, arbitrary choices about the composition of the initial list of options, or other contingencies. We might conclude that the ideal of equality of resources embraces a variety of different distributions, each of which satisfies the ideal, and that the ideal is therefore partially indeterminate. This would show limitations on the power of the ideal to discriminate between certain distributions, but would not for that reason show that the ideal is either incoherent or practically impotent. So it is worth trying to develop the idea of an equal auction as a test of the theoretical standing and power of the political ideal.

Second, a fully developed description of an equal auction, adequate for a more complex society, might provide a standard for judging actual institutions and distributions in the real world. Of course no complex, organic society would have, in its history, anything remotely comparable to an equal auction. But we can nevertheless ask, for any actual distribution, whether it falls within the class of distributions that might have been produced by such an auction over a defensible description of initial resources. Or, if it is not, how far it differs from or falls short of the closest distribution within this class. The device of the auction might provide, in other words, a standard for judging how far an actual distribution, however it has been achieved, approaches equality of resources at any particular time.

Third, the device might be useful in the design of actual political institutions. Under certain (perhaps very limited) circumstances, when the conditions for an equal auction are at least roughly met, then an actual auction might be the best means of reaching or preserving equality of resources in the real world. This will be true, particularly, when the results of such an auction are antecedently indeterminate in the way just described, so that any result the auction reaches will respect equality of resources even though it is not known, in advance, which result would be reached. In such a case it may be fairer to conduct an actual auction than to choose, through some other political means, one rather than another of the results that an auction might produce. Even in such a case it will rarely be possible or desirable to conduct an actual auction in the design our theoretical investigations recommend. But it may be possible to design an auction surrogate—an economic or political institution having sufficient of the characteristics of a theoretical equal auction so that the arguments of fairness recommending an actual auction were it feasible also recommend the surrogate. The economic markets of many countries can be interpreted, even as they stand, as forms of auctions. (So, too, can many forms of democratic political process.) Once we have developed a satisfactory model of an actual auction (to the extent we can) we can use that model to test these institutions, and reform them to bring them closer to the model.

Nevertheless our project is in the main, within the present essay, entirely theoretical. Our interest is primarily in the design of an ideal, and of a device to picture that ideal and test its coherence, completeness, and appeal. We shall therefore ignore practical difficulties, like problems of gathering information, which do not impeach these theoretical goals, and also make simplifying counterfactual assumptions which do not subvert them. But we should try to notice which simplifications we are making, because they will be of importance, particularly as to the third and most practical application of our projects, at any later stage, at which we consider second-best compromises of our ideal in the real world.

III. Luck and Insurance

If the auction is successful as described, then equality of resources holds for the moment among the immigrants. But perhaps only for the moment, because if they are left alone, once the auction is completed, to produce and trade as they wish, then the envy test will shortly fail. Some may be more skillful than others at producing what others want and will trade to get. Some may like to work, or to work in a way that will produce more to trade, while others like not to work or prefer to work at what will bring them less. Some will stay healthy while others fall sick, or lightning will strike the farms of others but avoid theirs. For any of these and dozens of other reasons some people will prefer the bundle others have in say, five years, to their own.

We must ask whether (or rather how far) such developments are consistent with equality of resources, and I shall begin by considering the character and impact of luck on the immigrants' post-auction fortunes. I shall distinguish, at least for the moment, between two kinds of luck. Option luck is a matter of how deliberate and calculated gambles turn out—whether someone gains or loses through accepting an isolated risk he or she should have anticipated and might have declined. Brute luck is a matter of how risks fall out that are not in that sense deliberate gambles. If I buy a stock on the exchange that rises, then my option luck is good. If I am hit by a falling meteorite whose course could not have been predicted, then my bad luck is brute (even though I could have moved just before it struck if I had any reason to know where it would strike). Obviously the difference between these two forms of luck can be represented as a matter of degree, and we may be uncertain how to describe a particular piece of bad luck. If someone develops cancer in the course of a normal life, and there is no particular decision to which we can point as a gamble risking the disease, then we will say that he has suffered brute bad luck. But if he smoked cigarettes heavily then we may prefer to say that he took an unsuccessful gamble.

Insurance, so far as it is available, provides a link between brute and option luck, because the decision to buy or reject catastrophe insurance is a calculated gamble. Of course, insurance does not erase the distinction. Someone who buys medical insurance and is hit by an unexpected meteorite still suffers brute bad luck, because he is worse off than if he had bought insurance and not needed it. But he has had better option luck than if he had not bought the insurance, because his situation is better in virtue of his not having run the gamble of refusing to insure.

Is it consistent with equality of resources that people should have different income or wealth in virtue of differing option luck? Suppose some of the immigrants plant valuable but risky crops while others play it safer, and that some of the former buy insurance against uncongenial weather while others do not. Skill will play a part in determining which of these various programs succeed, of course, and we shall consider the problems this raises later. But option luck will also play a part. Does its role threaten or invade equality of resources?

Consider, first, the differences in wealth between those who play it safe and those who gamble and succeed. Some people enjoy, while others hate, risks; but this particular difference in personality is comprehended in a more general difference between the kinds of lives that different people wish to lead. The life chosen by someone who gambles contains, as an element, the factor of risk; someone who chooses not to gamble has decided that he prefers a safer life. We have already decided that people should pay the price of the life they have decided to lead, measured in what others give up in order that they can do so. That was the point of the auction as a device to establish initial equality of resources. But the price of a safer life, measured in this way, is precisely forgoing any chance of the gains whose prospect induces others to gamble. So we have no reason to object, against the background of our earlier decisions, to a result in which those who decline to gamble have less than some of those who do not.

But we must also compare the situation of those who gamble and win with that of those who gamble and lose. We cannot say that the latter have chosen a different life and must sacrifice gains accordingly; for they have chosen the same lives as those who won. But we can say that the possibility of loss was part of the life they chose—that it was the fair price of the possibility of gain. For we might have designed our initial auction so that people could purchase (for example) lottery tickets with their clamshells. But the price of those tickets would have been some amount of other resources (fixed by the odds and the gambling preferences of others) that the shells would otherwise have bought, and which will be wholly forgone if the ticket does not win.

The same point can be made by considering the arguments for redistribution from winners to losers after the event. If winners were made to share their winnings with losers, then no one would gamble, as individuals, and the kind of life preferred by both those who in the end win and those who lose would be unavailable. Of course, it is not a good argument, against someone who urges redistribution in order to achieve equality of resources, that redistribution would make some forms of life less attractive or even impossible. For the demands of equality (we assume in this essay) are prior to other desiderata, including variety in the kinds of life available to people. (Equality will in any case make certain kinds of lives—a life of economic and political domination of others, for example—impossible.) In the present case, however, the difference is apparent. For the effect of redistribution from winners to losers in gambles would be to deprive both of lives they prefer, which indicates, not simply that this would produce an unwanted curtailment of available forms of life, but that it would deprive them of an equal voice in the construction of lots to be auctioned, like the man who hated both plovers' eggs and claret but was confronted only with bundles of both. They both want gambles to be in the mix, either originally or as represented by resources with which they can take risks later, and the chance of losing is the correct price, measured on the metric we have been using, of a life that includes gambles with a chance of gain.

We may, of course, have special reasons for forbidding certain forms of gambles. We may have paternalistic reasons for limiting how much any individual may risk, for example. We may also have reasons based in a theory of political equality for forbidding someone to gamble with his freedom or his religious or political rights. The present point is more limited. We have no general reason for forbidding gambles altogether in the bare fact that in the event winners will control more resources than losers, any more than in the fact that winners will have more than those who do not gamble at all. Our initial principle, that equality of resources requires that people pay the true cost of the lives that they lead, warrants rather than condemns these differences.

We may (if we wish) adjust our envy test to record that conclusion. We may say that in computing the extent of someone's resources over his life, for the purpose of asking whether anyone else envies those resources, any resources gained through a successful gamble should be represented by the opportunity to take the gamble at the odds in force, and comparable adjustments made to the resources of those who have lost through gambles. The main point of this artificial construction of the envy test, however, would be to remind us that the argument in favor of allowing differences in option luck to affect income and wealth assumes that everyone has in principle the same gambles available to him. Someone who never had the opportunity to run a similar risk, and would have taken the opportunity had it been available, will still envy some of those who did have it.

Nor does the argument yet confront the case of brute bad luck. If two people lead roughly the same lives, but one goes suddenly blind, then we cannot explain the resulting differences in their incomes either by saying that one took risks that the other chose not to take, or that we could not redistribute without denying both the lives they prefer. For the accident has (we assume) nothing to do with choices in the pertinent sense. It is not necessary to the life either has chosen that he run the risk of going blind without redistribution of funds from the other. This is a fortiori so if one is born blind and the other sighted.

But the possibility of insurance provides, as I suggested, a link between the two kinds of luck. For suppose insurance against blindness is available, in the initial auction, at whatever level of coverage the policy holder chooses to buy. And also suppose that two sighted people have, at the time of the auction, equal chance of suffering an accident that will blind them, and know that they have. Now if one chooses to spend part of his initial resources for such insurance and the other does not, or if one buys more coverage than the other, then this difference will reflect their different opinions about the relative value of different forms or components of their prospective lives. It may reflect the fact that one puts more value on sight than the other. Or, differently, that one would count monetary compensation for the loss of his sight as worthless in the face of such a tragedy while the other, more practical, would fix his mind on the aids and special training that such money might buy. Or simply that one minds or values risk differently from the other, and would, for example, rather try for a brilliant life that would collapse under catastrophe than a life guarded at the cost of resources necessary to make it brilliant.

But in any case the bare idea of equality of resources, apart from any paternalistic additions, would not argue for redistribution from the person who had insured to the person who had not if, horribly, they were both blinded in the same accident. For the availability of insurance would mean that, though they had both had brute bad luck, the difference between them was a matter of option luck, and the arguments we entertained against disturbing the results of option luck under conditions of equal antecedent risk hold here as well. But then the situation cannot be different if the person who decided not to insure is the only one to be blinded. For once again the difference is a difference in option luck against a background of equal opportunity to insure or not. If neither had been blinded, the man who had insured against blindness would have been the loser. His option luck would have been bad–though it seems bizarre to put it this way–because he spent resources that, as things turned out, would have been better spent otherwise. But he would have no claim, in that event, from the man who did not insure and also survived unhurt.

So if the condition just stated were met–if everyone had an equal risk of suffering some catastrophe that would leave him or her handicapped, and everyone knew roughly what the odds were and had ample opportunity to insure–then handicaps would pose no special problem for equality of resources. But of course that condition is not met. Some people are born with handicaps, or develop them before they have either sufficient knowledge or funds to insure on their own behalf. They cannot buy insurance after the event. Even handicaps that develop later in life, against which people do have the opportunity to insure, are not randomly distributed through the population, but follow genetic tracks, so that sophisticated insurers would charge some people higher premiums for the same coverage before the event. Nevertheless the idea of a market in insurance provides a counterfactual guide through which equality of resources might face the problem of handicaps in the real world.

Suppose we can make sense of and even give a rough answer to the following question. If (contrary to fact) everyone had at the appropriate age the same risk of developing physical or mental handicaps in the future (which assumes that no one has developed these yet) but that the total number of handicaps remained what it is, how much insurance coverage against these handicaps would the average member of the community purchase? We might then say that but for (uninsurable) brute luck that has altered these equal odds, the average person would have purchased insurance at that level, and compensate those who do develop handicaps accordingly, out of some fund collected by taxation or other compulsory process but designed to match the fund that would have been provided through premiums if the odds had been equal. Those who develop handicaps will then have more resources at their command than others, but the extent of their extra resources will be fixed by the market decisions that people would supposedly have made if circumstances had been more equal than they are. Of course, this argument does involve the fictitious assumption that everyone who suffers handicaps would have bought the average amount of insurance, and we may wish to refine the argument and the strategy so that that no longer holds.[6] But it does not seem an unreasonable assumption for this purpose as it stands.

Can we answer the counterfactual question with sufficient confidence to develop a program of compensation of that sort? We face a threshold difficulty of some importance. People can decide how much of their resources to devote to insurance against a particular catastrophe only with some idea of the life they hope to lead, because only then can they decide how serious a particular catastrophe would be, how far additional resources would alleviate the tragedy, and so forth. But people who are born with a particular handicap, or develop one in childhood, will of course take that circumstance into account in the plans they make. So in order to decide how much insurance such a person would have bought without the handicap we must decide what sort of life he would have planned in that case. But there may be no answer, even in principle, to that question.

We do not need, however, to make counterfactual judgments that are so personalized as to embarrass us for that reason. Even if people did all have equal risk of all catastrophes, and evaluated the value and importance of insurance differently entirely due to their different am-

6. The averaging assumption is a simplifying assumption only, made to provide a result in the absence of the detailed (and perhaps, for reasons described in the text, indeterminate) information that would enable us to decide how much each handicapped person would have purchased in the hypothetical market. If we had such full information, so that we could tailor compensation to what a particular individual in fact would have bought, the accuracy of the program would be improved. But in the absence of such information averaging is second best, or in any case better than nothing.

bitions and plans, the insurance market would nevertheless be structured through categories designating the risks against which most people would insure in a general way. After all, risks of most catastrophes are now regarded by the actual insurance market as randomly distributed, and so we might follow actual insurance practice, modified to remove the discriminations insurers make when they know that one group is more likely, perhaps for genetic reasons, to suffer a particular kind of brute bad luck. It would make sense to suppose, for example, that most people would make roughly the same assessment of the value of insurance against general handicaps, such as blindness or the loss of a limb, that affect a wide spectrum of different sorts of lives. (We might look to the actual market to discover the likelihood and the contours of more specialized insurance we might decide to use in more complex schemes, like the insurance of musicians against damage to their hands, and so forth.)

We would, in any case, pay great attention to matters of technology, and be ready to adjust our sums as technology changed. People purchase insurance against catastrophes, for example, against a background of assumptions about the remedial medical technology, or special training, or mechanical aids that are in fact available, and about the cost of these remedies. People would seek insurance at a higher level against blindness, for example, if the increased recovery would enable them to purchase a newly discovered sight-substitute technology, than they would if that increased recovery simply swelled a bank account they could not, in any case, use with much satisfaction.

Of course, any judgments that the officials of a community might make about the structure of the hypothetical insurance market would be speculative and open to a variety of objections. But there is no reason to think, certainly in advance, that a practice of compensating the handicapped on the basis of such speculation would be worse, in principle, than the alternatives, and it would have the merit of aiming in the direction of the theoretical solution most congenial to equality of resources.

We might now remind ourselves of what these alternatives are. I said in Part 1 of this essay that the regime of equality of welfare, contrary to initial impressions, does a poor job of either explaining or guiding our impulse to compensate the severely handicapped with extra resources. It provides, in particular, no upper bound to compensation so long as any further payment would improve the welfare of the wretched; but this is not, as it might seem, generous, because it leaves the standard for actual compensation to the politics of selfishness broken by sympathy, politics that we know will supply less than any defensible hypothetical insurance market would offer.

Consider another approach to the problem of handicaps under equality of resources. Suppose we say that any person's physical and mental powers must count as part of his resources, so that someone who is born handicapped starts with less by way of resources than others have, and should be allowed to catch up, by way of transfer payments, before what remains is auctioned off in any equal market. People's powers are indeed resources, because these are used, together with material resources, in making something valuable out of one's life. Physical powers are resources for that purpose in the way that aspects of one's personality, like one's conception of what is valuable in life, are not. Nevertheless the suggestion, that a design of equality of resources should provide for an initial compensation to alleviate differences in physical or mental resources, is troublesome in a variety of ways. It requires, for example, some standard of "normal" powers to serve as the benchmark for compensation.[8] But whose powers should be taken as normal for this purpose? It suffers, moreover, from the same defect as the parallel recommendation under equality of welfare. In fact, no amount of initial compensation could make someone born blind or mentally incompetent equal in physical or mental resources with someone taken to be "normal" in these ways. So the argument provides no upper bound to initial compensation, but must leave this to a political compromise likely to be less generous, again, than what the hypothetical insurance market would command.

Quite apart from these practical and theoretical inadequacies, the

7. Cf. Amartya Sen, "Equality of What?," *The Tanner Lectures on Human Values*, Vol. 1 (Salt Lake City: University of Utah Press, 1980), pp. 197, 218.

8. The hypothetical insurance approach does not require any stipulation of "normal" powers, because it allows the hypothetical market to determine which infirmities are compensable.

suggestion is troublesome for another reason. Though powers are resources, they should not be considered resources whose ownership is to be determined through politics in accordance with some interpretation of equality of resources. They are not, that is, resources for the theory of equality in exactly the sense in which ordinary material resources are. They cannot be manipulated or transferred, even so far as technology might permit. So in this way it misdescribes the problem of handicaps to say that equality of resources must strive to make people equal in physical and mental constitution so far as this is possible. The problem is, rather, one of determining how far the ownership of independent material resources should be affected by differences that exist in physical and mental powers, and the response of our theory should speak in that vocabulary.

It might be wise (if for no other reason than as a convenient summary of the argument from time to time) to bring our story of the immigrants up to date. By way of supplement to the auction, they now establish a hypothetical insurance market which they effectuate through compulsory insurance at a fixed premium for everyone based on speculations about what the average immigrant would have purchased by way of insurance had the antecedent risk of various handicaps been equal. (We choose for them, that is, one of the simpler possible forms of instituting the hypothetical insurance market. We shall see, when we discuss the problem of skills, that they might well choose a more complex scheme of the sort discussed there.)

But now a question arises. Does this decision place too much weight on the distinction between handicaps, which the immigrants treat in this compensatory way, and accidents touching preferences and ambitions (like the accident of what material resources are in fact available, and of how many other people share a particular person's taste)? The latter will also affect welfare, but they are not matters for compensation under our scheme. Would it not now be fair to treat as handicaps eccentric tastes, or tastes that are expensive or impossible to satisfy because of scarcity of some good that might have been common? We might compensate those who have these tastes by supposing that everyone had an equal chance of being in that position and then establishing a hypothetical insurance market against that possibility.

A short answer is available. Someone who is born with a serious handicap faces his life with what we concede to be fewer resources, just on that account, than others do. This justifies compensation, under a scheme devoted to equality of resources, and though the hypothetical insurance market does not right the balance—nothing can—it seeks to remedy one aspect of the resulting unfairness. But we cannot say that the person whose tastes are expensive, for whatever reason, therefore has fewer resources at his command. For we cannot state (without falling back on some version of equality of welfare) what equality in the distribution of tastes and preferences would be. Why is there less equality of resources when someone has an eccentric taste that makes goods cheaper for others, than when he shares a popular taste and so makes goods more expensive for them? The auction, bringing to bear information about the resources that actually exist and the competing preferences actually in play, is the only true measure of whether any particular person commands equal resources. If the auction has in fact been an equal auction, then the man of eccentric tastes has no less than equal material resources, and the argument that justifies a compensatory hypothetical auction in the case of handicaps has no occasion even to begin. It is true that this argument produces a certain view of the distinction between a person and his circumstances, and assigns his tastes and ambitions to his person, and his physical and mental powers to his circumstances. That is the view of a person I sketched in the introductory section, of someone who forms his ambitions with a sense of their cost to others against some presumed initial equality of economic power, and though this is different from the picture assumed by equality of welfare, it is a picture at the center of equality of resources.

In one way, however, my argument might well be thought to overstate the distinction between handicaps and at least certain sorts of what are often considered preferences. Suppose someone finds he has a craving (or obsession or lust or, in the words of an earlier psychology, a "drive") that he wishes he did not have, because it interferes with what he wants to do with his life and offers him frustration or even pain if it is not satisfied. This might indeed be some feature of his physical needs that other people would not consider a handicap at all: for example, a generous appetite for sex. But it is a "preference" (if that is the right word) that he does not want, and it makes perfect sense to say that he would be better off without it. For some people these unwanted tastes include tastes they have (perhaps unwittingly) themselves cultivated, such as a taste for a particular sport or for music of a sort difficult to obtain. They regret that they have these tastes, and believe they would be better off without them, but nevertheless find it painful to ignore them. These tastes are handicaps; though for other people they are rather an essential part of what gives value to their lives.

Now these cases do not present, for particular people, borderline cases between ambitions and handicaps (though no doubt other sorts of borderline cases could be found). The distinction required by equality of resources is the distinction between those beliefs and attitudes that define what a successful life would be like, which the ideal assigns to the person, and those features of body or mind or personality that provide means or impediments to that success, which the ideal assigns to the person's circumstances. Those who see their sexual desires or their taste for opera as unwanted disadvantages will class these features of their body or mind or personality firmly as the latter. These are, for them, handicaps, and are therefore suitable for the regime proposed for handicaps generally. We may imagine that everyone has an equal chance of acquiring such a craving by accident. (Of course, for each person the content of a craving that would have that consequence would be different. We are supposing here, not the risk of any particular craving, but the risk of whatever craving would interfere with set goals in that way.) We may then ask—with as much or as little intelligibility as in the case of blindness—whether people generally would purchase insurance against that risk, and if so at what premium and what level of coverage. It seems unlikely that many people would purchase such insurance, at the rates of premium likely to govern if they sought it, except in the case of cravings so severe and disabling as to fall under the category of mental disease. But that is a different matter. The important point, presently, is that the idea of an insurance market is available here, because we can imagine people who have such a craving not having it, without thereby imagining them to have a different conception of what they want from life than what in fact they do want. So the idea of the imaginary insurance auction provides at once a device for identifying cravings and distinguishing them from positive features of personality, and also for bringing these cravings within the general regime designed for handicaps.

IV. Labor and Wages

Equality of resources, once established by the auction, and corrected to provide for handicaps, would be disturbed by production and trade. If one of the immigrants, for example, was specially proficient at producing tomatoes, he might trade his surplus for more than anyone else could acquire, in which case others would begin to envy his bundle of resources. Suppose we wished to create a society in which the division of resources would be continuously equal, in spite of different kinds and degrees of production and trade. Can we adapt our auction so as to produce such a society?

We should begin by considering a different sequence after which people would envy each other's resources, and the division might be thought no longer to be equal. Suppose all the immigrants are in fact sufficiently equal in talent at the few modes of production that the resources allow so that each could produce roughly the same goods from the same set of resources. Nevertheless they wish to lead their lives in different ways, and they in fact acquire different bundles of resources in the initial auction and use them differently thereafter. Adrian chooses resources and works them with the single-minded ambition of producing as much of what others value as possible; and so, at the end of a year, his total stock of goods is larger than anyone else's. Each of the other immigrants would now prefer Adrian's stock to his own; but by hypothesis none of them would have been willing to lead his life so as to produce them. If we look for envy at particular points in time, then each envies Adrian's resources at the end of the year, and the division is therefore not equal. But if we look at envy differently, as a matter of resources over an entire life, and we include a person's occupation as part of the bundle of his goods, then no one envies Adrian's bundle, and the distribution cannot be said to be unequal on that account.

Surely we should take the second, synoptic, point of view. Our final aim is that an equal share of resources should be devoted to the lives of each person, and we have chosen the auction as the right way to measure the value of what is made available to a person, through his decision, for that purpose. If Bruce chooses to acquire land for use as a tennis court, then the question is raised how much his account should be charged, in the reckoning whether an equal share has been put to his use, in virtue of that choice, and it is right that his account should be charged the amount that others would have been willing to pay had the land been devoted to their purposes instead. The appeal of the auction, as a device for picturing equality of resources, is precisely that it enforces that metric. But this scheme will fail, and the device disappoint us, unless Adrian is able to bid a price for the same land that reflects his intention to work rather than play on it and so to acquire whatever gain would prompt him to make that decision. For unless this is permitted, those who want tomatoes and would pay Adrian his price for them will not be able to bid indirectly, through Adrian's decision, against Bruce, who will then secure his tennis court at a price that, because it is too low, defeats equality of resources. This is not, I should add, an argument from efficiency as distinct from fairness; but rather an argument that in the circumstances described, in which talents are equal, efficiency simply is fairness, at least as fairness is conceived under equality of resources. If Adrian is willing to spend his life at drudgery, in return for the profit he will make at prices that others will pay for what he produces, then the land on which he would drudge should not be used for a tennis court instead, unless its value as a tennis court is greater as measured by someone's willingness to invade an initially equal stock of abstract resources.

Now this is to look at the matter entirely from the standpoint of those who want Adrian's tomatoes, a standpoint that treats Adrian only as a means. But we reach the same conclusion if we look at the matter from his point of view as well. If someone chooses to have something inexpensive in his life, under a regime of equality of resources, then he will have more left over for the rest of what he wants. Someone who accepts Algerian wine may use it to wash down plovers' eggs. But a decision to produce one thing rather than another with land, or to use the land for leisure rather than production, is also the choice of something for one's life, and this may be inexpensive as well. Suppose Adrian is desperate for plovers' eggs but would rather work hard at tilling his land than settle for less than champagne. The total may be no more expensive, measured in terms of what his decisions cost others, than a life of leisure and grape juice. If he earns enough by working hard, or by working at work that no one else wants to do, to satisfy all his expensive tastes, then his choice for his own life costs the rest of the community no more than if his tastes were simpler and his industry less. So we have no more reason to deny him hard work and high consumption than less work and frugality. The choice should be indifferent under equality of resources, so long as no one envies the total package of work plus consumption that he chooses. So long as no one envies, that is, his life as a whole. Of course, Adrian might actually enjoy his hard work, so that he makes no sacrifice. He prefers working hard to anything else. But this cannot provide any argument, under equality of resources, that he should gain less in money or other goods by his work than if he hated every minute of it, any more than it argues against charging someone a low price for lettuce, which he actually prefers to truffles.

So we must apply the envy test diachronically: it requires that no one envy the bundle of occupation and resources at the disposal of anyone else over time, though someone may envy another's bundle at any particular time. It would therefore violate equality of resources if the community were to redistribute Adrian's wealth, say, at the end of each year. If everyone had equal talents (as we have been assuming just now), the initial auction would produce continuing equality of resources even though bank-account wealth became more and more unequal as years passed.

Is that unlikely condition—that everyone has equal talent—absolutely necessary to that conclusion? Would the auction produce continuing equality of resources if (as in the real world) talents for production differed sharply from person to person? Now the envy test would fail, even interpreted diachronically. Claude (who likes farming but has a black thumb) would not bid enough for farming land to take that land from Adrian. Or, if he did, he would have to settle for less in the rest of his life. But he would then envy the package of Adrian's occupation and wealth. If we interpret occupation in a manner sensitive to the joys of craft, then Adrian's occupation, which must then be described as skillful, craftsmanlike farming, is simply unavailable to Claude. If we interpret occupation in a more census-like fashion, then Claude may undertake Adrian's occupation, but he cannot have the further resources that Adrian has along with it. So if we continue to insist that the envy test is a necessary condition of equality of resources, then our initial auction will not insure continuing equality, in the real world of unequal talents for production.

But it may now be objected that we should not insist on the envy test at this point, even in principle, for the following reason. We are moving too close to a requirement that people must not envy each other, which is different from the requirement that they must not envy each other's bundles of resources. People may envy each other for a variety of reasons: some are physically more attractive, some more easily satisfied with their condition, some better liked by others, some more intelligent or able in different ways, and so on. Of course, under a regime of equality of welfare each of these differences would be taken into account, and transfers made to erase their welfare consequences so far as possible or feasible. But the point of equality of resources is fundamentally different: it is that people should have the same external resources at their command to make of them what, given these various features and talents, they can. That point is satisfied by an initial auction, but since people are different it is neither necessary nor desirable that resources should remain equal thereafter, and quite impossible that all envy should be eliminated by political distribution. If one person, by dint of superior effort or talent, uses his equal share to create more than another, he is entitled to profit thereby, because his gain is not made at the expense of someone else who does less with his share. We recognized that, just now, when we conceded that superior industry should be rewarded, so that Adrian, who worked hard, should be allowed to keep the rewards of his effort.

Now this objection harbors many mistakes, but they all come to this: it confuses equality of resources with the fundamentally different idea sometimes called equality of opportunity. It is not true, in the first place, that someone who does more with his initial share does not, in so doing, lessen the value of what others have. If Adrian were not so successful at agriculture, then Claude's own efforts would be rewarded more, because people would buy his inferior produce having no better alternative. If Adrian were not so successful and hence so rich he would not be able to pay so much for wine, and Claude, with his smaller fortune, would be able to buy more at a cheaper price. These are simply the most obvious consequences of the fact that the immigrants form one economy, after the initial auction, rather than a set of distinct economies. Of course these consequences also follow from the situation we discussed a moment ago. If Adrian and Bruce have the same talents, but Adrian chooses to work harder or differently and acquires more money, then this may also decrease the value of Claude's share to him. The difference between these two circumstances, if there is one, lies elsewhere; but it is important to reject the claim, instinct in some arguments for equality of opportunity, that if people start with equal shares the prosperity of one does no damage to the other.

Nor is it true that if we aim at a result in which those with less talent do not envy the circumstances of those with more talent we have destroyed the distinction between envying others and envying what they have. For Adrian has two things that Claude would prefer to have which belong to Adrian's circumstances rather than his person. The desires and needs of other people provide Adrian but not Claude with a satisfying occupation, and Adrian has more money than Claude can have. Perhaps nothing that can be done, by way of political structure or distribution, to erase these differences and remove the envy entirely. We cannot, for example, alter the tastes of other people by electrical means so as to make them value what Claude can produce more and what Adrian can produce less. But this provides no argument against other schemes, like schemes of education that would allow Claude to find satisfaction in his work or of taxation that would redistribute some of Adrian's wealth to him, and we could fairly describe these schemes as aiming to remove Claude's envy of what Adrian has rather than of what Adrian is.

Important as these points are, it is more important still to identify and correct another mistake that the present objection makes. It misunderstands our earlier conclusion, that when talents are roughly equal the auction provides continuing equality of resources, and so misses the important distinction between that case and the present argument. The objection supposes that we reached that conclusion because we accept, as the basis of equality of resources, what we might call the starting-gate theory of fairness: that if people start in the same circumstances, and do not cheat or steal from one another, then it is fair that people keep what they gain through their own skill. But the starting-gate theory of fairness is very far from equality of resources. Indeed it is hardly a coherent political theory at all.

The starting-gate theory holds that justice requires equal initial resources. But it also holds that justice requires laissez-faire thereafter, in accordance, presumably, with some version of the Lockean theory that people acquire property by mixing their labor with goods or something of that sort. But these two principles cannot live comfortably together. Equality can have no greater force in justifying initial equal holdings when the immigrants land–against the competing that all property should be available for Lockean acquisition at that time–than later in justifying redistributions when wealth becomes unequal because people's productive talents are different. The same point may be put the other way around. The theory of Lockean acquisition (or whatever other theory of justice in acquisition is supposed to justify the laissez-faire component in a starting-gate theory) can have no less force in governing the initial distribution than it has in justifying title through talent and effort later. If the theory is sound later, then why does it not command a Lockean process of acquisition in the first instance, rather than an equal distribution of all there is? The moment when the immigrants first land is, after all, an arbitrary point in their lives at which to locate any one-shot requirement that they each have an equal share of any available resources. If that requirement holds then, it must also hold on the tenth anniversary of that date, which is, in the words of the banal and important cliché, the first day in the rest of their lives. So if justice requires an equal auction when they land, it must require a fresh, equal auction from time to time thereafter; and if justice requires laissez-faire thereafter, it must require it when they land.

Suppose someone replies that there is an important difference between the initial distribution of resources and any later redistribution. When the immigrants land, no one owns any of the resources, and the principle of equality therefore dictates equal initial shares. But later, after the initial resources have been auctioned, they are each owned in some way by someone, so that the principle of equality is superceded by respect for people's rights in property or something of that sort. This reply begs the question straightway. For we are considering precisely the question whether a system of ownership should be established in the first instance that has that consequence, or, rather, whether a different system of ownership should be chosen that explicitly makes any acquisition subject to schemes of redistribution later. If the latter sort of system is chosen, at the outset, then no one can later complain that redistribution is ruled out by his property rights alone. I do not mean that no theory of justice can consistently distinguish between justice in initial acquisition and justice in transfer on the ground that anyone may do what he wants with property that is already his. Nozick's theory, for example, does just that. This is consistent, because his theory of justice in initial acquisition purports to justify a system of property rights which have that consequence: justice in transfer, that is, flows from the rights the theory of acquisition claims are acquired in acquiring property. But the theory of initial acquisition on which the starting-gate theory relies, which is equality of resources, does not even purport to justify a characterization of property that necessarily includes absolute control without limit of time thereafter.

So the starting-gate theory, that the immigrants should start off equal in resources but grow prosperous or lean through their own efforts thereafter, is an indefensible combination of very different theories of justice. Something like that combination makes sense in games, such as Monopoly, whose point is to allow luck and skill to play a highly circumscribed and, in the last analysis, arbitrary, role; but it cannot hold together a political theory. Our own principle, that if people of equal talent choose different lives it is unfair to redistribute halfway through those lives, makes no appeal to the starting-gate theory at all. It is based on the very different idea that the equality in question is equality of resources devoted to whole lives. This principle offers a clear answer to the question that embarrasses the present objection. Our theory does not suppose that an equal division of resources is appropriate at one moment in someone's life but not at any other. It argues only that resources available to him at any moment must be a function of resources available or consumed by him at others, so that the explanation of why someone has less money now may be that he has consumed expensive leisure earlier. Nothing like that explanation is available to explain why Claude, who has worked as hard and in the same way as Adrian, should have less in virtue of the fact that he is less skillful.

So we must reject the starting-gate theory, and recognize that the requirements of equality (in the real world at least) pull in opposite directions. On the one hand we must, on pain of violating equality, allow the distribution of resources at any particular moment to be (as we might say) ambition-sensitive. It must, that is, reflect the cost or benefit to others of the choices people make so that, for example, those who choose to invest rather than consume, or to consume less expensively rather than more, or to work in more rather than less profitable ways, must be permitted to retain the gains that flow from these decisions in an equal auction followed by free trade. But on the other hand, we must not allow the distribution of resources at any moment to be endowment-sensitive, that is, to be affected by differences in ability of the sort that produce income differences in a laissez-faire economy among people with the same ambitions. Can we devise some formula that offers a practical, or even a theoretical, compromise between these two, apparently competing, requirements?

We might mention, but only to dismiss, one possible response. Suppose we allow our initial auction to include, as resources to be auctioned, the labor of the immigrants themselves, so that each immigrant can bid for the right to control part or all of his own or other people's labor. Special skills would accrue to the benefit, not of the laborer himself, but of the community as a whole, like any other valuable resource the immigrants found when they landed. Except in unusual cases, since people begin with equal resources for bidding, each agent would bid enough to secure his own labor. But the result would be that each would have to spend his life in close to the commercially most profitable manner he could, or, at least if he is talented, suffer some very serious deprivation if he did not. For since Adrian, for example, is able to produce prodigious income from farming, others would be willing to bid a large amount to have the right to his labor and the vegetables thereof, and if he outbids them, but chooses to write indifferent poetry instead of farming full time, he will have spent a large part of his initial endowment on a right that will bring him little financial benefit. This is indeed the slavery of the talented.

We cannot permit this, but it is worth pausing to ask what grounds we have for barring it. Shall we say that since a person owns his own mind and body, he owns the talents that are only capacities thereof, and therefore owns the fruits of those talents? This is, of course, a series of nonsequiturs. It is also a familiar argument in favor of the laissez-faire labor market we have decided is a violation of equality of resources when people are unequal in talent. But we could not accept it in any case, because it uses the idea of pre-political entitlement based on something other than equality, and that is inconsistent with the premise of the scheme of equality of resources we have developed.

So we must look elsewhere for the ground of our objection to taking people's labor as a resource for the auction. We need not, in fact, look very far; for the principle that people should not be penalized for talent is simply part of the same principle we relied on in rejecting the apparently opposite idea, that people should be allowed to retain the benefits of superior talent. The envy test forbids both of these results. If Adrian is treated as owning whatever his talents enable him to produce, then Claude envies the package of resources, including occupation, that Adrian has over his life considered as a whole. But if Adrian is required to purchase leisure time or the right to a less productive occupation at the cost of other resources, then Adrian will envy Claude's package. If equality of resources is understood to include some plausible version of the envy test, as a necessary condition of an equal distribution, then the role of talent must be neutralized in a way that no simple addition to the stock of goods to be auctioned can accomplish.

We should turn, therefore, to a more familiar idea: the periodic redistribution of resources through some form of income tax.[9] We

9. Notice that our analysis of the problem that differential talents presents to equality of resources calls for an income tax, rather than either a wealth or a consumption tax. If people begin with equal resources, then we wish to tax to adjust for different skills so far as these produce different income, because it is only in that way that they threaten equality of resources. Someone's decision to spend rather than save what he has earned is precisely the kind of decision whose impact should be determined by the market uncorrected for tax under this analysis. Of course, there might be technical or other reasons why a society

want to develop a scheme of redistribution, so far as we are able, that will neutralize the effects of differential talents, yet preserve the consequences of one person choosing an occupation, in response to his sense of what he wants to do with his life, that is more expensive for the community than the choice another makes. An income tax is a plausible device for this purpose because it leaves intact the possibility of choosing a life in which sacrifices are constantly made and discipline steadily imposed for the sake of financial success and the further resources it brings, though of course it neither endorses nor condemns that choice. But it also recognizes the role of genetic luck in such a life. The accommodation it makes is a compromise; but it is a compromise of two requirements of equality, in the face of both practical and conceptual uncertainty how to satisfy these requirements, not a compromise of equality for the sake of some independent value such as efficiency.

But of course the appeal of a tax depends on our ability to fix rates of taxation that will make that compromise accurately. It might be helpful, in that aim, if we were able to find some way of identifying, in any person's wealth at any particular time, the component traceable to differential talents as distinguished from differential ambitions. We might then try to devise a tax that would recapture, for redistribution, just this component. But we cannot hope to identify such a component, even given perfect information about people's personalities. For we will be thwarted by the reciprocal influence that talents and ambitions exercise on each other. Talents are nurtured and developed, not discovered full-blown, and people choose which talents to develop in response to their beliefs about what sort of person it is best to be. But people also wish to develop and use the talents they have, not simply because they prefer a life of relative success, but because the exercise of talent is enjoyable and perhaps also out of a sense that an

dedicated to equality of welfare would introduce taxes other than income taxes. Such a society might want to encourage savings, for example. But these taxes would not be responses to the problem now under consideration. Should unearned (investment) income be taxed under the present argument? I assume that unearned income reflects skill in investment as well as preferences for later consumption, in which case that argument would extend to taxing such income. Since I am not considering, in this essay, the problem of later generations, I do not consider inheritance or estate taxes at all.

unused talent is a waste. Someone with a good eye or a skilled hand conceives a picture of what would make his life valuable that someone more clumsy would not.

So we cannot hope to fix the rates of our income tax so as to redistribute exactly that part of each person's income that is attributable to his talent as distinguished from his ambitions. Talents and ambitions are too closely intertwined. Can we do better by proceeding on a slightly different tack? Can we aim to fix rates so as to leave each person with the income he would have had if, counterfactually, talents for production had all been equal? No, because it is impossible to say, in any relevant way, what sort of world that would be. We should have to decide what sort and level of talent everyone would have equally, and then what income people exploiting those talents to different degrees of effort would reach. Should we stipulate that in that world everyone would have the talents that the most talented people in the real world now have? Do we mean, by "the most talented people," the people who are able to earn the most money in the actual world if they work single-mindedly for money? But in a world in which everyone could hit a high inside pitch, or play sexy roles in films, with equal authority, there would probably be no baseball or films; in any case no one would be paid much for exercising such talents. Nor would any other description of the talents everyone would be supposed to have in equal degree be any more help.

But though this crude counterfactual exercise must fail, it suggests a more promising exercise. Let us review our situation. We want to find some way to distinguish fair from unfair differences in wealth generated by differences in occupation. Unfair differences are those traceable to genetic luck, to talents that make some people prosperous but are denied to others who would exploit them to the full if they had them. But if this is right, then the problem of differential talents is in certain ways like the problem of handicaps we have already considered.

V. Underemployment Insurance

Though skills are different from handicaps, the difference can be understood as one of degree: we may say that someone who cannot play basketball like Wilt Chamberlain, paint like Piero, or make money like Geneen, suffers from an (especially common) handicap. This description emphasizes one aspect of skills, which is their genetic and, hence, luck component, at the expense of hiding the more intimate and reciprocal play we noticed between skills and ambitions. But it also points to one theoretical solution to the problem of identifying at least the minimum requirements of a fair redistribution policy responding to differences in skill. We may capitalize on the similarities between handicaps and relative lack of skill to propose that the level of compensation for the latter be fixed, in principle, by asking how much insurance someone would have bought, in an insurance sub-auction with initially equal resources, against the possibility of not having a particular level of some skill.

Of course, there is no actual insurance market against lack of what we ordinarily take to be skill, as there is an insurance market against catastrophes that result in handicap. For one thing, a person's level of skills is sufficiently fixed and known, at least roughly, before that person enters the insurance market, so that lack of skill is primarily a matter of history rather than future contingency. (There are other reasons as well, which we shall have to identify in a moment.) But let us nevertheless try to frame a hypothetical question something like the question we asked in the case of handicaps. Suppose an imaginary world in which, though the distribution of skills over the community were in the aggregate what it actually is, people for some reason all had the same antecedent chance of suffering the consequences of lacking any particular set of these skills, and were all in a position to buy insurance against these consequences at the same premium structure. How much insurance would each buy at what cost? If we can make sense of that question, and answer it even by fixing rough lower limits on average, then we shall have a device for fixing at least the lower bounds of a tax-and-redistribution program satisfying the demands of equality of resources.

There are several ways in which we might construct a hypothetical or imaginary insurance market of that sort. We might try to imagine, for example, that people are ignorant of the skills they actually have, though they know how many people will turn out to have each skill, and therefore what their own chances are. People might then be supposed to insure against turning out to lack some particular skill at some particular level, either a very precise skill like the ability to capture September light at dusk in oil, or a more general skill, like a very good memory or a quick way with numbers. This model would be very like the model we constructed for handicaps, and would therefore provide theoretical continuity for our theory as a whole. We might even propose to integrate the two hypothetical insurance markets by taking seriously the suggestion that the lack of some skill is just another handicap, and simply asking how many so-called skills would find their way into a general market for catastrophe insurance.

But this model for the hypothetical insurance market for skills is subject to a certain objection. We noticed, in considering the hypothetical insurance market for handicaps the following difficulty. There is a certain indeterminacy in the issue of what ambitions and tastes someone who is handicapped would have if he were not, and this indeterminacy infects the question of how much of what insurance he would then buy. The indeterminacy is manageable in the case of ordinary handicaps, because generalizations are nevertheless possible. But it would not be manageable in the case of skills, because if we suppose that no one has any idea what talents he has, we have stipulated away too much of his personality to leave any intelligible base for speculation about his ambitions, even in a general or average way. The connection between talents and ambitions, which I described earlier, is much closer than that between ambitions and handicaps—it is, for one thing, reciprocal—and much too close to permit that sort of counterfactual speculation.

So let us suppose, not that people are wholly ignorant of what talents they have, but rather that for some other reason they do not have any sound basis for predicting their economic rent—what income the talents they do have can produce. Or even whether the economic situation will be such that these talents will find any employment at all. There are, of course, many different ways of imagining such a state of affairs, and it does not much matter, for present purposes, which we select. So let us fall back on our immigrants once again. Suppose that, before the initial auction has begun, information about the tastes, ambitions, talents, and attitudes toward risk of each of the immigrants, as well as information about the raw materials and technology available, is delivered to a computer. It then predicts not only the results of the auction but also the projected income structure –the number of people earning each level of income–that will follow the auction once production and trade begin, on the assumption that there will be no income tax.

Now the computer is asked a further hypothetical question. Assume each immigrant knows the projected income structure but is ignorant of the computer's data base, except for its information about himself, and is therefore radically uncertain what income level his own talents would permit him to occupy. He supposes, in fact, that he has the same chance as anyone else of occupying any particular level of income in the economy, though he takes the number projected for that level into account. Assume that there is no monopoly in insurance, and that insurance firms offer policies of the following sort. Insurance is provided against failing to have an opportunity to earn whatever level of income, within the projected structure, the policy holder names, in which case the insurance company will pay the policy holder the difference between that coverage level and the income he does in fact have an opportunity to earn.[10] Premiums will vary with the level of coverage chosen, must be the same for everyone at any particular coverage level, and will be paid, not out of the policy holder's initial stock of resources (or clamshells) but rather from future earnings after the auction at fixed periods. How much of such insurance would the immigrants, on average, buy, at what specified level of income coverage, and at what cost?

That problem seems amenable, at least in principle, to the various types of analysis that economists devote to problems of decision making under uncertainty, and there is no reason to doubt that the computer could furnish an answer. Even without the computer's information and powers, we can make some general observations about what

10. Other forms for the insurance market we are imagining are possible, but those I have considered seem to produce roughly the same results. Amartya Sen has suggested to me, for example, that the insurer might offer a policy guaranteeing the named level of coverage to every policyholder, but making the premium depend on the economic rent people turn out to have. This would not, I think, produce different results from the arrangement I describe as elaborated in the next section, and I think it useful to consider, as I do there, why these elaborations would be necessary.

it is likely to predict. Economists make a rough distinction between two kinds of decisions under uncertainty. An insurance problem is posed when a small cost purchases reimbursement for an unlikely but serious loss. A gambling problem is posed when a small cost purchases a small chance of a large gain. Let us define a financially advantageous bet of either of these types as a bet such that the cost of the bet is less than the amount of the return if "successful"–if the covered risk eventuates or if the bet is won–discounted by the improbability of success. If I am offered insurance for $1 against a loss of $10 that is equally likely as not to occur, or a ten-to-one bet at any size that a coin flip will come up heads, these are both financially advantageous bets. A bet is financially disadvantageous if the cost of the bet exceeds the expected return so calculated. Let us say that someone is risk-neutral if he will accept any financially advantageous bet and reject any financially disadvantageous bet, no matter what the size or other character of the bet.

Commercial insurance companies and commercial bookmakers will offer only financially disadvantageous bets, of course, because their income must equal not only the expected return to the policy holders and bettors but also their costs, including opportunity costs. So if everyone were risk-neutral no one would buy insurance or bet on the numbers or pools. But almost no one is risk-neutral over the full range of his utility curve: for almost everyone the marginal utility of more money declines over at least part of the graph that pictures how his welfare behaves as a function of his income. It is fairly easy to see how this explains the phenomenon of commercial insurance (though of course any explanation of why rates for particular policies are what they are would require more detailed information about these utility functions and would be much more complex). Suppose there is a one-in-ten chance that my $50,000 house will burn down in the next year, and I am offered full insurance at a cost of $6,000. I am offered the choice, that is, between a certainty of $44,000 (if I purchase insurance) and a gamble with an expected return of $45,000 (if I do not). If the loss of my house would be more than nine times as serious as the loss of $6,000 (because, for example, I could not find or borrow enough money to build a suitable house, my marriage would dissolve, and my children become delinquent) then it is worth my while to buy the insurance, though it is a financially disadvantageous bet. It is much harder to explain gambles. (Kenneth Arrow, discussing gambling, quoted the preacher who reached a sticky point in theology and said that the problem was a very difficult one which the congregation should face firmly and then pass on.) It is perhaps necessary to suppose that gamblers either mistake the actual odds (because they think that luck is a lady) or attach value to uncertainty for its own sake; and though both these assumptions hold sometimes, it seems doubtful that they hold sufficiently often to explain how popular gambling is.

What can be said, against this background, about the two hypothetical insurance markets we have described? Our insurance market for handicaps is sufficiently like ordinary insurance markets and requires no special comment. But the hypothetical insurance market we just described for talents is different, in part because it seems, at first blush, to allow for decisions that look much more like gambles than insurance. For it might seem that many immigrants would leap at the chance to buy a policy that would protect them against not having the very highest income projected for the economy, and would pay them, if they do not, the difference between the great income and what they actually can earn. But in fact that policy would be a very poor wager indeed.[11] We take it as given that insurance at that level would be a financially disadvantageous bet. Otherwise it would not be offered by the insurance firm. So if it is a good bet, it is good on grounds of expected welfare rather than financial grounds, as is my insurance policy on my house. But the bet is much more likely to be silly than sound in welfare terms.

Since (unlike lottery tickets generally) the chances of "winning" are extremely high—very few immigrants will turn out to have that maximum earning power—the cost of the premium will be extremely high as well.[12] It will approach the value of the projected return if the

11. If I am wrong in this, the hypothetical insurance argument would insist on radical redistribution and substantial wealth equality. So the scheme would offer an argument for that consequence, on that assumption.

12. I have neglected the question of the technology that will be available for the insurance firms to decide and prove who has what level of talent, and the cost of that technology. I assume that the computer will have that information, as part of its technological data base, and will use it to predict the premium structure and other incidents of the contract of insurance. When I speak of

risk eventuates. So someone who buys this insurance faces an extremely high chance of gaining very little. Suppose he loses, however; suppose he is one of those who does have the maximum earning power. He is now in a much worse position than if he had never insured, because he must now work at close to his top earning capacity just to pay the high premium for his insurance on which he collected nothing—just, that is, to break even. He will be a slave to his maximum earning power.[13]

Now just how bad a bargain this is will depend upon facts not specified, including the question of how many people can be expected to have the talents necessary to earn at the highest level. But it is likely to be a very bad bargain in any case. It is very different from the situation that apparently tempts large numbers of people to make financially disadvantageous bets on vast lotteries, which is the prospect of a small chance of a large fortune in return for a very small certain cost. This insurance decision would be the very different financially disadvantageous bet of a very small chance of a very great loss in return for the very large chance of a very small gain, and nothing in the literature of the psychology of gambling (except perhaps the literature of Russian roulette) supports the idea that bets of that character would be popular.

Nor does the explanation of why people purchase ordinary financially disadvantageous insurance policies offer it any support either. I buy insurance on my house because the marginal utility loss of an uncompensated fire is so much greater than the utility cost of the premium. But considerations of marginal utility would, in anything, condemn rather than support any immigrant's bet that he would not have the skills necessary to earn the highest income. For that bet pits the almost certain prospect of a tiny and probably unnoticeable welfare gain against the tiny chance of an enormous welfare loss on financially disadvantageous terms. Of course, we make assumptions

"winning" or "losing" the insurance bet, I mean qualifying for compensation or failing to qualify under the incidents the computer predicts.

13. He may also run the even graver risk of losing under the tests specified in the policy and yet not, in fact, having the ability to earn the covered level. I do not emphasize this risk because it assumes failures of technology about which it is impossible to speculate. I therefore assume, arguendo, that no one will be in that perilous position.

in presuming that almost no one would have a utility curve that would make that bet sensible in welfare terms. But it does seem plausible that almost no one would.

Does this argument prove too much? Does it prove that insurance against lacking skills, on the model we described, would almost always be a bad buy for almost everyone? If so, it would seem to follow that the hypothetical insurance device could not, after all, provide reasonable guides for redistribution through income tax. Or, perhaps worse, it might suggest that no such redistribution would ever be justified. But in fact the argument does not have this consequence, because the lower the income level chosen as the covered risk the better the argument becomes that most people given the chance to buy insurance on equal terms would in fact buy at that level. The argument becomes compelling, I think, well above the level of income presently used to trigger transfer payments for unemployment or minimum wage levels in either Britain or the United States.

The argument becomes stronger, as the chosen income level declines, for the following reasons. First, as the level declines, the odds that any particular person will have the talents necessary to earn that income, at full stretch, improve and, for a substantial section of income levels in normal economies, improve faster than the rate of that decline. Many more than twice as many people have the abilities necessary to earn the amount earned in the fiftieth percentile than in the ninety-ninth percentile of a normal income distribution. So the premium falls, and falls, at least over a considerable range, at a rate faster than the rate of the coverage. So, correspondingly, do the odds against "winning," of course, but the situation grows steadily closer to the normal case of insurance, in which people incur a small certain loss to prevent an unlikely great loss whose marginal utility consequences are serious enough to justify on the welfare space a financially disadvantageous transaction. For even though the financial loss in falling from, say, the seventieth to the sixtieth percentile in income is vastly greater than the loss in falling from the fortieth to the thirtieth, the welfare consequences will probably, on average, be much worse for the latter drop.

As the income level covered drops, moreover, so the penalties of "losing" the insurance bet, by turning out to have the abilities to earn at least that income, diminish in importance and, once again, at what seems likely to be a faster rate. Someone who "loses" in this sense must work hard enough to cover his premium before he is free to make the tradeoffs between work and consumption he would have been free to make if he had not insured. If the level of coverage is high then this will enslave the insured, not simply because the premium is high, but because it is extremely unlikely that his talents will much surpass the level that he has chosen, which means that he must indeed work at full stretch, and that he will not have much choice about what kind of work to do. Only one form of work, and that full time, will be likely to produce the income needed to pay the premium that is now his albatross. So his penalty has special welfare disadvantages not measurable in ordinary financial terms. It is these that make it appropriate to speak of his enslavement.

But as the level of coverage and hence the premium drop, these special welfare disadvantages are not simply mitigated, but entirely fall away. For it becomes likely that anyone who has the qualifications needed to earn at, say, the thirtieth percentile level will also have the talents to earn at a higher level, and so would retain a considerable freedom of choice about the character of work, and the mix of work and labor and additional consumption, that he prefers. Even if he just barely "loses" the insurance bet, and has exactly the talents needed to earn the level of income he covered, but no more, he will still probably retain a great freedom of choice. The premium will be small enough to sustain even if he works at a lower level of income than he could, particularly if he is willing to sacrifice consumer goods. There are, moreover, a greater variety of types of jobs that will produce a lower level of income than a higher (at least in most complex economies at least over a range of income levels) so that someone committed to earning as much as he can will have more choice of work, nevertheless, at that lower level. Even if he must work flat out, and has no choice in his work, his situation is very little worse than if he had taken no insurance. For if the coverage level is so low that almost everyone must earn it to have a decent life, he would have worked that way and practically that hard anyway, and if the premium is very low, as it then would be, he would not have to work much harder to cover it. He is not much differently enslaved by his talent if he insures than he would be enslaved by his lack of talent if he did not.

The hypothetical insurance market might nevertheless produce apparent anomalies. Suppose two people, Deborah and Ernest, both purchase insurance at the sixtieth percentile level. Deborah is beautiful and could in fact earn at the ninetieth percentile as a movie star. They have otherwise the same talents and interests, and these other talents would not earn at the sixtieth level. Ernest recovers under his policy, but Deborah does not. She is faced with the choice of a movie career, which she detests, or trying to pay the premium and the other expenses of her life from whatever salary she could earn at jobs she and Ernest would both prefer. Ernest can have both such a job and compensation under his policy, and is therefore much better off.[14] Is this unfair? Deborah is, as it turns out, enslaved by her singular talent. But this is because she ran the risk described in the text by purchasing insurance at a coverage level commanding a high premium and such that few jobs could produce income approaching that level. Ernest ran the same risk, but had better option luck. The anomaly is therefore only a further (and more complex) example of the undesirable welfare risks of insurance at a high level. If Deborah and Ernest had purchased insurance at a lower level, the premium would have been lower and Deborah would have had a much better choice of jobs other than a film career. They would still fare differently, but the difference would be much less, and would (arguably) then be an appropriate mark of the fact that Deborah had an option Ernest did not. In any case, this unfairness, if it is unfairness, would disappear in any plausible translation of the hypothetical insurance market into an actual tax scheme of the sort described in the next section.

VI. Tax as Premium

Now let us assume that the computer fixes the average coverage level that would be reached in the hypothetical insurance market, and declares some premium to be the premium that level would command. The premiums would be a sufficiently small proportion of the coverage so that (for the average person) the expected welfare of insuring

14. Thomas Scanlon provided this example.

would be higher than the expected welfare of not doing so. Can we translate that hypothetical insurance structure into a tax scheme? Can we base tax rates on the assumed premium, and then redistribute by paying those who do not have the ability to earn at the assumed coverage level the difference between that level and what they can earn?

We might assume that any tax system constructed to model the hypothetical insurance market in that way would suffer from serious defects. First, it seems unfair that everyone, rich and poor alike, should pay the same tax, but this would seem to be the consequence of modeling tax rates on hypothetical premiums. Second, the requirement that both the incidence and amount of payments from the fund depend on what the recipient could earn if willing seems inefficient and troublesome in a variety of ways. It might be very expensive to enforce that requirement, and in practice the requirement will tempt some people to cheat by hiding their abilities under a bushel. In any case, even honest people cannot know what they might earn at a given occupation without trying, and in the case of some professions, trying is impossible without half a lifetime of preparation. So a battery of new tests to discover latent talent would be necessary, and these would be vulnerable to many sorts of mistakes.

But these objections make certain assumptions about what the hypothetical insurance market would be like, and these assumptions are both unjustified and probably false. For suppose the insurance firms had offered, in place of the flat rate premium for a given coverage that the objections assume, a premium fixed as an increasing percentage of the income the policy owner turns out to earn. The premium of someone who barely earns the average coverage amount would be less than the premium the insurance market would have fixed on a flat-rate basis, though the premium of someone who earns much more would be much greater. The insurance firms would have reason to offer this different scheme if the total premiums paid would be more, and the immigrants buying insurance would also have reason to accept it if the change increased their expected welfare under the conditions of equal risk we have stipulated. Since we assume declining marginal utility of money over the range in question, as part of the assumptions on which we speculate that insurance would be bought at all, these conditions will be met. Immigrants would prefer a "bet" under which the cost of the bet will be an increasing function of their income, and will prefer it by enough to provide profit for the insurance firms even counting the increased administrative costs of a progressive premium scheme.

The firms would also have reason to reduce what insurers call the "moral hazard" of such insurance—the risk that insurance will make the covered even more likely to occur or the level of recovery higher if it does—and to pass on part of their savings to policy holders as inducements to accept the necessary constraints. One technique insurers now use for that purpose is co-insurance. First-person automobile policies provide, for example, that the owner must assume the first several hundred dollars of damage before the insurer makes up the balance. Co-insurance in our story means that if one of the immigrants is unable to earn the average coverage amount, he will receive somewhat less than that amount as compensation. Of course people will buy insurance at a higher coverage level with co-insurance than without, though at a lower premium than the higher amount would otherwise attract, which means that the average coverage level would be higher than under a scheme with no co-insurance. But if substantial savings for the insurance firm would result from reducing the moral hazard (and popular resentment about welfare cheats under present welfare schemes in the United States, for example, supposes that they would), then the savings in premium along the range should also be substantial, which argues that the presumed coverage would indeed be higher under a co-insurance regime. How much co-insurance would obtain—by how much the payment to those who fail to earn the covered amount would fall short of that amount and what the effect on the premium structure and the presumed coverage level would be—depends only on information that has already been given to the computer.

The problem and the cost of accuracy in determining people's actual abilities to earn are made less pressing for the insurer by co-insurance. It sharply reduces the motives people have for not earning at least the covered amount in order to claim that they cannot earn it. But the insurer has another device for reducing the cost of accuracy enough to lower premiums and so make the device attractive to policy holders. Since policy holders will almost always have more information about their abilities and opportunities than the insurer will, there is room for joint savings by assigning the burden of proof to the policy holders. This burden will be more severe for coverage at the higher levels, I expect, than at the lower, because it is often difficult to prove that one could not have had a career for which special training or education or experience is necessary unless one has undertaken these. But if I am right that the average coverage level would be relatively low, we need not pursue that problem. At the lower levels the proof will be easily provided by attempts to find employment that have failed, or by evidence of less than average general physical and mental abilities, and so forth.

So the actual insurance profile the computer would predict is likely to be much more complex than the simple structure our defective tax system copied. If the immigrants translate this more complex insurance profile into a tax scheme they reach a more recognizable pattern of tax. They might establish a graduated income tax financing transfer payments in the amount of the difference between the average coverage level less the co-insurance factor and what an applicant can plausibly argue is the highest income he can in fact command. This exercise, of course, neither need nor should stop at this point. Further reflection about the hypothetical insurance market might develop further refinements or adjustments to the corresponding tax scheme. And we might decide that a tax scheme should differ from the best approximation of that hypothetical market for other reasons. We might decide that a tax scheme so closely modeled on that market is offensive to privacy, or too expensive in administrative costs, or too inefficient in other ways. We might decide, for these or other reasons, that a scheme that tied redistribution to actual earnings rather than to ability to earn, for example, was a better second-best approximation to the ideal of mimicking the insurance market than any other scheme we could develop.

But I want to put aside, for this essay, any further study of these issues, because we have carried them far enough, I think, to justify turning instead to the question waiting in the wings. Is the general approach sensible? Is a tax scheme constructed as a practical translation of a hypothetical insurance market, which assumes equal initial assets and equal risk, a proper response to the problem of differential talents under equality of resources?

It might be criticized from two standpoints: either that it does not justify enough redistribution, or that it justifies too much. The latter objection seems the weaker of the two. Recall the competing constraints we discovered. Equality requires that those who choose more expensive ways to live—which includes choosing less productive occupations measured by what others want—have less residual income in consequence. But it also requires that no one have less income simply in consequence of less native talent. Any objection that the transfer payments guided by the hypothetical insurance market are too great must show a lively danger that the first of these requirements has been given insufficient weight. But if the hypothetical insurance market justifies the selection of a particular level as the average assumed coverage, this argues strongly for the probability that any particular immigrant is ready to work at one of the occupations that could produce income at that level rather than have a lesser income, if the choice were available. Otherwise he would have run too great a risk by taking on a premium that makes sense only for someone ready to earn the income needed to support it, particularly if the average level is sufficiently low so that the odds were great that he would have to do exactly that. He may be ready to do so because a wide range of occupations, suiting very different personalities, would produce that level of income, or because that level is necessary to lead what popular culture considers an acceptable style of life, or, more probably, both together. Of course, this favorable counterfactual would not, in fact, hold in the case of everyone who lacks appropriate talents, or does not find a job in the employment market for some other reason. Some such immigrants would not have taken insurance at the average coverage level even if it were available. But this presents a question of valuing the relative importance of the two constraints on a theory of redistribution for differential talents. Taking the average coverage level as decisive, which we have done, is an appropriate way of weighting them equally.[15] It supposes that it is at least no worse to err on the

15. As in the case of handicaps, I have chosen to make premiums and therefore tax payments turn on the average coverage level as a simplifying device. I might, of course, have chosen either the median or the mode of coverage se-

side of redistributing to someone who would not have insured than to deny redistribution to one who would have. Someone may be able to show that this is wrong, and thereby justify taking, as the standard for redistribution through taxation, a lower figure such that the odds are very strong, possibly even overwhelming, that any particular person would have insured at that level. But I do not know of any arguments for that view. Perhaps the difference between the average coverage level and some lower level of that sort would not be very great. Perhaps, that is, utility curves of most people would look very much alike for the lower percentiles of possible coverage. But this is one of the legion of quasi-technical questions that I make no attempt to discuss here.

The opposite objection, that transfer payments based on the average coverage level are not enough, is more difficult to answer. It might be supported in two different ways. It might be said that the hypothetical insurance approach is the wrong approach through which to attempt to compromise the two requirements we discovered; or that the hypothetical insurance approach is the right approach but requires transfer payments at a higher level. The second argument presumably agrees with the objection just described, that the two requirements of equality in wage structure are not of equal weight, but insists that it is much worse to deny payments to someone who would have insured than to award them to someone who would not have done so. It insists, in other words, that the level of assumed coverage chosen should be some level such that it is very unlikely that anyone would have insured above that level. Once again, it is worth noticing that this level might not be much above the average coverage level. We saw, earlier, reasons why almost no one would insure at a very high income level in any case. But once again I shall not explore the technical issue raised, but simply set the suggestion aside until the substantive argument, in favor of special deference to one requirement of equality in wage structure over the other, has been made.

lected in the hypothetical market instead of the average or mean. It is an interesting question whether either of these would be better. I chose the average on the assumption that our assessment of the chance of error in particular cases (the chance that the "premium" extracted differs from what the individual in question actually would have paid in the hypothetical market) should reflect the amount as well as the fact of that difference.

But we must now state and consider the important argument that the hypothetical insurance market is altogether the wrong approach to the problem of reconciling these two requirements, because it undervalues the transfer payments that those whose talents are not in great demand should receive. The hypothetical insurance market approach aims to put such people in the position they would have been in had the risk of their fate been subjectively equally shared. But it does not make them as well-off in the end as those whose talents are in more demand, or those with similar talents lucky enough to find more profitable employment. Some people (movie stars and captains of industry and first basemen) in fact earn at a rate far beyond the rate of coverage any reasonable person would choose in an insurance market, as our inspection showed. The hypothetical insurance market approach is beside the point (it might be said) exactly because it provides no answer to someone who is unable to find a job, points to the movie star and declares, perfectly accurately, that he would do that work for that pay if asked. The fact that no one would buy coverage at a movie-star level in an equal insurance market simply underscores the injustice. The movie star had no need to buy that insurance. He won his life of luxury and glamor without it. The brute fact remains that some people have much more than others of what both desire, through no reason connected with choice. The envy test we once seemed to respect has been decisively defeated, and no defensible conception of equality can argue that equality recommends that result.

This is a powerful complaint, and there is no answer, I think, but to summarize and restate our earlier arguments to see if they can still persuade with that complaint ringing in our ears. Let us return to the immigrants. Claude cannot argue, on grounds of equality, for a world in which he has the movie star's income. The immigrants cannot create a world in which everyone who would be willing to work movie-star hours can have movie-star pay. If Claude is unhappy with his situation, even after the tax scheme is put into play, he must propose a world in which no one will have such an income and his income will be relatively (and perhaps absolutely) higher in consequence. But whichever such world he proposes will be changed not only for those who under our scheme would have more than he does, but for everyone else as well, including those who for one reason or another, including their preferences for work, leisure, and consumption, will have less. If, for example, no one can earn movie-star wages, people who wish to watch movies may perhaps find very different fare available which, rightly or wrongly, they will not regard as highly as what they now have. It is, of course, impossible to say in advance just what the consequences of any profound change in an economic system would be, and who would gain or lose in the long run. These changes could not be properly charted along any one simple dimension. They could not be measured simply in the funds or other "primary goods" available to one or another economic class, for example. For they also affect the prices and scarcity of different goods and opportunities that members of any particular class, even economic class, will value very differently from one another. That is exactly why the immigrants chose an auction, sensitive to what people in fact wanted for their lives, as their primary engine for achieving equality.

So though Claude may truly say that the difference between him and the movie star does not reflect any differences in tastes or ambitions or theories of the good, and so does not in itself implicate our first, ambition-sensitive requirement of equality in wage structure, he could not recommend any general change in relative economic positions that would not wreak wholesale and dramatic changes in the positions of others, changes which do implicate that requirement. Of course, this fact does not in itself rule out any changes that Claude might propose. On the contrary, the status quo achieved by laissez-faire production and trade from an equal start has no natural or privileged status, as I have been at pains to emphasize, particularly my argument against the "starting-gate" theory of equality. If Claude can show that a proper conception of equality of resources recommends some change, the fact that many people from all ranks would be then worse-off, given their particular tastes and ambitions, provides no objection, any more than the fact that Claude is worse-off without some change in itself provides an argument in favor of that change. I mean to emphasize only that Claude needs some argument in favor of the change he recommends which is independent of his own relative position. It is not enough for him to point to people, even those of the same ambitions and tastes as himself, who do better as things are.

The argument from the hypothetical insurance market is such an argument. It contrasts two worlds. In the first those who are relatively disadvantaged by the tastes and ambitions of others, vis-à-vis their own talents to produce, are known in advance and bear the full consequences of that disadvantage. In the second the same pattern of relative disadvantage holds, but everyone has subjectively an equal antecedent chance of suffering it, and so everyone has an equal opportunity of mitigating the disadvantage by insuring against it. The argument assumes that equality prefers the second world, because it is a world in which the resources of talent are in one important sense more evenly divided. The hypothetical insurance argument aims to reproduce the consequences of the second world, as nearly as it can, in an actual world. It answers those who would do better in the first world (who include, as I said, many of those who would have more money at their disposal in the second) by the simple proposition that the second is a world that, on grounds independent of how things happen to work out for them given their tastes and ambitions, is more nearly equal in resources.

The availability of that argument is no bar to the production of other arguments showing how some further change would improve equality of resources still further. Let me remind you, however, that it is hard to anticipate how great a motive we should have to search for further arguments if the hypothetical insurance argument were in fact accepted and enforced in, for example, our immigrant case. That would depend, among other things, on how high a level of income could be shown to be the average coverage level in that society. It might be that wealth disparities would be so greatly reduced by the features of the economy we have already described that we would be much less troubled than we might suspect in advance by the wealth inequalities that would remain. Indeed it might be that the costs in overall efficiency of even those features would be so great that those who are prepared to compromise equality of resources either for general utility or in service of some strategy of making the worst-off as well-off as possible, would argue that even that much equality would be condemned by their more embracing conception of justice.

Of course, many of the political philosophers and theorists who object to inequality are concerned, not simply with how poor those at the bottom are in absolute terms, but with what might be called the moral costs of a society with substantial wealth inequality, costs that remain, and indeed are sometimes exacerbated, when the position of the least well-off is sharply improved but the inequality remains. It would be a mistake, however, to suppose that the bizarre and mutually dependent attitudes about wealth that mark our own society—the ideas that the accumulation of wealth is a mark of a successful life and that someone who has arranged his life to acquire it is a proper object for envy rather than sympathy or concern—would find any footing in an economic system that is free of genuine poverty and that encourages people, as the initial auction encourages them, to see bank account wealth as simply one ingredient among others of what might make a life worth living. For in our world, these attitudes are sustained and nourished by the assumption that a life dedicated to the accumulation of wealth or to the consumption of luxuries—a major part of whose appeal lies just in the fact that they are reserved for the very rich—is a valuable life for people given only one chance to live. That proposition comes as close as any theory of the good life can to naked absurdity.

It is no doubt an important question for social psychology and intellectual history how that proposition finds a footing in any society. It has, after all, been condemned in all literature or any other form of art taken seriously for very long in even deeply capitalistic communities, and though I understand the possibility, that its rejection in art might be parasitic upon its unthinking acceptance in life, the protests of even the most popular forms of art nevertheless deepen the mystery. My present point is much more banal than any attempt to solve that mystery. It is simply this: that we are so ignorant about the complex genealogy of the implausible attitudes about wealth that we find among us, which those who point to the moral costs of the market system deplore, that we would do wrong to assume in advance that these same attitudes will rise in a market system whose very point is to encourage the kind of reflective examination about costs and gains under which these attitudes would seem most likely to shrivel and disappear.

But it is nevertheless important to try to discover arguments showing that equality of resources, as a distinct ideal, would recommend erasing even those wealth differentials that the hypothetical insurance argument would permit, and this project is not threatened by my uncertainty whether we should feel dismayed, or find our intuitions undermined, if we did not in fact discover any. I do not doubt that such arguments can be found, and it is part of my purpose to provoke them. But it is worth mentioning certain arguments that do not seem promising. It might be said, for example, that equality of resources would approve a different world still, in which people had in fact equal talents for production, more than either of the other two I described, so that we ought to strive to create a system in which wealth differences traceable to occupation were no greater than they would be in that world. There is an important point locked in that claim, which is that an egalitarian society ought, just in the name of equality, to devote special resources to training those whose talents, as things fall out, place them lower on the income scale. That is part of the larger question of an egalitarian theory of education, which I have not even attempted to take up here. But the more general point suffers from the fact that we could not even begin to replicate the wealth distribution that would hold in that different world without making assumptions about the mix of talents that everyone in that world would share in equal abundance, and no specification of the mix could be neutral amongst the various ambitions and tastes in the real world in which we attempt that replication.

Suppose someone says simply (and with creditable impatience) that equality of resources just *must* prefer a world in which people have more nearly equal wealth than they are likely to have in a world of free trade, even against a background of equal initial wealth and even as corrected by the hypothetical insurance market. To deny that (it might be said) is simply to prefer other values to equality, not to state an acceptable conception of equality itself. That is, of course, exactly what my arguments have been meant to challenge. Once we understand the importance, under equality of resources, of the requirement that any theory of distribution must be ambition-sensitive, and understand the wholesale effects of any scheme of distribution or redistribution on the lives which almost everyone in the community will want and be permitted to lead, we must regard with suspicion any flat statement that equality of resources just must be defined in a way that ignores these facts. Equality of resources is a complex ideal. It is probably (as the various arguments we have canvassed in this essay suggest) an indeterminate ideal that accepts, within a certain range, a variety of different distributions. But this much seems clear: any defensible conception of that ideal must attend to its different dimensions, and not reject out of hand the requirement that it be sensitive to the cost of one person's life to other people. The present suggestion, that genuine theories of equality must be concerned only with the quantity of disposable goods or liquid assets people command at a particular time, is a piece of pre-analytic dogma that does not, in fact, protect the boundaries of the concept of equality from confusion with other concepts, but rather thwarts the attempt to picture equality as an independent and powerful political ideal.

VII. Other Theories of Justice

It is hardly worth repeating how far the remarks here fall short of a full theory of equality of resources even under simple and artificial conditions like those of the immigrant society. I have said nothing, for example, about how far equality, properly understood, constrains people from giving to others what they are entitled to keep and use for themselves. That question includes, of course, the troublesome issue whether those who have amassed wealth through sacrifices in their own lives should be allowed to pass this on as extra wealth for their children. Nor have I said anything about what accommodation an equal distribution of resources should make for radical changes in people's minds about how they wish to spend their lives. Is someone entitled to a fresh stock of resources when he rejects his former life and wants a fresh start? Suppose he is a profligate who has wasted his initial endowment and now finds himself with less than he needs to provide even for basic needs in later life.

These questions are of great theoretical interest, and of central practical importance when we come to ask what the requirement of an equal start, which in our immigrant world could be satisfied by an equal initial auction, would mean for the real world. I have also set aside the entire issue of equality of political power, for another occasion, though as I noticed at the beginning it is quite illegitimate to regard political equality as an issue entirely distinct from economic equality. Nevertheless we have, I think, covered enough ground here so that it might be useful to contrast the direction in which we are traveling with those taken by certain prominent theories of justice now in the field.

It should be reasonably plain how the conception of equality defended here contrasts with equality of welfare, the theory considered in Part I of this essay. There is nothing in the idea of an equal initial auction, followed by trade and production constrained by taxation mimicking hypothetical insurance markets, that either aims at equality in any concept of welfare or makes convergence toward such equality likely. Indeed, there is no place in the theory, as developed so far, even for comparisons of the welfare levels of different people. The theory does make use of the idea of individual utility levels, for example in the calculations it recommends about how people would behave in certain hypothetical markets. But these calculations use only the rather antiseptic concept of utility proposed by von Neumann and Morgenstern, among others, rather than any of the more complex and judgmental conceptions of welfare that are necessary for interpersonal comparisons, whose shortcomings I discussed in Part I.

There might well be interesting connections between the theory described here and some form of utilitarianism, which commands the maximization of some concept of welfare overall rather than equality in its distribution. Of course our theory as a whole could not be expected to maximize any concept of welfare across society, except under special and quite extravagant assumptions about individual utility functions. The assumption that people should enter economic activity with equal initial resources, for example, would count as a dubious theorem for a utilitarian, rather than as a cardinal axiom. Nevertheless the idea of an equal auction for goods and services, from the base of an equal abstract distribution of economic power, might seem to suggest a utilitarian strain in the theory, because an auction would promote overall utility better than a more mechanical division of available goods into equal lots. I do not think that this mild similarity, insofar as it does exist, is entirely accidental. On the contrary utilitarianism owes part of its appeal, I think, to the fact that in certain circumstances a distribution that maximizes overall marginal utility from an intuitively fair basic distribution would also be recommended by the present conception of equality. This is even more plainly true of the wealth-maximization theory of justice, which is a cousin of utilitarianism, and is presently popular among academic lawyers. This theory argues that, at least in certain circumstances, a distribution that maximizes marginal wealth is fair. The circumstances in which the wealth-maximization theory seems intuitively plausible[16] are in fact just the circumstances in which our conception of equality would probably recommend the decision that in fact maximizes wealth. The overall fit between our conception of equality and the wealth-maximization theory, indeed, is likely to be closer than the fit of our conception of equality with utilitarianism.[17] But in both cases, so far as the present argument holds, the connection is one-way only. A distribution that fits the two theories is fair because equality, and not the maximization of either utility or wealth, recommends it.

There are also at least superficial connections between the theory of equality of resources suggested here and various forms of the Lockean theory of justice in private property, particularly in Robert Nozick's distinguished and influential version. Of course the differences, even on the surface, are more striking. There is no place in a theory like Nozick's for anything like the idea of an equal distribution of abstract economic power over all the goods under social control. But both Nozick's theory and equality of resources as described here give a prominent place to the idea of a market, and recommend the distribution that is achieved by a market suitably defined and constrained. It may be that those parts of Nozick's arguments that seem intuitively most persuasive are based on examples where the present theory would reach very similar results.

The famous Wilt Chamberlain example is a case in point. Nozick supposes an equal distribution of wealth, followed by uncoerced trades to mutual advantage in which each of many people pays a small sum

16. As it often does, I think, in hard cases at common law when those who would benefit and those who would lose by the introduction of any new rule of law are or must be assumed to be classes roughly equal in their command over resources. See R. Posner, "The Ethical and Political Basis of the Efficiency Norm in Common Law Adjudication," *Hofstra Law Review* 8 (1980): 487.

17. R. Dworkin, "Why Efficiency?" *Hofstra Law Review* 8 (1980): 563.

to watch Chamberlain play basketball, after which he grows rich and wealth is no longer equal. Equality of resources would not denounce that result, considered in itself. Chamberlain's wealth reflects the value to others of his leading his life as he does. His greater wealth, at the end of the process, is of course traceable mainly to his greater talent, and only in small part, we may assume, to the fact that he is willing to lead a life that others would not be. But almost no one would have purchased, in the hypothetical insurance market we described, insurance against not having talents that would provide such wealth. That insurance would be, for almost everyone, a strikingly irrational investment. So our discussion would not justify taxing any of Chamberlain's wealth for redistribution to others not so fortunate, if we attend only to the fact, as Nozick does, that others have much less wealth than he does.

But our discussion left open, as Nozick's did not, that arguments justifying such redistribution might be found in a more thorough study of the actual circumstances in which wealth like Chamberlain's is accumulated. Suppose Chamberlain plays his game, not in a community whose only wealth disparity lies in his enormous wealth against the equal wealth of all others, each of whom has only slightly less than the most equal distribution we can imagine, but in Philadelphia in the early 1970s. Now a great many people earn less than the average presumed coverage of a plausible hypothetical insurance market for that society, so even if we assume that the complex wealth differences we find are all traceable to lack of talent rather than lack of an equal start (which is absurd), we are still required to put in place a tax system for redistribution to them, and Chamberlain will be required to contribute to that system. Indeed, since our argument justified the conclusion that premiums in the hypothetical insurance market would lie at progressive rates, based on income realized, Chamberlain would be required to contribute more than anyone else, both absolutely and as a percentage of his income. When the discussion is broadened in this way, equality of resources travels very far from the boundaries of the nightwatchman state.

The difference between the use the two theories make of the market is therefore clear enough. For Nozick the role of the market in justifying distributions is both negative and contingent. If someone has justly acquired something, and chooses to exchange it with someone else in return for the latter's goods or services, then no objection can be taken, in the name of justice, to the distribution that results. The history of the transaction insulates it from attack and, in this negative way, certifies its moral pedigree. There is no room, in this theory, for hypothetical markets of any form, except in the special case of restitution for demonstrated injustice in the past. For Nozick does not use the market (as, for example, some wealth-maximizers do) simply to define another of what he calls "patterned" theories of justice. Justice consists, not in the distribution that a fair market of rational persons would reach, but in the distribution that has actually, as a matter of historical contingency, been reached by a process that might, but need not, include any market transactions at all.

Under equality of resources the market, when it enters, enters in a more positive but also more servile way. It enters because it is endorsed by equality, as the best means of enforcing, at least up to a point, the fundamental requirement that only an equal share of social resources be devoted to the lives of each of its members, as measured by the opportunity cost of such resources to others. But the value of actual market transactions ends at just that point, and the market must be abandoned or constrained when analysis shows, from any direction, that it has failed in this task, or that an entirely different theoretical or institutional device would do better. Hypothetical markets are plainly of comparable theoretical importance to actual markets for this purpose. We are less certain about their results, but have a great deal more flexibility in their design, and the objection that they have no historical validity is simply beside the point.

I shall try to say something, finally, about the connections and differences between our conception of equality of resources and John Rawls' theory of justice. That theory is sufficiently rich to provide a question of connection at two different levels. First, how far do the arguments in favor of equality of resources, as described, follow the structure of argument Rawls deploys? How far do they depend, that is, on the hypothesis that people in the original position Rawls described would choose the principles of equality of resources behind the veil? Second—and independently—how far are the requirements of equality of resources different from the two principles of justice that Rawls suggests people in that position would in fact choose?

It is obviously better to start with the second of these questions. The comparison in point is that between equality of resources and Rawls' second principle of justice, whose main component is the "difference" principle which requires no variation from absolute equality in "primary goods" save as works to the benefit of the worst-off economic class. (Rawls' first principle, which establishes what he calls the priority of liberty, has more to do with the topics I have set aside as belonging to political equality.) The difference principle, like our conception of equality of resources, works only contingently in the direction of equality of welfare on any conception of welfare. If we distinguish broadly between theories of equality of welfare and of resources, the difference principle is an interpretation of equality of resources.

But it is nevertheless a rather different interpretation than our conception. From the standpoint of our conception, the difference principle is not sufficiently fine-tuned in a variety of ways. There is a conceded degree of arbitrariness in the choice of any description of the worst-off group, and this is, in any case, a group whose fortunes can be charted only through some mythical average or representative member of that group. In particular, the structure seems insufficiently sensitive to the position of those with natural handicaps, physical or mental, who do not themselves constitute a worst-off group, because this is defined economically, and would not count as the representative or average member of any such group. Rawls calls attention to what he calls the principle of redress, which argues that compensation should be made to people so handicapped, as indeed it is, in the way I described, under our conception of equality. But he notes that the difference principle does not include the principle of redress, though it would tend in the same direction insofar as special training for the handicapped, for example, would work to the benefit of the economically worst-off class. But there is no reason to think that it would, at least in normal circumstances.

It has often been pointed out, moreover, that the difference principle is insufficiently sensitive to variations in distribution above the worst-off economic class. This complaint is sometimes illustrated with bizarre hypothetical questions. Suppose an existing economic system is in fact just. It meets the conditions of the difference principle because no further transfers of wealth to the worst-off class would in fact improve its position. Then some impending catastrophe (for example) presents officials with a choice. They can act so that the position of the representative member of the small worst-off class is worsened by a just noticeable amount or so that the position of everyone else is dramatically worsened and they become almost as poor as the worst-off. Does justice really require the much greater loss to everyone but the poorest in order to prevent a very small loss by them?

It may be a sufficient reply to such questions that circumstances of that sort are very unlikely to arise, and that in fact the fates of the various economic orders are or can easily be "chained" together so that improvements in the worst-off class will in fact be accompanied by improvement in at least the other classes just above them. But this reply does not remove the theoretical question whether, in all circumstances, it is really and exclusively the situation of the worst-off group that determines what is just.

Equality of resources, as described here, does not single out any group whose status has that position. It aims to provide a description (or rather a set of devices for aiming at) equality of resources person by person, and the considerations of each person's history that affect what he should have, in the name of equality, do not include his membership in any economic or social class. I do not mean that our theory, even as so far detailed, claims any impressive degree of accuracy for those devices. On the contrary, even in the artificially simple case we treated, we several times had to concede speculation and compromise, and sometimes even indeterminacy, in the statement of what equality would require in particular circumstances. But the theory nevertheless proposes that equality is in principle a matter of individual right rather than group position. Not, of course, in the sense that each has a predetermined share at his disposal regardless of what he does or what happens to others. On the contrary, the theory ties the fates of people together in the way that the dominant devices of actual and hypothetical markets are meant to describe. But in the different sense that the theory supposes that equality defines a relation among citizens that is individualized for each, and therefore can be seen to set entitlements as much from the point of view of each person as that of anyone else in the community. Even when our theory helps itself to the idea of an average utility curve, as it does in the construction of hypothetical insurance markets, it does so as a matter of probability judgments about particular people's particular tastes and ambitions, in the interests of giving them what they are, as individuals, entitled to have, rather than as part of any premise that equality is a matter of equality between groups. Rawls, on the other hand, assumes that the difference principle ties justice to a class, not as a matter of second-best practical accommodation to some deeper version of equality which is in principle more individualized, but because the choice in the original position, which defines what justice even at bottom is, would for practical reasons be framed in class terms from the start.[18]

It is impossible to say, a priori, whether the difference principle or equality of resources will work to achieve greater absolute equality in what Rawls calls primary goods. That would depend upon circumstances. Suppose, for example, that the tax necessary to provide the right coverage for handicaps and the unemployed has the long-term effect of discouraging investment and in this way reducing the primary-goods prospects of the representative member of the worst-off class. Certain individual members of the worst-off group who are handicapped or who are and will remain unemployed would be better off under the tax scheme (as would, we should notice, certain members of other classes as well) but the average or representative member of the worst-off class would be worse off. The difference principle, which looks to the worst-off group as a whole, would condemn the tax, but equality of resources would recommend it nevertheless.

In the circumstances of the familiar bizarre questions just described, when a just noticeable loss to the representative member of the worst-off class, from a just base, could be prevented only by very substantial losses to those better off, the difference principle is committed to preventing that small loss even at that cost. Equality of resources, on the contrary, would be sensitive to quantitative differences of just the sort that those who take objection to Rawls' theory

18. J. Rawls, *A Theory of Justice* (Cambridge, MA: Harvard University Press, 1971), p. 98.

on that account believe should matter. If the base is an equal division of resources, this means, not that any transfer to the worst-off group would work to the long-run loss of that group, which it might or might not, but that any such transfer would be an invasion of equality because it would be unfair to others. The fact that those at the bottom do not have more would not indicate that it is impossible to give them more, but rather that they have all that they are entitled to have. If some economic catastrophe is now threatened, a government that allows a much greater loss to fall on one citizen, in order to avert a much smaller loss to a second, would not be treating the former as an equal, because, since equality in itself requires no further special attention to the second, that government must have more concern with his fate than it has for the fate of others. So if the loss threatened to the financially worst-off is indeed really inconsequential to him, as the bizarre question assumes, then that is an end of the matter.

But it does not follow that equality of resources turns into utilitarianism in the face of examples like these. It is, in fact, sensitive to more, or at least different, quantitative information than either the difference principle or utilitarianism is. For suppose the impending catastrophe threatens the worst-off group, not with a trivial loss as in the original question, but with a substantial loss, though not as great, in aggregate, as the loss threatened to those better-off. Equality of resources must ask whether the calculations of the hypothetical insurance market, and of the tax scheme in force, took adequate account of the risk of the threat now about to materialize. It might not have. The possibility of a substantial loss from the unexpected quarter, if it had been anticipated, might have led the average buyer in that market to purchase either catastrophe or unemployment insurance at a higher level of coverage, and this fact might affect an official's present decision about how to distribute the coming loss. He might be persuaded, for example, that allowing the loss to fall on those better-off, in spite of the overall welfare loss, would reach a situation closer to the situation all would have been in had the tax scheme better reflected what people would have done in the hypothetical market with that additional information.

Such contrasts in the practical advice that the difference principle and equality of resources offer in particular circumstances are in fact myriad, and these examples are meant only to be suggestive of others. These contrasts are organized around the theoretical distinction I have already noticed. The difference principle is tuned to only one of the dimensions of equality that equality of resources recognizes. The former supposes that flat equality in primary goods, without regard to differences in ambition, taste, and occupation, or to differences in consumption, let alone differences in physical condition or handicap, is basic or true equality. Since (once the priority of liberty is satisfied) justice consists in equality, and since true equality is just this flat equality, any compromise or deviation can be justified only on the grounds that it is in the interests of the only people who might properly complain of the deviation.

This uni-dimensional analysis of equality would plainly be unsatisfactory if applied person by person. It would fall before the argument that it is not an equal division of social resources when someone who consumes more of what others want nevertheless has as much left over as someone who consumes less. Nor that someone who chooses to work at a more productive occupation, measured by what others want, should have no more resources in consequence than someone who prefers leisure. (It would fall before such arguments, that is, unless it were converted into a form of equality of welfare through the doubtful proposition that equality in primary goods, in spite of different consumption or occupational histories, is the best guarantee of equality of welfare.)

So (as Rawls, as I said, makes plain) the difference principle does not tie itself to groups rather than individuals as a second-best accommodation to some deeper vision of equality that is individualistic. Any such deeper vision would condemn the difference principle as inadequate. It ties itself to groups in principle, because the idea of equality among social groups, defined in economic terms, is especially congenial to the flat interpretation of equality. Indeed, the idea of equality as equality among economic groups permits no other interpretation. Since the members of any economic group will be widely diverse in tastes, ambitions, and conceptions of the good life, these must drop away from any principle stating what true equality among groups requires, and we are left with only the requirement that they must be equal in the only dimension on which they can, as groups, possibly differ. The tie between the difference principle and the group taken as its unit of social measure is close to definitional.

We must be wary in rushing from this fact to conclusions about Rawls' theory of justice as a whole. The first of his principles of justice is plainly meant to be individualistic in a way that the difference principle is not, and any evaluation of the role of the individual in the theory as a whole would require a careful analysis of that principle and of the manner in which the two principles might work in harness. But insofar as the difference principle is meant to express a theory of equality of resources, it expresses a theory different in its basic vocabulary and design from the theory sketched here. It might well be worthwhile to pursue that difference further, perhaps by elaborating and working out in more detail the differences between the consequences of the two theories in practical circumstances. But I shall turn instead to the first of the two issues of comparison I distinguished.

I have tried to show the appeal of equality of resources, as interpreted here, only by making plainer its motivation and defending its coherence and practical force. I have not tried to defend it in what might be considered a more direct way, by deducing it from more general and abstract political principles. So the question arises whether that sort of defense could be provided and, in particular, whether it could be found in Rawls' general method. The fact that equality of resources differs in various ways, some of them fundamental, from Rawls' own difference principle is not decisive against this possibility. For perhaps we might show that the people who inhabit Rawls' original position would choose, behind the veil of their ignorance, not his difference principle, but either equality of resources or some intermediate constitutional principles such that, when the veil was lifted, it would be discovered that equality of resources satisfied these principles better than the difference principle could.

I hope it is clear that I have not presented any such argument here. It is true that I have argued that an equal distribution is a distribution that would result from people's choices under certain circumstances, some of which, as in the case of the hypothetical insurance markets, require the counterfactual assumption that people are ignorant of what in fact they are very likely to know. But this argument is different from an argument from the original position in two ways. First, my arguments have been designed to permit people as much knowledge as it is possible to allow them without defeating the point of the exercise entirely. In particular, they allow people enough self-knowledge, as individuals, to keep relatively intact their sense of their own personality, and especially their theory of what is valuable in life, whereas it is central to the original position that this is exactly the knowledge people lack. Second, and more important, my arguments are constructed against the background of assumptions about what equality requires in principle. It is not intended, as Rawls' argument is intended, to establish that background. My arguments enforce rather than construct a basic design of justice, and that design must find support, if at all, elsewhere than in those arguments.

I do not mean to suggest, however, that I am simply agnostic about the project of supporting equality of resources as a political ideal by showing that people in the original position would choose it. I think that any such project must fail. Or rather that it is misconceived, because some theory of equality, like equality of resources, is necessary to explain why the original position is a useful device–or one among a number of useful devices–for considering what justice is. The project, as just described, would therefore be too self-sustaining. The device of an original position (as I have argued at length elsewhere)[19] cannot plausibly be taken as the starting point for political philosophy. It requires a deeper theory beneath it, a theory that explains why the original position has the features that it does and why the fact that people would choose particular principles in that position, if they would, certifies those principles as principles of justice. The force of the original position as a device for arguments for justice, or of any particular design of the original position for that purpose, depends, in my view, on the adequacy of an interpretation of equality of resources that supports it, not vice versa.

19. R. Dworkin, *Taking Rights Seriously* (Cambridge, MA: Harvard University Press, 1977), chap. 6.

CANADIAN JOURNAL OF PHILOSOPHY
Volume XIII, Number 1, March 1983

Egalitarianism

BRUCE M. LANDESMAN, University of Utah

Despite the popularity of equality as a political value, egalitarianism as a political theory has never, I think, been fully or successfully defended. I aim in this paper to begin the defense of such a view. The egalitarianism I have in mind has as its ideal a condition of *equal well-being for all persons at the highest possible level of well-being,* i.e. *maximum equal well-being.* Egalitarianism holds that society should be arranged so as to promote and maintain this state. Defending such a view involves, as I see it, three tasks. First, the ideal I have just mentioned must be made clearer and more specific and its implications for the distribution of particular goods such as material possessions and liberty must be revealed. Second, positive arguments must be given in support of an equal distribution of well-being as a requirement of morality and justice. And, thirdly, arguments to the effect that there are just or justified inequalities which seriously outweigh the claims of equality must be rebutted. This paper is largely devoted to the task of clarifying and showing the practical implications of the ideal.[1] This task of for-

1 This paper is intended to be the first of a series in which the other tasks I have mentioned are also undertaken.

mulation may seem modest in comparison to the second and third tasks of providing and assessing arguments, but there are knotty problems involved in clarifying the ideal which must be resolved for egalitarianism to emerge as a plausible theory.

I proceed as follows. In section I, I differentiate the egalitarianism I favor from two other views with which it contrasts in a natural way. In section II, I begin the task of clarification by examining different theories of the good or of well-being. Sections III and IV take up the question of the coordination of the ideal of equal well-being with the distribution of certain particular goods. Section V discusses the relation for an egalitarian between equality and other moral values. And in section VI, I discuss some problems raised by special needs, physically deprived individuals and scarcity. These discussions, taken together, reveal, I believe, the structure of a systematic and plausible egalitarian theory.

A remark about ideals and their specification: by an 'ideal' I mean a very abstract conception or 'moral picture' that strikes one as cogent and compelling. The ideal of equality I have mentioned seems to me to be the compelling conception at the root of egalitarianism. The specification and application of such an ideal, however, is not merely a matter of spelling out its meaning in more detail; it involves, in addition, interpreting the ideal in light both of the complexities of life and of other dimensions of morality. The result is a complex set of practical recommendations which will be different for different contexts and problems. The theory that emerges may seem to some to be more complicated and less straightforwardly (or simplistically) egalitarian than one might expect. I hope my discussion will bear out the reasonableness of this interpretive procedure.

I. Equality, Liberalism, and Uniformity

Many who call for equality object only to particular sorts of inequalities. In the seventeenth and eighteenth centuries, egalitarian reformers objected to inequalities based on feudal status. In the nineteenth century, with the rise of industrial capitalism, critics turned their attention to inequalities connected with economic class. While such inequalities are still with us, modern reformers have brought to attention and critcized inequalities based on race, ethnic and national status, and sex. In these cases critics object to inequalities based on what are held to be arbitrary or irrelevant characteristics. They have, however, often couched their objections by way of a positive demand for equality as such. It has been argued that they do not really favor general equality, but would condone much inequality so long as the arbitrary inequalities are removed. Thus H.J. McCloskey asserts that

> equality has rarely been favored for its own sake. It is particular equalities that have been demanded and defended, particular inequalities condemned. Talk about favoring equality is therefore extremely misleading....[2]

McCloskey is certainly correct with regard to many thinkers. There is a natural and rational human tendency to back up particular demands with general principles, which may be stronger than the goals aimed at. Such principles, further, may serve as more persuasive slogans than more limited goals and may, as Marx noted, elicit the support of segments of the community which are not intended beneficiaries of the reform.

The view which objects to the particular inequalities I have mentioned but is willing to allow other inequalities is inherent in modern welfare liberalism. This liberalism can be summarized as follows: it rules out the arbitrary inequalities of race, sex, ethnic status, religion and initial position in society. But it is willing to countenance other inequalities which would be justified on one or more of the following grounds: desert and contribution, liberty and respect for rights, conduciveness to general welfare, and necessity. Such inequalities would be allowed, however, only if three background conditions are met: a) there should be no individuals or groups which fall below a minimum floor of well-being; b) the permitted inequalities should not be 'too great'; and c) there should be a good deal of equal opportunity for everyone to achieve the advantaged positions. Different theories will fill out this schema in different ways.[3] Egalitarianism, in contrast, involves a commitment to equality

---

2 H.J. McCloskey, 'A Right to Equality? Re-Examining the Case for a Right to Equality,' *Canadian Journal of Philosophy*, **6** (1976) 632

3 Rawls' theory of justice is, I think, an example of this view; what is special to him is the specification of the floor as that which maximizes the welfare of the worst-off group, and the justification of inequalities not on the basis of the desert of those favored, but on the utility of rewarding them, that is, on the incentive value of inequalities in making everyone better off. See John Rawls, *A Theory of Justice* (Cambridge, MA: Harvard University Press 1971), especially Chapters II, III and V.

Other, though less clear, examples of this view can be found in W.K. Frankena, 'The Concept of Social Justice,' in R. Brandt, ed., *Social Justice* (Englewood Cliffs, NJ: Prentice-Hall 1962) 1-29; W.K. Frankena, 'Some Beliefs about Justice,' in K.E. Goodpaster, ed., *Perspectives on Morality* (Notre Dame,IN: University of Notre Dame Press 1976) 93-106; Gregory Vlastos, 'Justice and Equality,' in Brandt, *Social Justice*, 31-72; and N. Rescher, *Distributive Justice* (Indianapolis, IN: Bobbs-Merrill 1966) Ch. 5.

much stronger than this. It agrees with the liberal view in rejecting as arbitrary the particular inequalities mentioned above, but it does not condone the other inequalities the liberal finds justified.

The egalitarianism I favor should be differentiated not only from liberalism but from another egalitarian view which is formulated by Isaiah Berlin in his well-known essay, 'Equality.'[4] Berlin asserts that

> the ideal of complete social equality embodies the wish that *everything and everybody should be as similar as possible* to everything and everybody else....the demands for human equality which have been expressed both by philosophers and by men of action can best be represented as modifications of this *absolute* and perhaps *absurd* ideal....[The egalitarian] will tend to wish so to condition human beings that the highest degree of equality of natural properties is achieved, the greatest degree of mental and physical, that is to say, *total uniformity*...

Berlin doubts that

> extreme equality of this type – the maximum similarity of a body of all but indiscernible human beings – has ever been consciously been put forward as an ideal by any serious thinker.

Nevertheless, he holds that

> if we ask what kinds of equality have, in fact, been demanded, we shall see, I think, that they are specific modifications of this absolute ideal, and that it therefore possesses the central importance of an ideal limit or *idealized model at the heart of all egalitarian thought.*[5]

In sum, we have here the claim that the ideal at the 'heart' of the egalitarian tradition is a demand or wish for *absolute equality,* which in turn is understood as a demand or wish for *total similarity* and *uniformity*.

Berlin gives little argument for this proposition, taken either as the historical claim that egalitarians have embraced this ideal, or as the logical claim that they must embrace it. The ideal has been used recently by H.J. McCloskey and J.R. Lucas[6] as a basis for subjecting

4 I. Berlin, 'Equality,' *Proceedings of the Aristotelian Society,* **56** (1955-56) 301-26, reprinted in W.T. Blackstone, ed., *The Concept of Equality* (Minneapolis: Burgess Publishing Co. 1969) 14-34

5 'Equality,' in Blackstone, 22, 24, 25: italics added

6 H.J. McCloskey, 'Egalitarianism, Equality and Justice,' *Australasian Journal of Philosophy,* **44** (1966) 50-69; 'A Right to Equality?', 625-42. J.R. Lucas, 'Against

egalitarianism to easy ridicule. But they have attacked a 'straw man.' It is very implausible to think that an egalitarian has or must have uniformity as his fundamental aim. Why should he want or wish for uniformity? If he has a wish, it is that persons, all of them, do well, equally well, and it is a commonplace that equal well-being is at least logically compatible with the satisfaction of quite different preferences and the pursuit of different life-styles. At best the uniformity view can rest on the empirical claim that the only way to achieve equal well-being is to make people similar and treat them uniformly. On this understanding, however, uniformity is not part of the egalitarian *ideal,* but is at most a means, and a dubious one at that, to equal well-being.

II. Equality and Theories of the Good

The egalitarian should aim, I have said, for maximum equal well-being. But what is well-being? The question of what constitutes a person's well-being is a familiar one in moral and political philosophy. An answer to it is a 'theory of the good' or of 'the good life.' Let's call the view that maximum equal well-being should be sought the *egalitarian principle.* A complete egalitarian theory needs to conjoin the egalitarian principle with some theory of the good which specifies the meaning of well-being.

There are two broad types of theories of the good. First, there are those theories which understand a person's good in terms of the satisfaction of that person's actual desires or of the desires a person would have if he or she had correct information. (This is meant to rule out wanting something only because one has a mistaken factual belief about it, such as wanting to drink a glass of water, not knowing it is poisonous.) Such theories do not permit the assessment of actual or factually corrected desires, and I shall call them *subjective* theories of the good.[7] Theories

Equality,' *Philosophy,* **40** (1965) 296-307, reprinted in H.A. Bedau, ed., *Justice and Equality* (Englewood Cliffs, NJ: Prentice-Hall 1971) 138-51; 'Justice,' *Philosophy,* **40** (1972) 229-48; 'Against Equality Again,' *Philosophy,* **42** (1977) 255-80. I think the ideal also plays some role in H.A. Bedau's more sensitive criticism in 'Radical Egalitarianism,' in H.A. Bedau, ed., *Justice and Equality,* 168-80.

7 I borrow and modify here some terminology used by Thomas Scanlon in 'Preference and Urgency,' *Journal of Philosophy,* **72** (1975) 655-69. I draw the distinction between subjective and objective theories roughly, but I think adequately for my purposes; a more careful account of the two theories is certainly possible.

of this sort have been developed within the utilitarian tradition and modern sophisticated versions of them are the successors of the simple hedonism which identifies the good with pleasure. The second kind of theory, an *objective* theory of the good, holds that people's actual or informed desires are open to assessment, so that the satisfaction of such desires may not be good for a person. A person's good may be independent, at least to some extent, of what he happens to want, and he may come to want some things only because they are, independent of his actual or informed wants, good. The earliest philosophical example of an objective theory is Plato's.

Subjective theories are favored by many because they are neutral with respect to different substantive conceptions of the good. Objective theories are said to foist one person or group's view of the good on others, while subjective theories avoid this. For this reason, subjective theories are deeply engrained in the liberal, individualist tradition. I think, however, that subjective theories are not ultimately adequate, and that the view that objective theories must be intolerant or elitist can be avoided. I cannot, however, argue fully for these claims in this paper. In section III, I start with a subjective theory of the good but find some reasons for modifying it. In section IV, I simply assume an objective theory. I will make these assumptions explicit and explain their function in the appropriate places.

A second dimension on which theories of the good can be compared is what I shall call their content. That is, a theory of the good may put forward a list of particular sorts of things, e.g. wealth, freedom, happiness, etc., which it holds to be good for people. A subjective theory will defend its list on the ground that these are things people want or would want if correctly informed, while an objective theory will hold that such things are good for people on other grounds. In other words, a subjective or objective criterion provides the *justification* for the given content of a theory.

Consider now the view that the following are all basic elements of human well-being: material goods; individual liberty or the liberty to lead one's life as one sees fit; political liberty which includes the liberty to vote and participate in community affairs and equality before the law; self-development or the ability to develop one's powers and talents; and self-respect. Let's leave open the question of whether such a content is justified on subjective or objective grounds. The view that these are basic constituents of human good is inherent in the liberal tradition. Suppose, then, we call the theory which involves this list the liberal theory of the good.[8] The egalitarianism I want to consider and defend

8 Some may cavil at calling this a liberal theory. In his recent essay 'Liberalism' (in S. Hampshire, ed., *Public and Private Morality* [Cambridge: Cambridge Univer-

combines the egalitarian principle with this theory of the good. We might call the resulting structure *Liberal Egalitarianism.* One can be an egalitarian without being a liberal egalitarian. Consider here Dostoyevsky's Grand Inquisitor who favors equal security without liberty, or Plato's notion of equal self-development which also excludes liberty, or certain bureaucratic instantiations of Marxism which emphasize equal material well-being without freedom or self-development.

In this paper I do not argue for but presuppose the liberal theory ot the good. My aim is to sketch a comprehensive liberal egalitarian theory. in the next two sections I take up the question of what happens when the egalitarian principle and the liberal theory are put together. In particular, what distribution of these basic goods is required by the ideal of equal well-being? It might be thought that there is an obvious answer; the ideal requires an equal distribution. We shall see that this is not always so, and where it is so, it is not obviously so. There are genuine complexities involved in *coordinating* the ideal with distributions of goods.

III. Goods, Satisfactions and Intensity

In order to consider the coordination issue it is important to take note of the commonplace, suggested earlier in the discussion of Berlin, that sameness of treatment is not necessarily identical with equality of treatment. If people with different preferences are given a good which some want but which others are indifferent to, they may have been treated similarly but unequally. On the other hand, spending more resources on the sick or educationally deprived may be dissimilar but equal treatment. Obviously what underlies this is the very important distinction between the good things people may receive or possess – whether these be concrete material things or intangibles such as rights and opportunities – as contrasted with the satisfactions the possession or use

sity Press 1978] 133-43), Ronald Dworkin claims that the subjective theory of the good is at the heart of liberalism. It seems to me, however, that liberalism also involves the goods I have just mentioned and this content is difficult to justify solely on subjective grounds. There is thus a tension in liberalism between the content and the justification of the good. Recall here the tension between Mill's hedonism, on one hand, and his emphais on self-development and individuality, on the other hand. The first suggests a subjective, the second an objective, view of the good. Mill's distinction between higher and lower pleasures is an inadequate attempt at resolving this tension.

of these goods make possible. Giving more goods, then, to the sick or deprived is a way of bringing them to or near the same level of satisfactions as the healthy or privileged, and that is why it can be seen as equality of treatment.[9]

Let us look more carefully at the distinction between goods and satisfactions. At the root of the distinction is the fact that what a good 'does' for a person is a function of many things: whether or not he wants it, the intensity of his want, how much he wants it in comparison with other things he wants, the way its possession or use fits into his overall aims and activities, etc. By 'satisfactions' I do not mean feelings of satisfaction or pleasure – though I do not mean to exclude these – but the general contribution of the particular good to a person in light of the above factors. The crucial point for understanding the distinction is that the same level of good may make quite different contributions to different people because of their different preferences and aims. The question of how a distribution of goods affects the distribution of satisfactions will be a complex and contingent matter.

I said above that subjective theories of the good are favored by

---

9 The distinction between equality and sameness of treatment is clearly made by William Frankena in 'The Concept of Social Justice,' 11. The failure to make the distinction between good and satisfactions characterizes and vitiates H.J. McCloskey's attacks on egalitarianism. Consider this passage frome 'Egalitarianism, Equality and Justice,' 55: treating people equally

> would enjoin treating the weak and the strong, the stupid, the clever and the cunning as if they were identical. The deaf, dumb and blind would be treated in the *same* way as the person possessed of all his faculties. ... This equality of treatment would lead to gravely unequal states of affairs, to inequalities of power, status, wealth, privilege, etc. (italics added)

This conclusion follows only if equal treatment is defined as similar treatment but is not implied by an egalitarianism sensitive to the good-satisfaction distinction. An apparent failure to make this distinction also occurs, surprisingly, in Ronald Dworkin's contrast between equal treatment and treatment as an equal. See 'Liberalism,' 126, and *Taking Rights Seriously* (Cambridge, MA: Harvard University Press 1978) 227. For Dworkin 'equal treatment' seems to mean similar or identical treatment, 'the equal distribution of some opportunity or resource or burden.' 'Treatment as an equal,' on the other hand, means the receipt of equal concern and respect and may sometimes dictate equal treatment, sometimes not. Consider now equal treatment construed in terms of equality of satisfactions. This does not fall under Dworkin's concept of equal treatment. Does it come under treatment as an equal? Perhaps it does but that category can also be construed to involve significant inequalities of treatment. What Dworkin gives us, in effect is a contrast between sameness of treatment and a vague notion of equal concern which can be construed in a number of different ways: the notion of equality of satisfactions or well-being is simply overlooked.

many. Let us see, then, what happens if the ideal of equal well-being is interpreted in the light of such a theory. The subjective theory says that a person's well-being is constituted by his satisfactions; all satisfactions are elements of well-being, and the more satisfactions, the more well-being. This leads to the idea that egalitarianism should aim primarily at equality of satisfactions with goods distributed solely as a means to this end. I call this view the *primacy of satisfactions over goods*; the subjective theory seems to entail this. But given the diversity of people's preferences, this means that goods would have to be distributed in quite unequal amounts, i.e. the unequal amounts needed for equalizing satisfaction. So conjoining the ideal of equal well-being with the subjective theory of the good gives us this 'solution' to the coordination problem: unequal goods so as to promote equal satisfactions and well-being. And this will apply to the goods contained in the liberal theory of the good I am presupposing.

This result, however, leads to two problems. The first is a practical one. It seems unrealistic to think that society could have the detailed knowledge of individuals needed to distribute goods in the unequal manner required to attain equality of satisfactions. 'To each according to his needs or preferences' seems an unworkable criterion. There is, however, a device available to society which can go some way towards meeting this problem. Goods can be ranked in terms of their *convertibility* into other goods. Money, for example, is a highly convertible good because it is easily converted into, i.e. exchanged for, other things. It can be used by people in different ways to attain very different types of satisfactions. Liberty is also a highly convertible good in that people can use it to engage in quite different kinds of activities and projects, as suits them. Other goods, obviously, are of low convertibility. Basic social and economic institutions are geared to affecting the distribution of highly convertible goods such as money and liberty. It is thus not implausible to think that an equal distribution of such goods, given their capacity to be used in ways which meet individual wants and needs, would go some way towards producing a rough equality of satisfactions. Of course, there are important exceptions to this, as in the case of underprivileged or handicapped individuals or of those with unusually intense desires, and I shall discuss these cases in the following paragraphs. It does seem, however, that, with some exceptions, the egalitarian holding a subjective theory of the good can support an equal distribution of economic goods and liberty as a practicable means of approaching an equality of satisfactions.

The second problem, the problem of intense preferences, is more difficult. Suppose that A's needs and wants are few and easily satisfied, while B's are many and intense and he suffers deep disappointment at their frustration. And suppose, because of this, that B needs twice as much income as A to achieve equal satisfactions. In some contexts this may strike us as quite unfair, i.e. equality of goods may seem more just than equality of satisfactions.

A problem like this might lead one to conclude that the only reasonable thing to do is to aim for a certain distribution of goods and leave the satisfactions received as a concern only of the individuals involved. This underlies Rawls' decision in *A Theory of Justice* to formulate principles of justice in terms of the distribution of primary goods rather than of satisfactions. He justifies this as follows:

> It may be objected that expectations should not be defined as an index of primary goods anyway but rather as the satisfactions to be expected when plans are executed using these goods. After all it is in the fulfilment of these plans that men gain happiness.... Justice as fairness, however, takes a different view. For it does not look behind the use which persons make of the rights and opportunities available to them in order to measure, much less to maximize, the satisfactions they receive.... it is assumed that members of society are rational persons able to adjust their conceptions of the good to their situation.[10]

Rawls' point is that the satisfactions people receive from goods are 'up to them,' a function of their own 'plans of life' which are to an important extent under their voluntary control

I think that Rawls' point has cogency, but I do not think it justifies the strong conclusion that justice should be concerned only with the distribution of goods. (In effect Rawls replaces the primacy of satisfactions with the primacy of goods.) I shall sketch my reasons both for agreeing with Rawls' point and for disagreeing with his conclusion and then I will show the bearing of the result on egalitarianism and the coordination issue.

Let's recall the point that the satisfactions a person gets from a good are determined by such features as his wants, their intensity, how they rank vis-à-vis other wants, and his overall goals. To make this vivid we might note that a person's satisfactions are also influenced by these other features of his psychological 'make-up': his emotional responses; his temperament or basic disposition to be pleased or displeased by things; and his adaptability which involves the willingness to try out new activities, be open to new experiences, put in effort, put up with dissatisfactions and adjust to disappointments. Given this I want to

10 Rawls, 94. Rawls has clarified and reinforced this view in some of his essays that have appeared since *A Theory of Justice*. See 'A Kantian Conception of Equality,' *Cambridge Review*, **96** (1975) 97; 'Fairness to Goodness,' *Philosophical Review*, **84** (1975) 553-4; and 'Reply to Alexander and Musgrave,' *Quarterly Journal of Economics*, 88 (1974) 641-3.

argue as follows. Suppose that a person is not afflicted by some special *liability* such as a physical or mental handicap or a social handicap such as being educationally underprivileged, poor or oppressed. And suppose that society has secured an equal distribution of highly convertible goods such as wealth and liberty. In such cases it seems to me a morally sound claim to hold that it is a person's responsibility to coordinate and adjust his preferences with the available goods in a satisfying way. Moreover I think it can be argued that it is part of a person's good to do this, an aspect of his autonomy and self-determination. If this argument is correct, it would not be society's responsibility, in the case I have mentioned, to equalize satisfactions; in fact it would be morally inappropriate, even if possible, for society to do that which ought to be the product of self-determination. Despite the varying intensity of preferences, then, this would give the egalitarian a moral reason (and not just a practical reason) for supporting in these conditions an equal distribution of (highly convertible) goods and, as Rawls says, leaving it up to individuals how best to use them.

This argument for equal goods applies only in the conditions I have mentioned. When people have special handicaps or have undergone social deprivation or oppression, then I think the egalitarian should favor an unequal distribution of goods in order to equalize well-being.[11] In other words, the argument distinguishes between two causes of inequality of satisfactions: when such inequality is caused only by the special features of people's psychological make-ups, then equality of goods is in order; but, when the inequality is a function of special liabilities, attributable either to nature or society, then unequal goods are required to remove these. Thus Rawls is partly right in his emphasis on the distribution of goods, but his exclusive concern with this results in an inadequate treatment of the significant liabilities I have mentioned.[12]

---

11 An obvious application of this point is the justification of preferential treatment to overcome the liabilities caused by discrimination.

12 The point that Rawls' exclusive emphasis on primary good overlooks special physical needs was made by Brian Barry, *The Liberal Theory of Justice* (Oxford: Oxford University Press 1973) 55-6. My argument extends this criticism to the claim that social liabilities are also overlooked. In a later paper Rawls responds to a criticism of Barry's sort by saying that for the development of principles of justice, he assumes that 'everyone has normal physical needs so that the problem of special health care does not arise,' 'Some Reasons for the Maximin Criterion,' *American Economic Review*, **64** (1974) 142. I suppose he might also assume that the social liabilities I have mentioned don't exist either. This would be equivalent to assuming as the norm for principles of justice the special condi-

Bruce M. Landesman

I cannot claim to have given a fully developed argument here. In particular the distinction between the two causes of inequality I have mentioned needs more careful probing, and since space and my overall aims do not permit fuller treatment here, I will rest with this sketch.[13] But suppose the argument or something like it is sound. What does it tell us about the ideal of equal well-being? I think that it represents a move away from the subjective theory of the good. If attempting to adjust one's aims to the circumstances is an element of the good, then it is good even for one who does not want it. Consider further one who fails to some degree in this struggle and is dissatisfied and compare this person with one who has not struggled but has achieved a relatively effortless contentment. Who is better off? (Compare: Who is better off, a pig satisfied or Socrates dissatisfied?) I do not think there is a clear answer. One has contentment which is an element of well-being, while the other has developed and exercised capacities in order to meet a specifically human challenge, which is also an element of well-being. Each has something the other lacks. Well-being is, this suggests, a rather complex concept whose elements may be in opposition. Equality of satisfactions is thus a one-sided interpretation of well-being because it leaves out other elements. Equality of goods, then, in the circumstances I have mentioned is a reasonable compromise between these two elements, giving everyone a chance to attain both the satisfactions and the exercise of responsibility which are each important elements of well-being.[14]

---

tions in which I have held that equality of good is appropriate. But it seems to me to be highly arbitrary to develop principles of justice which overlook these crucial social and physical handicaps, and to say nothing about how society should respond to them.

13 I have found very valuable Thomas Scanlon's discussion of the intensity issue in his 'Preference and Urgency.' My conclusions, I think, are similar to his but he rejects the responsibility argument as sufficient. He holds that the argument presupposes an objective criterion of well-being according to which certain interests of persons are more urgent or important than others independent of how they feel about them; what matters is not the intensity but the urgency or importance of their desires. Scanlon's argument thus constitutes another set of reasons for maintaining equality of goods in the circumstances I have mentioned: such equality meets equally the equally important wants of persons, this being a more appropriate specification of equal well-being than equal satisfactions. Space does not permit me to show why I persist in the independent validity of the responsibility argument in the face of Scanlon's point, but I intend to treat this whole set of issues in detail in another paper.

14 A recent discussion which casts considerable doubt on the adequacy of subjective theories of the good based on satisfactions is Amartya Sen's 'Utilitarianism

If what I have argued is correct, I would like to draw the following moral. It would be a mistake for the egalitarian to understand equality of well-being exclusively in terms either of goods or of satisfactions. The egalitarian will have to choose between these two specifications of equality in different contexts. Some may take this as showing that egalitarianism is incoherent since it involves conflicting ideals of equality, but I take it as showing only that it is complicated. It is not a fanatical commitment to one simple idea, but a more complex commitment to the ideal of equal well-being which, in the process of being clarified and applied, will require that arguments be given in particular contexts for one way of applying the ideal rather than another – arguments in terms of different theories of the good and of the complexities of human life.

IV. Overall Equality and the Possibility of Trade-Offs

In the section just completed I considered some distributional questions raised by the goods-satisfactions distinction and the subjective theory of the good. In this section I consider another distributional issue, the possibility of what I shall call trade-offs between basic goods of different categories. To make this clearer, let's first note a distinction we can make between *particular* and *overall* equality. A particular equality exists when people have the same amounts of a particular good or satisfaction. Overall equality exists when the particular goods or satisfactions people have 'sum up' or 'balance out' so that people come out equally 'overall'; such equality is compatible with many particular inequalities. For the ideal of equal well-being, it would seem that what is fundamentally important is overall equality, and that particular inequalities would not be objectionable if they balanced out so as to produce overall equality of well-being. In fact some particular inequalities might be required if they were the *only* way to achieve such overall equality.[15]

---

and Welfarism,' *Journal of Philosophy*, **76** (1979) 463-89. Sen criticizes what he calls 'welfarism': 'the judgment of the relative goodness of alternative states of affairs must be based exclusively on, and taken as an increasing function of the respective collection of individual utilities in these states' (468). Welfarism involves the subjective theory of the good.

15 We have already seen some cases in which the distinction between particular and overall equality could have been used, e.g. giving those physically or socially deprived more resources than others is a particular inequality meant to promote overall equality. It is often assumed that the egalitarian must require par-

This doctrine of what we might call the 'primacy of overall equality' leads to another distributional problem. Consider the following goods which are part of the liberal theory of well-being; material possessions, individual and political liberty, and self-respect. Let's also add power to this list. The primacy of overall equality implies that unequal distributions of these goods would be justified (or required) if the inequalities compensate for or off-set each other in such a way that overall equality is achieved. For example, those with less liberty might have this balanced by more money, or those with less self-respect might be compensated by more power. But these trade-offs seem quite fanciful, just plain wrong. What, then, should an egalitarian say about this possibility which seems to be allowed by his theory?

In this section I will give some reasons why, contrary to appearances, the egalitarian should not support such trade-offs. I will try to make plausible the claim that the best way to achieve equal well-being is not to permit trade-offs between these goods, but, in general, to secure their equal distribution.[16] The appropriate place for trade-offs, I will suggest, is within certain categories of the goods. The arguments I give here frequently assume an objective theory of the good, for they involve grounds for the liberal goods independent to some extent of people's desires for them. My account thus depends on a fuller defense of such an objective theory.

---

ticular equality for every good. Cf. J.R. Lucas:

> We can secure Equality in certain respects between members of certain classes for certain purposes and under certain conditions; but never and necessarily never, Equality in all respects for all men for all purposes and under all conditions. The egalitarian is doomed to a life not only of grumbling and everlasting envy, but of endless and inevitable disappointment.
>
> ('Against Equality,' 150)

The rhetorical force of this claim depends I think on assuming that the egalitarian must favor complete particular equality.

16 It is possible to approach the issue of equality by distinguishing different types of equality, such as economic, political, social, moral, etc., and treating each of these separately in terms of moral arguments held appropriate for the particular case. In doing this one resists the idea, central to my outlook, of an overall good or well-being of which the goods connected to the particular types of equality are elements; and one resists the idea that those goods should be distributed, at least to some extent, in terms of their effect on the overall good. Part of my aim in this section is to show that the intuitions of the 'separatists' can be accommodated within the 'overall equality' framework I put forward.

1. *Economic Goods and Political Liberty.* Let's focus on these two goods and consider the possibility that overall equality could be produced by counterbalancing inequalities of them. Such a situation is not unimaginable: consider, for example, a society divided into two groups, money-makers who are debarred from political participation, and rulers who do not engage in economic activity. Such a division into artisans and rulers is familiar to us from Plato's *Republic*. One of the things wrong with this situation is that the two goods at issue, economic goods and political participation (or the right to it) are in a certain way incommensurable; they meet quite different fundamental human interests and promote qualitatively different aspects of human well-being. Roughly, economic goods enable us to satisfy our 'personal' preferences[17] while political participation facilitates our need for community. The good life, we might say, requires fair doses of both these goods, and this means that their qualitative differences are more important than any quantitative dimension in accord with which they might be compared or traded off. I don't want to claim, however, that these goods are never commensurable. At levels of severe economic deprivation a sacrifice of political liberty makes sense in return for economic well-being; while at levels of affluence, political participation may be more important than increased economic well-being. In the cases of need and luxury, then, these goods have different degrees of importance and trade-offs make sense. But in the ordinary range of social conditions, they are equally important, satisfying as they do quite different interests. Equal well-being and overall equality will thus best be served in those conditions if these goods are not traded off, but distributed equally.

2. *Economic Goods and Individual Liberty.* Are counterbalancing inequalities here plausible as a way of promoting equal well-being? Again I want to say no, but for a different reason. As background for the argument, let us note that there are two ways of understanding individual liberty. On the *formalistic* account, people are equally free to perform an act if no legal or conventional rules forbid them from doing it. Their equal liberty remains intact even if some cannot do it because they lack an appropriate means, e.g., income. On a *substantive account,* equal liberty requires not only the absence of prohibitive rules but equality of the relevant means, i.e., people have equal liberty only when they are equally able to perform an act.[18] Let us now consider the possibility of

17 Cf. the distinction between personal and external preferences put forth by Ronald Dworkin in *Taking Rights Seriously*, 231-8, 275-6; and 'liberalism,' 134-6

18 I make this distinction roughly, in a way adequate for my purposes but I do not deny it needs to be qualified. Those holding the substantive view may not re-

trade-offs between economic goods and individual liberty according to the substantive definition. The obvious point to make here is that these goods are to a high degree connected, that is the more economic goods one has, the more one can do and therefore the more liberty one has, and vice versa. Liberty and economic goods vary directly, not inversely, with each other. The idea of counterbalancing inequalities in this case is therefore not realistic. Equality of well-being is more likely to be reached if these goods are not traded off in the way contemplated, but are each distributed equally.[19]

3. *Power.* It is often assumed that since power is a good the egalitarian must favor its equal distribution.[20] But unequal power seems essential for political and social organization; moreover, it may be part of the concept of power that it must be possessed, at any given time, unequally. So the egalitarian is in a spot if he must call for such equality. I am inclined to argue, however, that power is the sort of good which is an element in the well-being of some people but not of others, and depends to a large extent, perhaps completely, on people's desire for it. (I do not consider myself worse off than someone else solely on the ground that he has power I neither have nor want very much.) In this regard I take power to be unlike economic goods and political and individual liberty.[21] Thus the unequal distribution of power can be compatible with overall equality just as the unequal distribution of skiing equipment need not detract from equal well-being. Of course power, as a matter of fact, tends to be strongly connected with other basic goods, so that while inequality of power may itself be benign, its consequences can be

quire complete equality of means, nor will those holding the formalistic account hold that means are entirely irrelevant. The standard defense of the formalistic account of liberty is Isaiah Berlin's, 'Two Concepts of Liberty,' *Four Essays on Liberty* (Oxford: Oxford University Press 1969), Ch. III. See also Freidrich Hayek, *The Constitution of Liberty* (South Bend, IN: Gateway Editions 1960) Chapters 1-2. A well-known criticism of Berlins view is Gerald MacCallum's 'Negative and Positive Freedom,' *Philosophical Review* 76 (1967) 312-34. A sensitive discussion of this issue, defending what I call the substantive conception of liberty, is Richard Norman, 'Does Equality Destroy Liberty,' in Keith Graham, ed., *Contemporary Political Philosophy* (Cambridge: Cambridge University Press 1982), 83-109.

19 If liberty is construed formalistically, then I think trade-offs with economic goods can be rejected on the sort of grounds of incommensurability appealed to for the rejection of trade-offs between economic goods and political liberty.

20 Cf. the discussion in J.R. Lucas, 'Against Equality,' 147-9

21 Rawls also distinguishes power from these other goods by leaving it off his list of primary goods. He thus denies that power is the sort of good a rational person would want whatever else he wants and this seems to be correct. See Rawls, 'Fairness to Goodness,' fn. 8, 542-3.

highly inegalitarian. For this reason the egalitarian should support institutional means both to weaken these connections and to produce a wider distribution of power.

What about the question of trade-offs, those having more power having less of something else? Since I have argued that more power does not necessarily make a person better off, there may be no inequality of the sort that needs to be balanced off, i.e. there is no problem for which trade-offs are the solution. So while in the case of economic goods and individual and political liberty, trade-offs seemed either undesirable or impossible, here they just seem unnecessary.[22]

4. *Self-respect.* I conclude this section with a few remarks about self-respect and then I summarize the argument. Self-respect is a fundamental good for 'when self-respect is lacking we feel our ends are not worth pursuing, and nothing has much value.'[23] Society, however, cannot distribute self-respect itself, for it is a good like satisfaction dependent not only on external conditions, but also on our inner psychological make-up and perhaps to some extent on our own efforts. Society can, however, influence the distribution of what Rawls refers to as the bases of self-respect. What are these bases? Surely the liberal goods we have been discussing – material possessions, individual and political liberty – are among the most important, and equal self-respect is therefore most likely to be achieved if these goods are distributed equally. But this means, further, that possible trade-offs between these goods and self-respect so as to attain overall equality of well-being are unrealistic, i.e. it is unreasonable to think we could compensate less income, for example, with more self-respect, or have less self-respect but more liberty. This is much like the case of trade-offs between economic goods and individual liberty: self-respect varies directly, not inversely, with the other liberal goods, and this makes trade-offs unworkable.

In an egalitarian society the grounds for equal self-respect will also include the public conviction of the equality of all persons. But there is another basis of self-respect that poses a problem. To some extent self-respect depends on the respect and admiration shown us by others in light of our development of particular talents and achievements. But since some will always be more talented or achieve more than others, it seems impossible that this basis be possessed equally and to that extent

22 I have overlooked in this section the fact that not all who want power and for whom it would be a good can have it. So inequality of power can sometimes be an inequality of well-being. This is an instance of the general problem of scarcity which I simply do not treat in this paper.

23 Rawls, 'Reply to Alexander and Musgrave,' 641

equal self-respect seems unattainable. Concerning this problem I simply want to point out a way in which the distinction between particular and overall equality can be useful. Not everyone can develop the *same* skills and achieve the respect of others in the *same* way, but a society which both encourages self-development and removes or neutralizes the social and physical obstacles to such development may bring it about that (almost) everyone is good at something which validly[24] elicits the admiration and esteem of others. In such a case particular inequalities of self-respect will balance out and help provide part of the basis for overall equality of self-respect.[25] I do not, however, want to rest too much on this argument. It is likely that even in the best of circumstances some will not excel in anything or will excel only in activities which have low social prestige. My main point is that this is a case in which particular inequalities may be structured so as to produce or approach overall equality.

In this section I have tried to make plausible[26] the view that the egalitarian committed to equal well-being will not want to allow trade-offs among the basic liberal goods, but will instead support an equal distribution of these goods. The doctrine of overall equality is compatible with such equality of goods and does not have the counter-intuitive implications it first seemed to have. At the same time I have suggested that balancing inequalities to achieve overall equality is plausible *within* the categories of self-respect, power, and – as suggested by the discus-

24 The use of the word 'validly' is meant to rule out 'gimmicked' self-respect, as when a person comes to respect a self not worthy of respect because he is misinformed or deceived into thinking some characteristic of his is admirable. All sorts of examples of this abound – see especially Thomas Hill, 'Servility and Self-Respect,' *The Monist*, **57** (1973) 87-104 – and the question of when self-respect is genuine is a difficult one.

25 Consider Lucas's remarks:

> Two inequalities are better than one. It is better to have a society in which there are a number of pecking orders, so that a person who comes in low according to one can nevertheless rate highly according to another. ... So long as we have plenty of different inequalities, nobody need be absolutely inferior. ('Against Equality Again,' 268)

An egalitarian can agree with this though I find Lucas's language insensitive to the difference between real and spurious self-respect referred to in the previous footnote.

26 I say 'make plausible' because I am aware that my arguments need more development, a task incompatible with my overall aim in this paper of exhibiting the different parts of a plausible egalitarian perspective.

sion in the previous section – economic goods. The overall solution of the coordination problem treated in the last two sections is that the egalitarian should support an equal distribution of basic goods except in the case of special physical and social liabilities where the concern for satisfactions instead of goods dictates an unequal distribution. When such liabilities are absent and goods are distributed equally, a person must take responsibility for how he adjusts his plans and aims to achieve a level of satisfaction. This completes the bulk of my discussion of the coordination problem, although there are a few additional remarks in the concluding section.

V. The Structure of Egalitarianism

Berlin says that the egalitarian seeks *absolute equality*, and he understands this as total uniformity. We have seen many reasons to doubt that an egalitarian must seek uniformity, but must he seek absolute equality understood in some other way? To answer this we must consider what could be meant by 'absolute' here and I will suggest two claims that might be intended, first that the egalitarian is committed to only one value – equality – and can recognize no others, and second that an egalitarian must support an absolute or exceptionless obligation that people be treated equally, in contrast to a prima facie or 'other things being equal' obligation. I will argue that both of these claims are false.

Regarding the first claim, it is easy to show that an egalitarian need not be committed to just one value. As we have seen the egalitarian will believe that certain things are good, and he wants goods to be distributed in a way that equalizes well-being. But these things are good independent of their equal distribution, that is they are values the egalitarian adopts in addition to equality. So he is not committed to just one value. We can put this in another way, using some standard philosophical vocabulary, by saying that egalitarianism (strictly: the egalitarian principle) is a theory of the right or of obligation, which needs to be supplemented by a theory of the good. (In this regard it is just like utilitarianism.) Equality may be its only *distributive value*, but it recognizes values other than distributive values, i.e. the goods to be distributed.[27]

27 Suppose an 'egalitarian' were to think that there is nothing of value besides equality, not well-being, not liberty, not self-respect, etc. Finding nothing of

A more plausible interpretation of the claim that egalitarianism is committed to absolute equality is the view that it can admit no distributive values other than equality, and allow no exceptions to this value. In other words, egalitarianism involves an absolute, rather than a prima facie, obligation that people be treated equally. Let me explain how I will use these familiar terms. An obligation to do an act of kind X is absolute if X must be done on every occasion on which it is an alternative; there are no exceptions to absolute obligations. An obligation to do an act of kind X is prima facie if there is always a moral reason for doing acts of kind X, but this reason may be overriden in particular cases when there are stronger moral reasons for not doing X; exceptions, in other words, are possible.[28]

I do not think that egalitarianism should involve an absolute obligation of equal treatment.[29] My reasons for this are simple and general. I feel sure that for any moral principle – except for a vacuous one like 'Always do what is right' – it is always possible that particular circumstances occur in which putting it into effect will be seriously wrong from the moral point of view. Morality is too complex, too pluralistic to permit any principle to be taken in an absolute manner. The same is true for the egalitarian principle: cases could arise in which an equal distribution will be, all things considered, morally inappropriate. I therefore think that a plausible egalitarian theory must involve a prima facie rather than an absolute obligation of equal treatment. (The first part of defending egalitarianism is providing arguments for such an obligation.)

---

value to distribute, he might then be led to the view that equality means nothing but identical or uniform treatment; this would be the only *content* he could give to equal treatment. This may explain how some get to the uniformity view of egalitarianism: they assume that the egalitarian can embrace only one value and needs no 'independent' theory of the good.

28 I have tried to clarify in more detail the notions of absolute and prima facie obligation, and have distinguished different types of prima facie obligation, in 'The Obligation to Obey the Law,' *Social Theory and Practice*, **2** (1972) 67-84.

29 I shall not discuss the important question of upon whom the obligation falls. I assume it falls on 'society' but it needs to be made clear what this means and what its implications are, and how this obligation accords with the obligations of particular individuals, in both public and private capacities. My overall view is that society needs to provide and maintain an egalitarian 'basic structure' within which individuals will be free to act on nonegalitarian reasons. But this needs more clarification than I can give it here, though some of my remarks concerning what I call the 'morality condition' below are relevant. For the notion of a 'basic structure' see Rawls, *A Theory of Justice*, sections 2, 11-17, 41-43; and 'The Basic Structure as Subject,' *American Philosophical Quarterly* **14** (1977) 159-65.

This conclusion may seem, however, to constitute a fatal weakening of the egalitarian position. An absolute obligation of equal treatment may be too strong, but a prima facie one may seem too weak: such an obligation may be easily overriden by opposing moral considerations, so that very little equality, all things considered, is required. But surely an egalitarian must hold that equality is the right outcome quite a lot of the time, if not 'always,' at least 'almost always.' But a prima facie obligation cannot guarantee this.

To meet this objection, it must be shown that the egalitarian prima facie obligation is strong enough to produce the outcome just mentioned. Let's say that the *strength* of a prima facie obligation is a function of how it fares against other obligations, how easily it overrides or is overriden. A weak obligation is easily overriden, a strong one overrides many others. The egalitarian must hold that the obligation to promote equal well-being is a strong one, in fact strong enough to require actual equality in most situations. In the remainder of this section I will show how it is possible for the egalitarian obligation to have such strength.

Let's call the requirement that the egalitarian obligation must as a matter of moral judgment almost always triumph over other distributive moral considerations *the morality condition.* I want to bring out two different ways in which the morality condition might be met. First, it might be held that the relation between the egalitarian obligation and other obligations is that the others come into play only in extreme or unusual conditions. The egalitarian obligation is the only one or the dominant one for the normal range of cases. What I have in mind is similar to a view often expressed about the concept of rights. If one has a right to do something, then one's liberty to do it may not be restricted solely on the grounds that doing so would promote the general good or the general interest. Rights, as Ronald Dworkin has maintained, are trumps people hold against ordinary utilitarian or general welfare considerations.[30] Nevertheless, on this view, if the exercise of a right would have extremely bad consequences, then it may be restricted. But this justification for restriction is applicable only in extreme cases, and the competing moral considerations override the right only in such cases. If such a view could be made out for the egalitarian obligation, then we could express the content of the obligation by saying that people have a *right to equality.* I shall call this way of meeting the morality condition the *rights solution.*

The second way of meeting the morality condition involves allowing other distributive considerations to have moral force in ordinary cir-

---

30 Cf. Ronald Dworkin, *Taking Rights Seriously,* xi.

cumstances, but holds that considerations of equality have a strong kind of priority with regard to these other considerations. I have in mind here something akin to what Rawls means when he speaks in one essay of the 'absolute weight' of justice in regard to utility, or when he says that his principle of equal liberty has lexical priority over the difference principle.[31] My idea is that the establishment of an egalitarian pattern or 'basic structure' is the dominant moral consideration and other moral considerations are permitted to play a role only in ways that are compatible with this pattern and do not upset it. Suppose, for example, that people with achievements and talents receive rewards or recognition it is held that they deserve. This will be acceptable only if such rewards do not upset the overall structure of equality.[32] As I suggested in discussing self-respect this may be possible if there are competing inequalities of reward so that everyone benefits to some degree. In such a case, then, other considerations are admitted as genuinely moral, and relevant in normal cases, but they have a kind of second-class citizenship and are allowed to play a role only in cooperation with equality. Since other moral considerations are not cancelled but contained here, I will call this second way of meeting the morality condition the *containment solution*.

I want to suggest a second condition – in addition to the morality condition – which I believe the egalitarian obligation should meet. This is the requirement that a good deal of equality not only be right as a matter of moral judgment, but that it be to a large degree practically realizable. I shall call this the *reality condition*. To make this clearer, let us suppose it has been successfully argued that there is a prima facie obligation of equal treatment of the strength implied by the morality condition. But suppose it is impossible to meet the obligation to any great extent; the facts of nature and human nature makes a great deal of equality impossible. In such a case the obligation is not overriden but, I shall say, *suspended* by necessity. We admit certain inequalities as necessary inequalities and strive for what we sadly agree is only second best. The reality condition requires that the amount of necessary in-

---

31 The idea of the absolute weight' of justice in regard to utility is found in 'Legal Obligation and the Duty of Fair Play,' in S. Hook, ed., *Law and Philosophy* (New York: New York University Press 1964) 13-14. For the notion of lexical priority see *A Theory of Justice*, sections 8, 14, 46, 82.

32 For a discussion of the sort of desert considerations which may be containable within an egalitarian structure, see Joel Feinberg, 'Justice and Personal Desert,' in *Doing and Deserving* (Princeton: Princeton University Press 1970), Ch. 4.

qualities be small or, if there are many such inequalities, that they be unimportant ones.[33]

A full discussion of the reality condition would raise difficult issues about the relations between moral judgments and empirical facts. On one hand, moral judgments have some independence of the facts so that an outcome may be right even though it cannot be brought about; on the other hand, a moral theory totally divorced from the possible is a utopia which, though perhaps intrinsically appealing, can provide little guidance for policy decisions in the real world. I cannot go in detail into the question of the relation between facts and moral judgments here, so my defense of the reality condition is somewhat ad hoc: I am interested in developing and defending a theory of justice which is not only theoretically appealing but has a chance of being satisfied in practice, something one can work for as well as think about, and has implications for current and forthcoming social controversies. The reality condition is sugested by this aim.[34]

An egalitarian theory which meets the morality anu reality conditions has a certain structure. It involves a prima facie obligation to promote and maintain equal well-being; it describes the role other moral considerations are permitted to play within a basic egalitarian framework; and it delineates the special conditions in which the obligation can be overridden or suspended. The prima facie obligation is strong enough so that the resulting theory is genuinely egalitarian, but it is not so strong that the theory becomes implausible. In other words, the resulting theory expresses the dominance of the value of equality in social life, but it does not embrace the 'absolute' or monolithic equality Berlin envisages, and it makes room for the complexity of moral deliberation.

33 For a discussion of necessary inequalities see H. Bedau, 'Radical Egalitarianism,' 175-6.

34 If a claim of necessity can overcome or suspend a right, then the reality condition could be seen as a special case of the rights solution of the morality condition; alternatively, it is a solution with respect to necessity analogous to the rights solution with respect to competing moral considerations.

Bruce M. Landesman

VI. Equality, Pareto Improvements, and Expensive Needs

If the moral structure described in the last section were filled in with the account of well-being defended in previous sections, we would have a fairly complete formulation of egalitarianism.[35] There are, however, two further problems about the meaning of 'maximum equal well-being' which need to be addressed. The first might be called the problem of 'pareto improvements.' Suppose it is possible to move from a condition of equality to one in which at least one person is better off and the rest remain the same (and no redistribution to restore equality at a higher level is possible). Or suppose we can move from equality to a condition involving inequality in which *everyone* is better off (and, again, no egalitarian redistribution is possible). These are both pareto improvements on equality. Should an egalitarian favor or oppose them? If he favors them, he seems to allow inequalities, but opposition in the name of equality, seems to mean a gratuitious sacrifice of well-being. Egalitarianism seems to imply opposing the pareto improvement, but that seems to be the morally wrong outcome.

Let me try to put the problem in more theoretical terms before dealing with it. The egalitarian ideal of maximum equal well-being expresses two moral concerns or values – one is a (comparative) concern that people do equally well, that their relations be characterized by equality, the other is a (non-comparative) concern for people's well-being, a wish that they do as well as they can. The egalitarian ideal thus involves two 'strands,' 'equal well-being' or the *equality strand,* and 'the maximum well-being each person is capable of' or the *humanitarian strand.* The ideal puts these together by qualifying the humanitarian strand in light of the equality strand and the result can be put as follows: given available resources, *the highest possible well-being each person is capable of compatible with a similar well-being for others.* This synthesis of the two strands leads to the pareto improvement problem in cases in which the fulfilment of the equality strand severely restricts the humanitarian strand, i.e. equal well-being seems to require the needless sacrifice of

35 The formulation is somewhat limited in scope because of two assumptions I have implicitly made. First, I have taken as my unit a single society, rather than the whole world; and secondly, I have tried to aim at picturing an egalitarian society at a single instant and have not considered what equality requires over time, especially in regard to future generations. The wider perspective involves both more people and fewer resources per capita and thus needs to be considered for a fully adequate treatment of the subject.

higher well-being for some, or even for all. In effect, while the synthesis gives the equality strand dominance, its plausibility is based on the assumption that each strand can be fulfilled to a high degree compatible with the other strand being equally fulfilled. The pareto improvement problem casts doubt on this. There are three phenomena which pose the problem most acutely: the existence of mentally or physically handicapped people, differences among ordinary person's in 'natural' capacities for well-being, and the possibility that social inequalities might benefit everyone. I will examine each of these in turn.

The existence of persons who are mentally or physically handicapped in such a way that their capacity for well-being is less than that of the ordinary non-deprived person raises the pareto improvement problem vividly. Suppose that such persons are brought up to their maximum level. The egalitarian ideal seems to require that no one else should be permitted to attain higher levels of well-being, i.e. that no one may reach a level of well-being higher than that attainable by a person with the lowest capacity for well-being. This is an unwelcome and inhumane conclusion. This is a case in which the egalitarian 'synthesis' I mentioned above falls apart. Instead of the two strands each contributing a fair amount to the outcome, the equality strand almost totally defeats the humanitarian strand. The egalitarian must admit here, I think, that his major idea does not work out right and he should conclude that, once the handicapped have been brought up to their maximum level, ordinary people may justifiably do better, i.e. he should admit the justifiability of this pareto improvement on equality. Suppose, however, that the egalitarian admits this? What implications does this have for the rest of his view? It might be thought that it is a fatal admission, that it will lead, in all consistency, to pareto and other sorts of departures from equality in so many other cases that egalitarianism has in effect been given up. I will show that this is not so, that the admission is a very limited and minor one, applicable only to this particular case. After showing that the admission does not have fatal implications, I will return to this case – the handicapped and the average – and put it in its proper light.

The obvious generalization from the case of the handicapped is this: differences in capacity for well-being exist not only between the average and the handicapped, but among normal persons as well. Allowing normal persons to do better than the handicapped seems to imply that normal persons with higher capacities should do better than those with lower capacities, once the latter have attained their maximum level. This suggests the following version of what Rawls calls the 'lexical difference principle' to govern differences in 'natural' capacities for well-being: first, maximize the well-being of persons with the lowest capacity; then, those with the next lowest, etc.[36] It might be thought that the egalitarian, admitting inequalities in regard to the handicapped and the average, must also admit them here and must accept this principle rather than a more egalitarian one. But this is not so. People have different capacities for well-being largely because they have different capacities for what I have called 'satisfactions.' I argued in section III, however, that the appropriate content of 'equal well-being' in the absence of special physical and social liabilities is *equality of goods*, leaving the satisfactions people receive as their responsibility. In fact, I argued that part of their well-being is constituted by dealing with the challenge of managing their resources in a satisfying and beneficial way. This means, then, that goods are not to be distributed in terms of intensity of preference or capacity for satisfactions. But since differences in capacity for well-being are a function of such differences in intensity and satisfactions, it means also that goods should not be distributed in accord with the lexical difference principle either. Instead goods are, for the egalitarian, to be distributed equally.

The point just made could be put slightly differently by saying that the arguments in section III concerning satisfactions and responsibility show in effect that the morally relevant capacities for well-being of average persons are, for all practical purposes, equal. I think this point could be argued more directly. It is difficult to see what facts would be appealed to in order to defend the claim of differential capacities for well-being – the fact that some have greater talents or are more temperamentally 'optimistic' than others won't be sufficient. But I cannot pursue this here; I will simply point out that if the argument is correct and the capacities for well-being of average persons are, in effect, equal, this means that there is *no need* for the lexical principle as an alternative to straight equality, and there is no problem-causing pareto departure from equality to consider. In other words, considering ordinary persons in light of their natural capacities for well-being, equal distribution poses no pareto improvement problem.

I turn now to social inequalities. It is often claimed that social inequalities provide incentives which draw out the efforts of people in a way which makes everyone better off, better off than they would be under conditions of equality. Given this 'incentive principle,' some

---

36 Rawls discusses the lexical difference principle in *A Theory of Justice*, 82-3. My version differs from his in that his is meant to cover both social inequalities and inequalities due to natural characteristics, while mine is intended to cover only the latter. The possibility of such a principle for social inequalities is discussed, and rejected below. I am indebted to Rolf Sartorius for suggesting the objections to my view that this section attempts to deal with.

social inequalities would be pareto improvements on equality. What should the egalitarian say here? I think that the egalitarian must argue against the incentive principle and hold that little, if any, social inequality is necessary or justified. Thus if social conditions are such that social inequalities would make everyone better off, the egalitarian view would be that the conditions which set the background for this possibility can and thus should be altered, and that the inequalities should not be allowed.

But how can the 'incentive principle' be attacked? Before addressing this question directly, let's note the use of the principle in Rawls' theory of justice. According to the difference principle, inequalities are justified when and only when they maximize the position of those who are worst off. While the difference principle does not justify inequalities per se, it lays down conditions under which they would be justified. Rawls clearly believes that these conditions hold. He says that there are inequalities in the basic structure of society which are 'especially deep,' 'pervasive' and 'presumably inevitable.'[37] He holds that better prospects for entrepreneurs may act as incentives to economic efficiency and innovation, with the resulting material benefits spreading 'throughout the system and to the least advantaged.' While Rawls says that he will 'not consider how far these things are true,'[38] his assumption that they are true is deeply embedded in his perspective and influences the way he treats many issues. But he never really probes the assumption or gives significant argument for it. In effect he simply takes society stratified along economic lines as a given and proposes the difference principle to modify and justify the inequalities so contained. It seems to me, however, that a theory of justice which relies so heavily on the claim of the necessity and benevolence of social inequalities must give that claim careful and substantial support.

I cannot in this paper subject the principle to a thorough criticism but I'll suggest several lines of attack. It seems to me that the incentive principle as commonly understood involves three claims:

(a) People will develop socially beneficial skills only if they are rewarded for doing so.

(b) The needed rewards must be economic ones.

---

37 Rawls, *A Theory of Justice*, 7

38 Ibid., 78

(c) The resulting reward structure must be an inegalitarian one.

It may be that these three claims are true of a capitalist economy and of the sorts of persons socialized in such an economy. But a society does not necessarily have to be organized along capitalist lines. An egalitarian's typical preference for socialism can thus be seen as reflecting the belief that in such an economy the incentive principle and the ensuing inequalities it brings out does not apply. But, it might be said, capitalist economies are the best and most efficient and make everyone better off than they would be under an egalitarian socialist system. Much can be said for and much against this. The point, though, is that the defense of the incentive principle in this way turns into a defense of capitalism as the best economic system. And it certainly cannot be said that the evidence is all in on that matter.

Let us then look at the incentive principle as a claim about 'human nature' independent of economic systems. The first part of the principle is very plausible. It is probably essential that people in general be rewarded for the development of talents, at least in the form of recognition and praise. Some social 'validation' of one's merits may indeed be necessary for the development of self-respect and self-worth. But it is a far cry from this to hold that talents must receive economic rewards. People can be motivated to develop talents in many ways, not only for the obvious goods of prestige, power and fame, but to achieve the admiration and respect of friends, relatives, and co-workers. And of course there are intrisically rewarding occupations in which people are moved to develop skills for their own sakes. How far work in general can be restructured so as to make this tendency more widespread is another big question, and most egalitarians would hold that improvements are possible in this direction.

It might be countered, however, that if other rewards are substituted for economic ones, inequality will still result, an inequality of prestige or power, instead of income or wealth.[39] In other words, even if part (b) of the incentive principles can be shown to be false, part (c) may still hold: an inegalitarian reward structure *of some sort* is necessary. Against this I shall simply refer to a possibility raised earlier in the discussion of self-respect, that unequal rewards in the distribution of recognition may be compatible with an overall equality of well-being. Suppose A develops a talent to a greater extent than B. He will therefore receive more recogni-

---

39 See J.R. Lucas, 'Against Equality,' 148-9.

tion, more 'psychic income' for that talent than B. But B may receive more recognition that A for the development of some other talent. Both can then receive the recognition they require for the motivation to develop important skills without any inequality of overall well-being. But it might be said that overall inequalities of prestige attached to different occupations must emerge. Suppose this were so in a society in which a) there is a public conviction of the equality of all persons, b) income and liberty are equal, and c) all are able to develop worthwhile talents. In such a context inequalities in prestige or power may amount to very little, if any inequality of well-being.

My point then, is that even if part (a) of the incentive principle is correct – and in some 'minimal' sense I am inclined to accept it – parts (b) and (c) do not follow, and (c) even if correct may be correct in a sense not incompatible with the egalitarian idea. Of course a lot more needs to be said to support these points. But I believe enough has been said to show that the incentive principle cannot be taken for granted and the desirability of socially-based pareto improvements on equality can be questioned.

I return now to the case of the handicapped and the ordinary. We see now that the admission of a pareto improvement on equality in this case does not require giving up equality among ordinary persons, nor does it imply the justification of social inequalities. It is in fact a special case in which the obligation of equal treatment is overridden by paretian considerations (the humanitarian strand overrides the equality strand of the egalitarian ideal here, we might say). I find such an exception unimportant, of a kind allowed by the morality condition mentioned in the previous section. The need for the exception arises solely because of the *profound* and *physically-based* differences in capacity for well-being involved. But these features, I have argued, are confined to this case.

At the beginning of this section I said that I would address two problems: the second problem may be called the problem of very expensive needs. Consider the handicapped again. I have already argued (pp. 33-5, 37) that when an equal distribution of goods would not make the same contribution for the well-being of ordinary and handicapped persons, an unequal distribution in favor of the handicapped is favored on egalitarian grounds; in such a case, equality of goods gives way, I said to the promotion of equality of satisfactions. But suppose that the amount of resources needed to bring the handicapped to their maximum level is so high that providing it means that normal persons must achieve levels of well-being far below what they are capable of. In other words, scarce resources flow so heavily to the handicapped that comparatively little is left for the rest. Similar issues could arise in the provision of expensive exotic medical technology.[40] Should an egalitarian support such a provision for expensive needs? I believe that equality indeed requires such a provision, but that there is a point at which the sacrifice required of others outweighs the prima facie obligation of equal well-being; in effect, considerations of utility or welfare override equality here, although the exact point of overriding would have to be determined separately for each case. I think once again, however, that this is the sort of exception to equality permitted by the morality condition. Special people and special needs confront the egalitarian (and other moral theorists) with special problems, but they do not defeat the overall theory and its adequacy for typical cases. Such, at least, has been the argument of this section.

I have tried in this paper to do a number of things: to clarify the ideal of maximum equal well-being, to show what implications it has for the distribution of basic goods, to explicate the moral structure of a non-absolutist egalitarian theory, and to achieve, as a result, a clear and initially appealing version of egalitarianism. Along the way, I hope to have laid to well-deserved rest the uniformity conception of equality and to have shown that egalitarianism need not be the simplistic, monolithic view it is sometimes taken to be, but is both more complex and sensible and has usually overlooked argumentative resources available to it. I have, of course, provided substantive arguments neither for a strong prima facie obligation of equal treatment, nor against the objections that are typically brought against egalitarianism on ground of desert, necessity, efficiency and rights. But I hope that my discussion sets the stage for a fruitful consideration of these issues, unimpeded both by confusions as to what egalitarianism is and by facile arguments directed only at misconceptions of it.[41]

March 1981

---

40 Note that in both these cases the expensive need is not a function of specially intense preferences, which I have argued, do not require special moral treatment.

41 I am indebted to the following for extremely valuable feedback on earlier versions of this paper: Margaret Battin, James Bogen, Leslie P. Francis, Charles Landesman, Richard Norman and Rolf Sartorius.

[9]

RICHARD J. ARNESON

EQUALITY AND EQUAL OPPORTUNITY FOR WELFARE

(Received 2 January, 1988)

Insofar as we care for equality as a distributive ideal, what is it exactly that we prize? Many persons are troubled by the gap between the living standards of rich people and poor people in modern societies or by the gap between the average standard of living in rich societies and that prevalent in poor societies. To some extent at any rate it is the gap itself that is troublesome, not just the low absolute level of the standard of living of the poor. But it is not easy to decide what measure of the "standard of living" it is appropriate to employ to give content to the ideal of distributive equality. Recent discussions by John Rawls[1] and Ronald Dworkin[2] have debated the merits of versions of equality of welfare and equality of resources taken as interpretations of the egalitarian ideal. In this paper I shall argue that the idea of equal opportunity for welfare is the best interpretation of the ideal of distributive equality.

Consider a distributive agency that has at its disposal a stock of goods that individuals want to own and use. We need not assume that each good is useful for every person, just that each good is useful for someone. Each good is homogeneous in quality and can be divided as finely as you choose. The problem to be considered is: How to divide the goods in order to meet an appropriate standard of equality. This discussion assumes that some goods are legitimately available for distribution in this fashion, hence that the entitlements and deserts of individuals do not predetermine the proper ownership of all resources. No argument is provided for this assumption, so in this sense my article is addressed to egalitarians, not their opponents.

I. EQUALITY OF RESOURCES

The norm of equality of resources stipulates that to achieve equality the

Philosophical Studies **56**: 77–93, 1989.

agency ought to give everybody a share of goods that is exactly identical to everyone else's and that exhausts all available resources to be distributed. A straightforward objection to equality of resources so understood is that if Smith and Jones have similar tastes and abilities except that Smith has a severe physical handicap remediable with the help of expensive crutches, then if the two are accorded equal resources, Smith must spend the bulk of his resources on crutches whereas Jones can use his resource share to fulfill his aims to a far greater extent. It seems forced to claim that any notion of equality of condition that is worth caring about prevails between Smith and Jones in this case.

At least two responses to this objection are worth noting. One, pursued by Dworkin,[3] is that in the example the cut between the individual and the resources at his disposal was made at the wrong place. Smith's defective legs and Jones's healthy legs should be considered among their resources, so that only if Smith is assigned a gadget that renders his legs fully serviceable in addition to a resource share that is otherwise identical with Jones's can we say that equality of resources prevails. The example then suggests that an equality of resources ethic should count personal talents among the resources to be distributed. This line of response swiftly encounters difficulties. It is impossible for a distributive agency to supply educational and technological aid that will offset inborn differences of talent so that all persons are blessed with the same talents. Nor is it obvious how much compensation is owed to those who are disadvantaged by low talent. The worth to individuals of their talents varies depending on the nature of their life plans. An heroic resolution of this difficulty is to assign every individual an equal share of ownership of everybody's talents in the distribution of resources.[4] Under this procedure each of the N persons in society begins adult life owning a tradeable 1/N share of everybody's talents. We can regard this share as amounting to ownership of a block of time during which the owner can dictate how the partially owned person is to deploy his talent. Dworkin himself has noticed a flaw in this proposal, which he has aptly named "the slavery of the talented."[5] The flaw is that under this equal distribution of talent scheme the person with high talent is put at a disadvantage relative to her low-talent fellows. If we assume that each person strongly wants liberty in the sense of ownership over his own time (that is, ownership over his own body for his entire lifetime), the high-talent person finds that his taste for liberty is very expensive, as his time is socially valuable and very much in demand, whereas the low-talent person finds that his taste for liberty is cheap, as his time is less valuable and less in demand. Under this version of equality of resources, if two persons are identical in all respects except that one is more talented than the other, the more talented will find she is far less able to achieve her life plan than her less talented counterpart. Again, once its implications are exhibited, equality of resources appears an unattractive interpretation of the ideal of equality.

A second response asserts that given an equal distribution of resources, persons should be held responsible for forming and perhaps reforming their own preferences, in the light of their resource share and their personal characteristics and likely circumstances.[6] The level of overall preference satisfaction that each person attains is then a matter of individual responsibility, not a social problem. That I have nil singing talent is a given, but that I have developed an aspiration to become a professional opera singer and have formed my life around this ambition is a further development that was to some extent within my control and for which I must bear responsibility.

The difficulty with this response is that even if it is accepted it falls short of defending equality of resources. Surely social and biological factors influence preference formation, so if we can properly be held responsible only for what lies within our control, then we can at most be held to be partially responsible for our preferences. For instance, it would be wildly implausible to claim that a person without the use of his legs should be held responsible for developing a full set of aims and values toward the satisfaction of which leglessness is no hindrance. Acceptance of the claim that we are sometimes to an extent responsible for our preferences leaves the initial objection against equality of resources fully intact. For if we are sometimes responsible we are sometimes not responsible.

The claim that "we are responsible for our preferences" is ambiguous. It could mean that our preferences have developed to their present state due to factors that lay entirely within our control. Alternatively, it could mean that our present preferences, even if they have arisen through processes largely beyond our power to control, are now within our control in the sense that we could now undertake actions, at greater or lesser cost, that would change our preferences in ways that we can foresee. If responsibility for preferences on the first construal held true, this would indeed defeat the presumption that our resource share should be augmented because it satisfies our preferences to a lesser extent than the resource shares of others permit them to satisfy their preferences. However, on the first construal, the claim that we are responsible for our preferences is certainly always false. But on the second, weaker construal, the claim that we are responsible for our preferences is compatible with the claim that an appropriate norm of equal distribution should compensate people for their hard-to-satisfy preferences at least up to the point at which by taking appropriate adaptive measures now, people could reach the same preference satisfaction level as others.

The defense of equality of resources by appeal to the claim that persons are responsible for their preferences admits of yet another interpretation. Without claiming that people have caused their preferences to become what they are or that people could cause their preferences to change, we might hold that people can take responsibility for their fundamental preferences in the sense of identifying with them and regarding these preferences as their own, not as alien intrusions on the self. T. M. Scanlon has suggested the example of religious preferences in this spirit.[7] That a person was raised in one religious tradition rather than another may predictably affect his lifetime expectation of preference satisfaction. Yet we would regard it as absurd to insist upon compensation in the name of distributive equality for having been raised fundamentalist Protestant rather than atheist or Catholic (a matter that of course does not lie within the individual's power to control). Provided that a fair (equal) distribution of the resources of religious liberty is maintained, the amount of utility that individuals can expect from their religious upbringings is "specifically not an object of public policy."[8]

The example of compensation for religious preferences is complex, and I will return to it in section II below. Here it suffices to note that even if in some cases we do deem it inappropriate to insist on such compensation in the name of equality, it does not follow that equality of resources is an adequate rendering of the egalitarian ideal. Differences among people including sometimes differences in their upbringing may render resource equality nugatory. For example, a person raised in a closed fundamentalist community such as the Amish who then loses his faith and moves to the city may feel at a loss as to how to satisfy ordinary secular preferences, so that equal treatment of this rube and city sophisticates may require extra compensation for the rube beyond resource equality. Had the person's fundamental values not altered, such compensation would not be in order. I am not proposing compensation as a feasible government policy, merely pointing out that the fact that people might in some cases regard it as crass to ask for indemnification of their satisfaction-reducing upbringing does not show that in principle it makes sense for people to assume responsibility (act as though they were responsible) for what does not lie within their control. Any policy that attempted to ameliorate these discrepancies would predictably inflict wounds on innocent parents and guardians far out of proportion to any gain that could be realized for the norm of distributive equality. So even if we all agree that in such cases a policy of compensation is inappropriate, all things considered, it does not follow that so far as distributive equality is concerned (one among the several values we cherish), compensation should not be forthcoming.

Finally, it is far from clear why assuming responsibility for one's preferences and values in the sense of affirming them and identifying them as essential to one's self precludes demanding or accepting compensation for these preferences in the name of distributive equality. Suppose the government has accepted an obligation to subsidize the members of two native tribes who are badly off, low in welfare. The two tribes happen to be identical except that one is strongly committed to traditional religious ceremonies involving a psychedelic made from the peyote cactus while the other tribe is similarly committed to its traditional rituals involving an alcoholic drink made from a different cactus. If the market price of the psychedelic should suddenly rise dramatically while the price of the cactus drink stays cheap, members of the first tribe might well claim that equity requires an increase in their subsidy to compensate for the greatly increased price of the wherewithal for their ceremonies. Advancing such a claim, so far as I can see, is fully compatible with continuing to affirm and identify with one's preferences and in this sense to take personal responsibility for them.

In practise, many laws and other public policies differentiate roughly between preferences that we think are deeply entrenched in people, alterable if at all only at great personal cost, and very widespread in the population, versus preferences that for most of us are alterable at moderate cost should we choose to try to change them and thinly and erratically spread throughout the population. Laws and public policies commonly take account of the former and ignore the latter. For example, the law caters to people's deeply felt aversion to public nudity but does not cater to people's aversion to the sight of tastelessly dressed strollers in public spaces. Of course, current American laws and policies are not designed to achieve any strongly egalitarian ideal, whether resource-based or not. But in appealing to commmon sense as embodied in current practises in order to determine what sort of equality we care about insofar as we do care about equality, one would go badly astray in claiming support in these practises for the contention that equality of resources captures the ideal of equality. We need to search further.

II. EQUALITY OF WELFARE

According to equality of welfare, goods are distributed equally among a group of persons to the degree that the distribution brings it about that each person enjoys the same welfare. (The norm thus presupposes the possibility of cardinal interpersonal welfare comparisons.) The considerations mentioned seven paragraphs back already dispose of the idea that the distributive equality worth caring about is equality of welfare. To bring this point home more must be said to clarify what "welfare" means in this context.

I take welfare to be preference satisfaction. The more an individual's preferences are satisfied, as weighted by their importance to that very individual, the higher her welfare. The preferences that figure in the calculation of a person's welfare are limited to self-interested preferences — what the individual prefers insofar as she seeks her own advantage. One may prefer something for its own sake or as a means to further ends; this discussion is confined to preferences of the former sort.

The preferences that most plausibly serve as the measure of the individual's welfare are hypothetical preferences. Consider this familiar account: The extent to which a person's life goes well is the degree to which his ideally considered preferences are satisfied.[9] My ideally considered preferences are those I would have if I were to engage in thoroughgoing deliberation about my preferences with full pertinent information, in a calm mood, while thinking clearly and making no reasoning errors. (We can also call these ideally considered preferences "rational preferences.")

To avoid a difficulty, we should think of the full information that is pertinent to ideally considered preferences as split into two stages corresponding to "first-best" and "second-best" rational preferences. At the first stage one is imagined to be considering full information relevant to choice on the assumption that the results of this ideal deliberation process can costlessly correct one's actual preferences. At the second stage one is imagined to be considering also information regarding (a) one's actual resistance to advice regarding the rationality of one's preferences, (b) the costs of an educational program that would break down this resistance, and (c) the likelihood that anything approaching this educational program will actually be implemented in one's lifetime. What it is reasonable to prefer is then refigured in the light of these costs. For example, suppose that low-life preferences for cheap thrills have a large place in my actual conception of the good, but no place in my first-best rational preferences. But suppose it is certain that these low-life preferences are firmly fixed in my character. Then my second-best preferences are those I would have if I were to deliberate in ideal fashion about my preferences in the light of full knowledge about my actual preferences and their resistance to change. If you are giving me a birthday present, and your sole goal is to advance my welfare as much as possible, you are probably advised to give me, say, a bottle of jug wine rather than a volume of Shelley's poetry even though it is the poetry experience that would satisfy my first-best rational preference.[10]

On this understanding of welfare, equality of welfare is a poor ideal. Individuals can arrive at different welfare levels due to choices they make for which they alone should be held responsible. A simple example would be to imagine two persons of identical tastes and abilities who are assigned equal resources by an agency charged to maintain distributive equality. The two then voluntarily engage in high-stakes gambling, from which one emerges rich (with high expectation of welfare) and the other poor (with low welfare expectation). For another example, consider two persons similarly situated, so they could attain identical welfare levels with the same effort, but one chooses to pursue personal welfare zealously while the other pursues an aspirational preference (e.g., saving the whales), and so attains lesser fulfillment of self-interested preferences. In a third example, one person may voluntarily cultivate an expensive preference (not cognitively superior to the preference it supplants), while another person does not. In all three examples it would be inappropriate to insist upon equality of welfare when welfare inequality arises through the voluntary choice of the person who gets lesser welfare. Notice that in all three examples as described, there need be no grounds for finding fault with any aims or actions of any of the individuals mentioned. No imperative of practical reason commands us to devote our lives to the maximal pursuit of (self-interested) preference satisfaction. Divergence from equality of welfare arising in these ways need not signal any fault imputable to individuals or to "society" understood as responsible for maintaining distributive equality.

This line of thought suggests taking equal opportunity for welfare to be the appropriate norm of distributive equality.

In the light of the foregoing discussion, consider again the example of compensation for one's religious upbringing regarded as affecting one's lifetime preference satisfaction expectation. This example is urged as a reductio ad absurdum of the norm of equality of welfare, which may seem to yield the counterintuitive implication that such differences do constitute legitimate grounds for redistributing people's resource shares, in the name of distributive equality. As I mentioned, the example is tricky; we should not allow it to stampede us toward resource-based construals of distributive equality. Two comments on the example indicate something of its trickiness.

First, if a person changes her values in the light of deliberation that bring her closer to the ideal of deliberative rationality, we should credit the person's conviction that satisfying the new values counts for more than satisfying the old ones, now discarded. The old values should be counted at a discount due to their presumed greater distance from deliberative rationality. So if I was a Buddhist, then become a Hindu, and correctly regard the new religious preference as cognitively superior to the old, it is not the case that a straight equality of welfare standard must register my welfare as declining even if my new religious values are less easily achievable than the ones they supplant.

Secondly, the example might motivate acceptance of equal opportunity for welfare over straight equality of welfare rather than rejection of subjectivist conceptions of equality altogether. If equal opportunity for welfare obtains between Smith and Jones, and Jones subsequently undergoes religious conversion that lowers his welfare prospects, it may be that we will take Jones's conversion either to be a voluntarily chosen act or a prudentially negligent act for which he should be held responsible. (Consider the norm: Other things equal, it is bad if some people are worse off than others through no voluntary choice or fault of their own.) This train of thought also motivates an examination of equal opportunity for welfare.

III. EQUAL OPPORTUNITY FOR WELFARE

An opportunity is a chance of getting a good if one seeks it. For equal opportunity for welfare to obtain among a number of persons, each must face an array of options that is equivalent to every other person's in terms of the prospects for preference satisfaction it offers. The preferences involved in this calculation are ideally considered second-best preferences (where these differ from first-best preferences). Think of two persons entering their majority and facing various life choices, each action one might choose being associated with its possible outcomes. In the simplest case, imagine that we know the probability of each outcome conditional on the agent's choice of an action that might lead to it. Given that one or another choice is made and one or another outcome realized, the agent would then face another array of choices, then another, and so on. We construct a decision tree that gives an individual's possible complete life-histories. We then add up the preference satisfaction expectation for each possible life history. In doing this we take into account the preferences that people have regarding being confronted with the particular range of options given at each decision point. Equal opportunity for welfare obtains among persons when all of them face equivalent decision trees — the expected value of each person's best (= most prudent[11]) choice of options, second-best, . . . nth-best is the same. The opportunities persons encounter are ranked by the prospects for welfare they afford.

The criterion for equal opportunity for welfare stated above is incomplete. People might face an equivalent array of options, as above, yet differ in their awareness of these options, their ability to choose reasonably among them, and the strength of character that enables a person to persist in carrying out a chosen option. Further conditions are needed. We can summarize these conditions by stipulating that a number of persons face *effectively* equivalent options just in case one of the following is true: (1) the options are equivalent and the persons are on a par in their ability to "negotiate" these options, or (2) the options are nonequivalent in such a way as to counterbalance exactly any inequalities in people's negotiating abilities, or (3) the options are equivalent and any inequalities in people's negotiating abilities are due to causes for which it is proper to hold the individuals themselves personally responsible. Equal opportunity for welfare obtains when all persons face effectively equivalent arrays of options.

Whether or not two persons enjoy equal opportunity for welfare at a time depends only on whether they face effectively equivalent arrays of options at that time. Suppose that Smith and Jones share equal opportunity for welfare on Monday, but on Tuesday Smith voluntarily chooses or negligently behaves so that from then on Jones has greater welfare opportunities. We may say that in an extended sense people share equal opportunity for welfare just in case there is some time at which their opportunities are equal and if any inequalities in their opportunities at later times are due to their voluntary choice or differentially negligent behavior for which they are rightly deemed personally responsible.

When persons enjoy equal opportunity for welfare in the extended sense, any actual inequality of welfare in the positions they reach is due to factors that lie within each individual's control. Thus, any such inequality will be nonproblematic from the standpoint of distributive equality. The norm of equal opportunity for welfare is distinct from equality of welfare only if some version of soft determinism or indeterminism is correct. If hard determinism is true, the two interpretations of equality come to the same.

In actual political life under modern conditions, distributive agencies will be staggeringly ignorant of the facts that would have to be known in order to pinpoint what level of opportunity for welfare different persons have had. To some extent it is technically unfeasible or even physically impossible to collect the needed information, and to some extent we do not trust governments with the authority to collect the needed information, due to worries that such authority will be subject to abuse. Nonetheless, I suppose that the idea is clear in principle, and that in practise it is often feasible to make reliable rough-and-ready judgments to the effect that some people face very grim prospects for welfare compared to what others enjoy.

In comparing the merits of a Rawlsian conception of distributive equality as equal shares of primary goods and a Dworkinian conception of equality of resources with the norm of equality of opportunity for welfare, we run into the problem that in the real world, with imperfect information available to citizens and policymakers, and imperfect willingness on the part of citizens and officials to carry out conscientiously whatever norm is chosen, the practical implications of these conflicting principles may be hard to discern, and may not diverge much in practise. Familiar information-gathering and information-using problems will make us unwilling to authorize government agencies to determine people's distributive shares on the basis of their preference satisfaction prospects, which will often be unknowable for all practical purposes. We may insist that governments have regard to primary good share equality or resource equality as rough proxies for the welfarist equality that we are unable to calculate. To test our allegiance to the rival doctrines of equality we may need to consider real or hypothetical examples of situations in which we do have good information regarding welfare prospects and opportunities for welfare, and consider whether this information affects our judgments as to what counts as egalitarian policy. We also need to consider cases in which we gain new evidence that a particular resource-based standard is a much more inaccurate proxy for welfare equality than we might have thought, and much less accurate than another standard now available. Indifference to these considerations would mark allegiance to a resourcist interpretation of distributive equality in principle, not merely as a handy rough-and-ready approximation.

IV. STRAIGHT EQUALITY VERSUS EQUAL OPPORTUNITY; WELFARE VERSUS RESOURCES

The discussion to this point has explored two independent distinctions: (1) straight equality versus equal opportunity and (2) welfare versus resources as the appropriate basis for measuring distributive shares. Hence there are four positions to consider. On the issue of whether an egalitarian should regard welfare or resources as the appropriate standard of distributive equality, it is important to compare like with like, rather than, for instance, just to compare equal opportunity for resources with straight equality of welfare. (In my opinion Ronald Dworkin's otherwise magisterial treatment of the issue in his two-part discussion of "What Is Equality?" is marred by a failure to bring these four distinct positions clearly into focus.[12])

The argument for equal opportunity rather than straight equality is simply that it is morally fitting to hold individuals responsible for the foreseeable consequences of their voluntary choices, and in particular for that portion of these consequences that involves their own achievement of welfare or gain or loss of resources. If accepted, this argument leaves it entirely open whether we as egalitarians ought to support equal opportunity for welfare or equal opportunity for resources.

For equal opportunity for resources to obtain among a number of persons, the range of lotteries with resources as prizes available to each of them must be effectively the same. The range of lotteries available to two persons is effectively the same whenever it is the case that, for any lottery the first can gain access to, there is an identical lottery that the second person can gain access to by comparable effort. (So if Smith can gain access to a lucrative lottery by walking across the street, and Jones cannot gain a similar lottery except by a long hard trek across a desert, to this extent their opportunities for resources are unequal.) We may say that equal opportunity for resources in an extended sense obtains among a number of persons just in case there is a time at which their opportunities are equal and any later inequalities in the resource opportunities they face are due to voluntary choices or differentially negligent behavior on their part for which they are rightly deemed personally responsible.

I would not claim that the interpretation of equal opportunity for resources presented here is the only plausible construal of the concept. However, on any plausible construal, the norm of equal opportunity for resources is vulnerable to the "slavery of the talented" problem that proved troublesome for equality of resources. Supposing that personal talents should be included among the resources to be distributed (for reasons given in section I), we find that moving from a regime of equality of resources to a regime that enforces equal opportunity for resources does not change the fact that a resource-based approach causes the person of high talent to be predictably and (it would seem) unfairly worse off in welfare prospects than her counterpart with lesser talent.[13] If opportunities for resources are equally distributed among more and less talented persons, then each person regardless of her native talent endowment will have comparable access to identical lotteries for resources that include time slices of the labor power of all persons. Each person's expected ownership of talent, should he seek it, will be the same. Other things equal, if all persons strongly desire personal liberty or initial ownership of one's own lifetime labor power, this good will turn out to be a luxury commodity for the talented, and a cheap bargain for the untalented.

A possible objection to the foregoing reasoning is that it relies on a vaguely specified idea of how to measure resource shares that is shown to be dubious by the very fact that it leads back to the slavery of the talented problem. Perhaps by taking personal liberty as a separate resource this result can be avoided. But waiving any other difficulties with this objection, we note that the assumption that any measure of resource equality must be unacceptable if applying it leads to unacceptable results for the distribution of welfare amounts to smuggling in a welfarist standard by the back door.

Notice that the welfare distribution implications of equal opportunity for resources will count as intuitively unacceptable only on the assumption that people cannot be deemed to have chosen voluntarily the preferences that are frustrated or satisfied by the talent pooling that a resourcist interpretation of equal opportunity enforces. Of course it is strictly nonvoluntary that one is born with a particular body and cannot be separated from it, so if others hold ownership rights in one's labor power one's individual liberty is thereby curtailed. But in principle one's self-interested preferences could be concerned no more with what happens to one's own body than with what happens to the bodies of others. To the extent that you have strong self-interested hankerings that your neighbors try their hand at, say, farming, and less intense desires regarding the occupations you yourself pursue, to that extent the fact that under talent pooling your own labor power is a luxury commodity will not adversely affect your welfare. As an empirical matter, I submit that it is just false to hold that in modern society whether any given individual does or does not care about retaining her own personal liberty is due to that person's voluntarily choosing one or the other preference. The expensive preference of the talented person for personal liberty cannot be assimilated to the class of expensive preferences that people might voluntarily cultivate.[14] On plausible empirical assumptions, equal opportunity for welfare will often find tastes compensable, including the talented person's taste for the personal liberty to command her own labor power. Being born with high talent cannot then be a curse under equal opportunity for welfare (it cannot be a blessing either).

V. SEN'S CAPABILITIES APPROACH

The equal opportunity for welfare construal of equality that I am espousing is similar to a "capabilities" approach recently defended by Amartya Sen.[15] I shall now briefly sketch and endorse Sen's criticisms of Rawls's primary social goods standard and indicate a residual welfarist disagreement with Sen.

Rawls's primary social goods proposal recommends that society should be concerned with the distribution of certain basic social resources, so his position is a variant of a resource-based understanding of how to measure people's standard of living. Sen holds that the distribution of resources should be evaluated in terms of its contribution to individual capabilities to function in various ways deemed to be objectively important or valuable. That is, what counts is not the food one gets, but the contribution it can make to one's nutritional needs, not the educational expenditures lavished, but the contribution they make to one's knowledge and cognitive skills. Sen objects to taking primary social goods measurements to be fundamental on the ground that persons vary enormously from one another in the rates at which they transform primary social goods into capabilities to function in key ways. Surely we care about resource shares because we care what people are enabled to be and do with their resource shares, and insofar as we care about equality it is the latter that should be our concern.

So far, I agree. Moreover, Sen identifies a person's well-being with the doings and beings or "functionings" that he achieves, and distinguishes these functionings from the person's capabilities to function or "well-being freedom."[16] Equality of capability is then a notion within the family of equality of opportunity views, a family that also includes the idea of equal opportunity for welfare that I have been attempting to defend. So I agree with Sen to a large extent.

But given that there are indefinitely many kinds of things that persons can do or become, how are we supposed to sum an individual's various capability scores into an overall index? If we cannot construct such an index, then it would seem that equality of capability cannot qualify as a candidate conception of distributive equality. The indexing problem that is known to plague Rawls's primary goods proposal also afflicts Sen's capabilities approach.[17]

Sen is aware of the indexing problem and untroubled by it. The grand theme of his lectures on "Well-being, Agency and Freedom" is informational value pluralism: We should incorporate in our principles all moral information that is relevant to the choice of actions and policies even if that information complicates the articulation of principles and precludes attainment of a set of principles that completely rank-orders the available alternative actions in any possible set of circumstances. "Incompleteness is *not* an embarrassment," Sen declares.[18] I agree that principles of decision should not ignore morally pertinent matters but I doubt that the full set of my functioning capabilities does matter for the assessment of my position. Whether or not my capabilities include the capability to trek to the South Pole, eat a meal at the most expensive restaurant in Omsk, scratch my neighbor's dog at the precise moment of its daily maximal itch, matters not one bit to me, because I neither have nor have the slightest reason to anticipate I ever will have any desire to do any of these and myriad other things. Presumably only a small subset of my functioning capabilities matter for moral assessment, but which ones?

We may doubt whether there are any objectively decidable grounds by which the value of a person's capabilities can be judged apart from the person's (ideally considered) preferences regarding those capabilities. On what ground do we hold that it is valuable for a person to have a capability that she herself values at naught with full deliberative rationality? If a person's having a capability is deemed valuable on grounds independent of the person's own preferences in the matter, the excess valuation would seem to presuppose the adequacy of an as yet unspecified perfectionist doctrine the like of which has certainly not yet been defended and in my opinion is indefensible.[19] In the absence of such a defense of perfectionism, equal opportunity for welfare looks to be an attractive interpretation of distributive equality.

NOTES

[1] John Rawls, 'Social Unity and Primary Goods,' in Amartya Sen and Bernard Williams, eds., *Utilitarianism and Beyond* (Cambridge: Cambridge University Press, 1982), pp. 159–185.

[2] Ronald Dworkin, 'What Is Equality? Part 1: Equality of Welfare,' *Philosophy and Public Affairs* **10** (1981): 185–246; and 'What Is Equality? Part 2: Equality of Resources,' *Philosophy and Public Affairs* **10** (1981): 283–345. See also Thomas Scanlon, 'Preference and Urgency,' *Journal of Philosophy* **72** (1975): 655–669.

[3] Dworkin, 'Equality of Resources.'

[4] Hal Varian discusses this mechanism of equal distribution, followed by trade to equilibrium, in 'Equity, Envy, and Efficiency,' *Journal of Economic Theory* **9** (1974): 63–91. See also John Roemer, 'Equality of Talent,' *Economics and Philosophy* **1** (1985): 151–186; and 'Equality of Resources Implies Equality of Welfare,' *Quarterly Journal of Economics* **101** (1986): 751–784.

[5] Dworkin, 'Equality of Resources,' p. 312.

It should be noted that the defender of resource-based construals of distributive equality has a reply to the slavery of the talented problem that I do not consider in this paper. According to this reply, what the slavery of the talented problem reveals is not the imperative of distributing so as to equalize welfare but rather the moral inappropriateness of considering all resources as fully alienable. It may be that equality of resources should require that persons be compensated for their below-par talents, but such compensation should not take the form of assigning individuals full private ownership rights in other people's talents, which should be treated as at most partially alienable. See Margaret Jane Radin, 'Market-Inalienability,' *Harvard Law Review* **100** (1987): 1849–1937.

[6] Rawls, 'Social Unity and Primary Goods,' pp. 167–170.

[7] Thomas Scanlon, 'Equality of Resources and Equality of Welfare: A Forced Marriage?', *Ethics* **97** (1986): 111–118; see esp. pp. 115–117.

[8] Scanlon, 'Equality of Resources and Equality of Welfare,' p. 116.

[9] See, e.g., John Rawls, *A Theory of Justice* (Cambridge, MA: Harvard University Press, 1971), pp. 416–424; Richard Brandt, *A Theory of the Good and the Right* (Oxford: Oxford University Press, 1979), pp. 110–129; David Gauthier, *Morals by Agreement* (Oxford: Oxford University Press, 1986), pp. 29–38; and Derek Parfit, *Reasons and Persons* (Oxford: Oxford University Press, 1984), pp. 493–499.

[10] In this paragraph I attempt to solve a difficulty noted by James Griffin in 'Modern Utilitarianism,' *Revue Internationale de Philosophie* **36** (1982): 331–375; esp. pp. 334–335. See also Amartya Sen and Bernard Williams, 'Introduction' to *Utilitarianism and Beyond*, p. 10.

[11] Here the most prudent choice cannot be identified with the choice that maximizes lifelong expected preference satisfaction, due to complications arising from the phenomenon of preference change. The prudent choice as I conceive it is tied to one's actual preferences in ways I will not try to describe here.

[12] See the articles cited in note 2. Dworkin's account of equality of resources is complex, but without entering into its detail I can observe that Dworkin is discussing a version of what I call "equal opportunity for resources." By itself, the name chosen matters not a bit. But confusion enters because Dworkin neglects altogether the rival doctrine of equal opportunity for welfare. For a criticism of Dworkin's objections against a welfarist conception of equality that do not depend on this confusion, see my 'Liberalism, Distributive Subjectivism, and Equal Opportunity for Welfare.'

[13] Roemer notes that the person with high talent is cursed with an involuntary expensive preference for personal liberty. See Roemer, 'Equality of Talent.'

[14] As Rawls writes, ". . . those with less expensive tastes have presumably adjusted their likes and dislikes over the course of their lives to the income and wealth they could reasonably expect; and it is regarded as unfair that they now should have less in order to spare others from the consequences of their lack of foresight or self-discipline." See 'Social Unity and Primary Goods,' p. 169.

[15] Amartya Sen, 'Well-being, Agency and Freedom: The Dewey Lectures 1984,' *Journal of Philosophy* **82** (1985): 169–221; esp. pp. 185–203. See also Sen, 'Equality of What?', in his *Choice, Welfare and Measurement* (Oxford: Basil Blackwell, 1982), pp. 353–369.

[16] Sen, 'Well-being, Agency and Freedom,' p. 201.

[17] See Allan Gibbard, 'Disparate Goods and Rawls' Difference Principle: A Social Choice Theoretic Treatment,' *Theory and Decision* **11** (1979): 267–288; see esp. pp. 268–269.

[18] Sen, 'Well-being, Agency and Freedom,' p. 200.

[19] However, it should be noted that filling out a preference-satisfaction approach to distributive equality would seem to require a normative account of healthy preference formation that is not itself preference-based. A perfectionist component may thus be needed in a broadly welfarist egalitarianism. For this reason it would be misguided to foreclose too swiftly the question of the possible value of a capability that is valued at naught by the person who has it. The development and exercise of various capacities might be an important aspect of healthy preference formation, and have value in this way even though this value does not register at all in the person's preference satisfaction prospects.

Department of Philosophy,
University of California, San Diego,
La Jolla, CA 92093,
U.S.A.

Part III
Equality of Opportunity

[10]

Competitive Equality of Opportunity

D. A. LLOYD THOMAS

The first purpose of this discussion is to distinguish between a number of distinct ideas which can be conveyed by the expression 'equality of opportunity'. Detailed attention will be given to only one of them: 'competitive' equality of opportunity. Other conceptions of equality of opportunity will be mentioned as part of a process of clarification, but then put aside. After an account of competitive equality of opportunity, and a discussion of its relationships to determinism and manipulability, two principal objections will be urged against it: that (given certain empirical beliefs) it leads to an incoherent account of what is a just distribution, and that the greater the extent to which competitive equality of opportunity is realized, the less (in one respect) are the opportunities available to persons to make autonomous choices.

Opportunities

First let us consider briefly what is an opportunity, with examples in mind such as 'She had an opportunity to take a job in California', 'He had many opportunities to get married', and 'There was no opportunity to escape'. One has an opportunity to do something or to have something provided that one can do it or have it if one chooses. One has no opportunity to do something or to have something if one cannot do it or have it even if one wishes to. But not everything one can do or have if one chooses is an opportunity. For something to be regarded as an opportunity it must, in addition, be seen as to some extent good. It is bitter sarcasm to say that someone had the opportunity to lose his life pointlessly in the Battle of the Somme. Opportunities may be good, wonderful, or not very good, but they cannot be regarded as in no respect good. Often opportunities are regarded as good both by the person who has them and by others. But it can be that A does not see a certain option as in any sense good, and hence does not regard it as an opportunity, while B does regard A's option as good, and thinks of A as having an opportunity.

What could people have in mind when they favour *equal* opportunity? Obviously it cannot be that everyone has the *same* opportunity (as when, for example, two people both have the opportunity to teach at Berkeley for a year). It is necessarily true that it cannot be a matter of the same opportunity, for in many cases A having a certain opportunity at a certain time to do something (for example, to marry B) precludes others from having the same opportunity.

'Ideal' Equality of Opportunity

Perhaps, then, for all to have equal opportunity would be for all to have an equally good (though different) set of opportunities. For example, A has the opportunity to join the Foreign Office, but not to be a Fellow of New College; B has the opportunity to be a Fellow of New College, but not to join the Foreign Office. A and B have different opportunities, but some would say that they were equally good opportunities. This view of equality of opportunity would hold in the following situation. Over P1's life there have been certain opportunities O1, O2 etc.; over P2's life there have been certain opportunities Oa, Ob etc., and so on for all persons. There is equality of opportunity if, by reference to some scale for evaluating opportunities, the score for P1's package is much the same as that for P2's package, and so on for everybody. For convenience this idea will be called 'ideal' equality of opportunity, but the label is not intended to suggest that there is anything ideal about it other than in the sense that it is a limit case.

For such a conception of equality of opportunity to be applicable (much less desirable), there would have to be an agreed interpersonal standard of comparison for the goodness of opportunities. Sometimes defenders of equality of opportunity appear to suppose that there is such a scale: for example, that opportunities can be compared by reducing them in some way to a common denominator of power, wealth and status. But it may be doubted whether there really is such a scale. For a certain option may appear good to one person and not to another, and it is not obvious that we can say, in all cases, that one of these judgements of goodness is simply mistaken. Some accept a conception of the goodness of opportunities related to power, wealth and status, while others reject this. Arguments can be given for both views, and it is doubtful whether either is simply mistaken. As Professor Nozick says '... life is not a race in which we all compete for a prize which someone has established; there is no unified race, with some person judging swiftness'.[1]

Non-competitive Equality of Opportunity

Let us make another attempt to get closer to a plausible account of equality of opportunity. The idea need not be that everyone should end up with an equally valuable cluster of opportunities. It could be that there shall be *some* opportunities that are present in everybody's cluster (at the appropriate stage of life). For example, there is the 'commitment to full employment': the attempt to make available to everyone the opportunity to do some kind of work, or again, the idea that everyone ought to have the opportunity to receive some kind of tertiary education. These ideas can be represented as follows. For every person P1, P2 etc. there ought to appear amongst his or her cluster of opportunities certain 'standard' opportunities, such as the opportunity to work (Ow), or the opportunity to receive some form of tertiary education (Ot). Thus:

For P1: O1 O2 *Ow Ot* O3 . . . etc.
For P2: Oa Ob *Ow Ot* Oc . . . etc.

This conception of equality of opportunity does not suffer from the difficulty with 'ideal' equality of opportunity. As it is not required that everyone shall have an equally valuable cluster of opportunities, the question of a common set of criteria for assessing clusters of opportunities does not arise. However, it is assumed that everyone has reason to want the opportunities to be made available in his or her cluster. Perhaps the justification would be (after Rawls) that they are opportunities everyone has reason to want whatever other opportunities they might want, i.e. whatever their rational plan of life might be. On this conception of equality of opportunity it is also supposed that there is some authority (nearly always assumed to be the state) which can so manipulate things as to make such opportunities available in everyone's cluster. It must be further supposed that it is legitimate for that authority to create such opportunities by, if need be, compulsorily reducing the opportunities others already have. It is commonly assumed that the state has the right to do this, and it is interesting to raise the question of where this right might come from. It does

1 Robert Nozick, *Anarchy, State, and Utopia*, Basil Blackwell, Oxford, 1974, p. 235.

not come from it merely being morally desirable that there should be such opportunities, even given that it can be shown that it is morally desirable. It may be morally desirable that my friend should make available equal opportunities to all of his children, in so far as he can, and not give them all to his favourite child. But that does not give me any right to interfere, if he does not.

The mere existence of a morally unacceptable situation does not, as such, give anyone the right to interfere. Hence arguments which move directly from premises about some state of affairs being morally unacceptable to a conclusion that the state has the right to put that state of affairs in order are incomplete. This is not to deny (to return to the case of the children) that certain forms of mistreatment provide grounds for a right to interfere. Though now the point turns on the ill-treated child having a claim against others, if those immediately responsible fail to discharge their obligations, rather than on the child being *unequally* treated. One might cite the case of parents who equally neglect all of their children.

My own view concerning this type of equality of opportunity is that such claims are justified (i) when they reasonably can be represented as moral claims which hold against the community generally, and (ii), when they are moral claims of such a kind that the state can legitimately compel their satisfaction. Perhaps, therefore, a claim for equal opportunity to receive medical attention is justified, while a claim for equal opportunity to receive some kind of tertiary education is not. The issues involved here are complex, and warrant separate discussion. It must be emphasized that in now setting aside non-competitive equality of opportunity, I intend *only* to set it aside. It is not supposed that it has been shown to lack justification.

The label 'non-competitive' has been used for this conception of equality of opportunity because the opportunities are seen as ones that ought to be available to anybody if they wish to take them up, not as ones that have to be striven for in competition with others. So we now have, on the one hand, 'ideal' equality of opportunity, which is an implausibly strong demand, and on the other, 'non-competitive' equality of opportunity, which many would regard as acceptable for certain types of opportunity, but not as the whole story. So we come to 'competitive' equality of opportunity.

Competitive Equality of Opportunity

The example to be taken is the English 'UCCA' selection process for filling student places at universities. We have persons (university departments) who have authority to distribute certain opportunities. They are to be distributed by reference to general criteria, those best satisfying the criteria, in competition with others, getting the opportunity. It will be supposed that there are not as many opportunities available as there are persons who desire them. Thus the position is clearly different from non-competitive equality of opportunity. Under that conception anyone who wished could take up the opportunity, whereas under the competitive conception not everyone who likes can have the opportunity. Indeed there could not be both non-competitive and competitive equality of opportunity in exactly the same respect, as the former presupposes that the opportunity is available to anyone who wants it, while the latter presupposes that not everyone who wants it can have it. It would seem, then, that competitive equality of opportunity has to do with the way in which these competitions are conducted. Perhaps, considering that opportunities are not going to be equal in the end, it would be better to call it 'fair competition for scarce opportunities'. For strictly speaking (as has already been suggested), if a person has an opportunity to do something, then he can do it if he chooses, but in the situation we are now considering it will not be the case that all of those who can be said to enjoy 'equal' opportunity will in fact get the opportunity. There is an interesting contrast between having an opportunity and some instances of having a right. One may have a right to vote (in law), but not be in a position to exercise that right (because one has been kidnapped). In the case of opportunities the connection between having the opportunity and the 'material' possibility of exercising it is closer. If being kidnapped prevents one from voting, then one does not have the opportunity to vote on that occasion. Thus linguistic propriety could lead one to say that what has been called 'non-competitive equality of opportunity' is the only form of equality of *opportunity*, properly speaking. Alternatively, the opportunity in question here could be thought of as the opportunity to compete, which each can have if he chooses, and not the opportunity to enjoy that which is competed for (i.e. the student places).

We now need to lay down the criteria for this kind of equality of opportunity. What would it be for there not to be equality of opportunity in this sense? There would not be equality of opportunity if A were selected and B were not because B was a Jew (or female, or of a certain nationality, or not rich, or not a relative of a member of the department etc.). What is the principle operative here?

First Principle:

The criteria of selection employed must include only characteristics relevant to the utilization of the opportunity to be made available.

Thus, with regard to the example in hand, being a Catholic is not a relevant characteristic, while being clever is. Clearly there can be disputes about what are the relevant characteristics, and also, if these are agreed, about what are reliable indications of their presence. But there is no need to go into details on these questions.

It would be widely allowed that this is a necessary condition for competitive equality of opportunity, but is it sufficient? Let us consider the following case. A and B are applicants for the last remaining place in a university department. A has impressive 'A level' results, performs well at interview, and has an enthusiastic reference. B has mediocre 'A level' results, is not forthcoming at interview, and has an indifferent reference. On this basis A is given the place. The first principle would appear to be satisfied, anyway, so long as any doubts about the connection between facts of this sort concerning A and B and their likely capacity to make good use of the opportunity are ignored. Has B enjoyed equal opportunity? Suppose that A has come from a rich, well-educated family, and has been to one of the academically best public schools in the country. B has come from a poor, ill-educated family, and has been to a comprehensive school with a not very well taught sixth form. Has B enjoyed equal opportunity with A? Some would say 'No', and justify their view in the following way. 'Compare A's past opportunities with B's. At various crucial points in his career A has had amongst his total set of opportunities better ones than those available to B. For example, at the time of starting secondary education A may have had the opportunity to go to several excellent public schools as well as to the local comprehensive. B only had the opportunity to go to the local comprehensive. These past opportunities have considerably influenced their respective chances of getting the new opportunity —the place in the department. Only if those past opportunities had been equally good (with regard to the effects that they have on factors relevant to the present competition) would there be "true" or "real" equality of opportunity.'

The chances of a person securing a new opportunity is seen as resting upon a pyramid of past opportunities which have already been enjoyed. But it is not only past opportunities which will affect a person's present chances of securing new opportunities: inherited potentialities and talents, early environment and so on, will also have this kind of effect. Having made the point more general, we can now state a second principle for competitive equality of opportunity.

Second Principle:

Equality of opportunity in a certain competition C exists only if the possession of all the factors (*F*1, *F*2 . . . *etc.*) *which affect success at C, and the possession of which is open to human manipulation, are so manipulated as to ensure that all competitors in C possess those factors to an equal extent.*

Here is an example of the operation of the second principle. The quality of secondary education a person receives affects his chances of success at the university entrance competition. It is also open to human manipulation. Therefore, if there is to be equality of opportunity in this competition, the quality of secondary education received will have to be manipulated so that all competitors have the benefit of an education of the same quality. From the second principle we can also see what factors might become relevant to equality of opportunity, though they are not now. Let us suppose that intelligence is at least partly dependent on genetic factors, and let us suppose further that we come to know how to manipulate these factors. Then, by the second principle, lack of equal possession of these factors would establish that there was not full equality of opportunity for those competitions in which intelligence is relevant.

Competitive Equality of Opportunity and Determinism

The second principle has an interesting consequence if it is conjoined with a belief in determinism. (Determinism is taken to be the doctrine that for any event, including any human act or performance, there is some set of antecedent events, and some set of covering general laws, such that the occurrence of the event in question is fully explicable by reference to these.) Let us suppose that determinism is true, and that it covers 'moral' factors relevant to doing well at competitions, such as trying hard and persevering against difficulties; that is, such 'moral' factors, as much as any other, are held to be determined by antecedent circumstances. Let us further suppose that our 'technology' has reached the point where any factor of relevance to success in a certain competition can be manipulated, and has been, so as to ensure equality of opportunity by satisfaction of the second principle.

If these conditions were satisfied the outcome of every competition would be a tie between all the contestants. For suppose that in some competition one contestant does better than the others. Then, according to determinism, there must be some antecedent event or events which account for this difference. Thus there is a factor relevant to success at the competition which could (by hypothesis) have been manipulated but was not. Therefore the second principle was not satisfied. Therefore in any competition in which the principle is satisfied the result must be a tie.[1] In these circumstances the demand for equality of opportunity becomes absurd. For the point of the competition was to decide on the allocation of scarce benefits by a contest in which some were more successful than others. But if the result of every competition is a tie, it is useless for making this allocation. Hence in the circumstances supposed there is an inconsistency between the demand for competitive equality of opportunity and the very presumption necessary to make that demand pointful, namely that the competition will resolve the allocation of scarce benefits.

The consequences of assuming complete manipulability are

1 It is not true that in every *particular* 'contest' or 'game' it follows from equal possession of the factors relevant to success that the outcome will be a tie. But such particular 'contests' or 'games' would be constituents of a 'competition' as that term is being understood here. If there were a general tendency for P1 to do better than P2 over a series of particular contests making up a competition, then it could not be said that the second principle had been satisfied in the circumstances supposed. For either there is some relevant factor which has not been equalized and which accounts for the general tendency, or else we can 'decide' who will win (for example, by deciding who goes first), in which case the truth of determinism is called into question. (I am grateful to John Baker for drawing my attention to this point.)

discussed by Bernard Williams, who says 'In these circumstances, where everything about a person is controllable, equality of opportunity and absolute equality seem to coincide . . .'.[1] It is true that in these circumstances there would be no reason for giving any one competitor a greater benefit at the end of the competition than any other. But this does not amount to a coincidence of competitive equality of opportunity and absolute equality. It is the end of equality of opportunity, for that notion had its point in the context of a competition where scarce benefits are to be distributed.

It is true that the distribution of benefits could be made arbitrarily, say, by drawing lots. And if in the odd particular contest chance is used as a tie-breaker, it is not unreasonable to think that the letter rather than the spirit of the first principle has been broken. But what is at issue here is a situation in which all of the factors relevant to success have been manipulated so as to ensure that all have equal chance of success, and therefore where all competitions (that is, series of 'contests' or 'games') ought to end in a tie between all competitors. In *that* case competitive equality of opportunity becomes absurd, for the competitions are now essentially lotteries, and more than the letter of the first principle has been violated.

Let us consider some of the alternatives open to those who hold the second principle, but who want to get themselves off this hook.

1. While allowing that determinism is true, it could be pointed out that manipulation is far from possible in respect of all factors relevant to success at competitions. Therefore unmanipulable differences will remain, such as (at present) differences in 'natural' intelligence and talent. These differences will result in some competitors doing better than others, and hence the competitions will not be pointless.

2. While it is doubtful whether determinism could be wholly rejected (for then, how could the factors which affect success at competitions be ascertained?), it could be rejected in part. One possibility here is that determinism holds of the possession of natural abilities, and of some acquired characteristics relevant to

1 'The Idea of Equality' in Peter Laslett and W. G. Runciman (eds.), *Philosophy, Politics and Society, Second Series*, Basil Blackwell, Oxford, 1962, pp. 128–129.

success at competitions, but not in general to the moral qualities relevant to success, or to the acquired abilities which have in part been acquired in virtue of the presence of certain moral qualities.

Now let us consider the position if it is accepted (a) that complete manipulability is not in fact possible, and (b), that determinism does not hold for all moral qualities. Then if there were complete equality of opportunity by reference to the second principle, the outcomes of competitions would depend upon differences in natural abilities, and differences in the application of moral qualities to the development of such natural abilities. We have arrived at the formula used by Michael Young's meritocrats: 'Intelligence and effort together make up merit (I+E = M)'.[1]

Some will be unhappy that this should be the outcome of competitive equality of opportunity. For the conception we have arrived at will amount to a state of affairs in which those who win have greater natural aptitudes and more sterling moral qualities. Now if one is (as we are now supposing) non-deterministic about at least some moral qualities, one might be prepared to allow that differential rewards on this basis are deserved by the person who gains them. But surely one's natural abilities are not due to any credit on one's own part: they are not deserved, so why should people receive different rewards on that basis?

The tracks that lead away from the contention that in a just society only deserved differences (at most) ought to be differentially rewarded will be ignored. Let us merely note where we get to if we suppose that any difference in rewards on grounds of natural ability is not justified, and that greater natural abilities ought to be suppressed in some way, or counteracted, for example, by deliberately worsening the opportunities of those with superior natural abilities. This would make the outcome of competitions wholly dependent on moral qualities. In a society with complete equality of opportunity as now understood, we would have the morally superior receiving the greater benefits. In so far as these benefits included power and status, it would be an elite of the morally virtuous, of the 'elect'. It scarcely needs emphasizing that this is a secular version of some forms of seventeenth-century Protestantism.

1 Michael Young, *The Rise of Meritocracy 1870–2033*, Thames and Hudson, London, 1958, p. 74.

Competitive Equality of Opportunity and Justice

The discussion of competitive equality of opportunity in relation to the assumptions of determinism and complete manipulability has not given rise to any conclusive objection to that conception of equal opportunity. For complete determinism may be denied, and complete manipulability does not, in any case, hold. An advocate of competitive equality of opportunity might be worried by future prospects, but there would appear to be no difficulty in his at present advocating this conception. However, when we come to consider the view of justice to which he is committed, there would appear to be an incoherence in his idea of equality of opportunity.

While complete manipulability may be science fiction, the implicit assumption that what is manipulable legitimately may be manipulated in order to satisfy the second principle, has been left unquestioned. But on reflection we may wish to constrain the whole-hearted pursuit of the requirements of the second principle up to the limits of our 'technological' capabilities. An example of such a constraint is one we may wish to impose to protect the family. For example, let us take two persons who start from an approximately equal position in circumstances of equality of opportunity. One succeeds in acquiring a position needing considerable education and skill, while the other has a routine job requiring little education. Now consider their children. So far as factors relevant to success in their competitions are concerned, they will not be equal. Thus we have a problem for those who advocate competitive equality of opportunity: the prizes won in the competitions of the first generation will tend to defeat the requirements of equality of opportunity for the next. Now someone sufficiently fanatical about equality of opportunity may say that such inequalities as arise from the competitions of one generation should be corrected before the competitions of the next begin. 'The virtues of the fathers shall not be visited upon the sons.' And it is possible to suggest how this might be attempted: children could be removed from their parents at an early age and put into state nurseries intended to provide a similar early environment for all children.

It is now possible to formulate an incoherence in the notion of distributive justice contained within the idea of competitive equality of opportunity. On the basis of the second principle, equality of opportunity requires equalization of the factors ((F1 . . . Fn) relevant to success. Unless at t1 all competitors possess F1 . . . Fn equally (where this is manipulatively possible) equality of opportunity has not been achieved. Thus, in accordance with the second principle, the following situation must hold, so far as manipulable factors are concerned:

At t1:	A		B		C
	F1 . . . Fn	=	F1 . . . Fn	=	F1 . . . Fn

The competition between A, B and C now proceeds, and we suppose that they have differing degrees of success, due, let us suppose, to differential possession of unmanipulable natural abilities. Let us suppose that later the distribution of benefits by the competition is as follows:

At t2:	A	B	C
	3b	2b	b

By reference to the idea of competitive equality of opportunity the distribution at t2 is just, given that the first principle also has been observed during the competition.

Now let us suppose that these differences in benefits received have an effect on the extent to which F1. . . Fn are possessed by members of the next generation (A1, B1, and C1) when their time comes to engage in a similar competition. This effect may be through the passing on of wealth, or of especially favourable environments, and so on. It is assumed, for the purpose of this argument, that the benefits received by the first generation do affect, and affect *favourably* the possession of factors relevant to success in the next generation. This is, of course, a contingent assumption, and it may not hold invariably. The position at t3 will then be as follows:

At t3:	A1		B1		C1
	F1 . . . Fn	>	F1 . . . Fn	>	F1 . . . Fn

Thus there emerges the following incoherence in the conception of justice contained within the idea of competitive equality of opportunity. At t2 the distribution of benefits is just because it is based on competitive performance under the two principles of equality of opportunity. At t2 the distribution of benefits is not just because it causes a distribution of the factors F1 . . . Fn at t3 contrary to that required by the second principle of equality of opportunity. Such an incoherence must remain so long as the distribution of benefits at t2 has a certain kind of causal influence on the possession of factors F1 . . . Fn in the competition following upon t3. This amounts to saying that the incoherence will remain so long as there are social relations between members of one generation and the next. The considerations are wider than the continued existence of inheritance and the family. For as John Charvet has pointed out,[1] even state nurseries could not guarantee equal treatment, and hence the doctrine must end up by not permitting persons to be dependent for their self-development on others at all; in other words, the doctrine is incompatible with the existence of any society at all.

This argument against competitive equality of opportunity does not bring up the problems which arise with the ideas of 'compensatory justice' or 'positive discrimination'. For those ideas have application in a context in which competitive equality of opportunity is assumed *not* to have existed, and the question then arises 'What is to be done about those for whom the chance of equal opportunity has now been lost for good, and who are at a disadvantage as a result?'. But as the difficulty just raised for competitive equality of opportunity assumes that its requirements *are* being universally adhered to (so far as this is possible), the problems of compensatory justice are beside the point.

Competitive Equality of Opportunity and Liberty

The idea of state nurseries and so on would generally be regarded as taking the demands of equality of opportunity too far, if only because of the gross infringement of the liberty of parents. (We will pass over the risk of no more Mozarts because not all can benefit equally from having Leopold for a father.) Those who are just concerned about liberty may come to see the situation like this. The pursuit of equality of opportunity is justified within the bounds of certain constraints; constraints imposed not only by the limits of our manipulative capacities, but also by respect for certain liberties, which will not always allow us to manipulate everything that we could. But it might be thought that if this much is granted, pursuit of equality of opportunity can be conducted within a framework of respect for liberty. I will conclude

1 'The Idea of Equality as a Substantive Principle in Society', in A. de Crespigny and A. Wertheimer (eds.), *Contemporary Political Theory*, Nelson's University Paperbacks, London, 1971, pp. 157–158.

by trying to show that the conflict between equality of opportunity and liberty is more serious than this. It is not that the pursuit of such equality only sometimes may impinge on liberty: its pursuit will inevitably conflict with liberty.

It is not possible to lessen any inequality of opportunity without to some degree compromising autonomous choice. The opportunities that any particular person enjoys do not grow on trees: they depend upon the choices made by others. (It is not claimed, of course, that this is true of all opportunities. Some are provided by nature, for example, the opportunity to sow early, but it is true of the opportunities relevant to competitive equality of opportunity.) Consider the case of employment opportunities. One person's opportunity to do a certain job is a matter of some other person or persons choosing to offer him that opportunity. Now given the differences between persons so far as desirable characteristics for doing a certain job are concerned, if the choice of those to whom jobs are to be offered is made autonomously, some will have many opportunities and others few. Thus if one were to demand that people should have nearer to equal opportunities, the only way in which this could be done would be to place restrictions on the autonomy of choice of those who have opportunities at their disposal. Thus movement in the direction of equality of opportunity will inevitably be a constraint on liberty. To this a qualification may be allowed. If those who have opportunities at their disposal *themselves* choose to distribute those opportunities more equally (rather than being constrained by some authority to do so), the autonomy of choice need not be compromised. But in general, greater equality of opportunity must bring into existence an elaborate, coercive administrative apparatus constantly attempting to re-impose equal opportunity on (what will appear to the bureaucrats as) the anarchy of individual choice, and the unequal opportunities that arise from it.

Now it may be that one is quite happy to see some liberty sacrificed for the sake of equality of opportunity in a field such as employment. But the same point, that the opportunities of some are the choices of others, applies (one might say, applies paradigmatically) in the case of personal relationships. The opportunities for friendship or for marriage of one person are the choices others make to be willing to be friends or to enter marriage. Here too the only way in which greater equality of opportunity could be brought about would be by restricting the autonomy of individual choices. Now it may be that one is not so happy about accepting the constraints necessary for greater equality of opportunity in this area. If one is not so happy about it, what is the difference between it and the economic cases?

But the argument just given is open to objection.

> 'Often the person entitled to transfer a holding has no special desire to transfer it to a particular person; this contrasts with a bequest to a child or a gift to a particular person. He chooses to transfer to someone who satisfies a certain condition (for example, who can provide him with a certain good or service in exchange, who can do a certain job, who can pay a certain salary), and he would be equally willing to transfer to anyone else who satisfied that condition. Isn't it unfair for one party to receive the transfer, rather than another who had less opportunity to satisfy the condition the transferrer used? Since the giver doesn't care to whom he transfers, provided the recipient satisfies a certain general condition, equality of opportunity to be a recipient in such circumstances would violate no entitlement of the giver.'[1]

So the persons awarding the prizes will not be concerned to bestow opportunities on particular persons, but on anyone who happens to satisfy best certain general characteristics deemed to be relevant in this competition. Indeed, if the persons awarding prizes do not do this, they will be violating the first principle for equality of opportunity. Then how could equality of opportunity amongst the competitors conflict with the autonomy of choice of those awarding the prizes? For example, it does not infringe upon the autonomy of choice of university selectors, who are looking for certain general characteristics, if those amongst whom they are choosing have enjoyed equality of opportunity.

To consider this objection, let us think in terms of a sequence of competitions, those who do best at C1 go on to C2, and so on. Then the autonomy of the choice made between the contestants in the last competition of the series is not compromised by the contestants in that competition having enjoyed equality of opportunity. But in order for them to have equality of opportunity at *that* point, it is required that the *earlier* relevant opportunities of the contestants should have been equalized (in virtue of the

1 Robert Nozick, op. cit. p. 236. The argument in this section is indebted to ideas developed by Professor Nozick in chapters 7 and 8 of that work.

second principle). Therefore the choices of those who made the earlier decisions must have been constrained if equality of opportunity is to be maintained. For example, those selecting pupils for good schools could not be allowed to make their choices without constraint, for if they did, then by the time these pupils got to the university selection stage, those who had been selected would have enjoyed better opportunities than those who were rejected. Therefore there would not be equality of opportunity in the subsequent competitions, such as university selection. It may be replied that this still allows for autonomous choice in the last competition of any series. But this will not allow so much autonomy of choice as might at first appear. If we are only concerned with equality of opportunity *up until* the choice is made, then the choice may be made autonomously. But if we are also concerned with equality of opportunity *after* the choice has been made, then the choice will have to be constrained so as to ensure that.

As an example of this last point, consider a company official trying to choose the best person for an executive training scheme. Having a university education makes an applicant more desirable for the job. Therefore those who have not had a university education do not enjoy equal opportunity with those who have. If they were to enjoy equal opportunity, they, too, ought to have had a university education. Therefore, those who choose who is to receive a university education cannot be allowed an autonomous choice, for if they are, equality of opportunity at a later stage may be affected. But similarly, the company official's choice may be regarded as affecting equality of opportunity at a later stage. For if those who have been selected for the executive training scheme have a much better opportunity of reaching top positions than those who came into the company in some other way, then the company official's decision will affect equality of opportunity at that later stage. Thus equality of opportunity is compatible with making autonomous choices only if either (a) there is to be no further concern about equality of opportunity beyond the point at which the choice is made, or (b), those making the choice themselves decide to allocate opportunities with an eye to ensuring equality of opportunity in later competitions.

The argument has attempted to show that there is in one respect an inevitable conflict between the pursuit of competitive equality of opportunity and the exercise of autonomous choice.

Of course it can also be claimed that the existence of equality of opportunity in any sphere increases the options actually available to many, and hence that it furthers the ideal of liberty (or, if not of liberty, then, at any rate, of *something* valuable). The problems of evaluating the resultant of these conflicting tendencies will not be entered into here.[1]

BEDFORD COLLEGE, UNIVERSITY OF LONDON

1 My thanks are due to Wes Cooper, Trudy Govier and Anne Lloyd Thomas, who commented in detail on an earlier draft of this paper. They are also due to members of the audiences when this paper was read in 1975 at the University of Calgary and at the University of Alberta, for their illuminating remarks in discussion.

[11]

The Concept of Equal Opportunity*

Peter Westen

The concept of equal opportunity represents something of a paradox for Americans. We profess to believe in equal opportunity, yet we allow unequal opportunity to abound. Some observers may conclude, with R. H. Tawney, that we are simply hypocritical—that, while we pay "homage" to equal opportunity, we also "resist most strenuously attempts to apply it."[1] I believe that the paradox of equal opportunity is more complex and the solution more interesting. I believe, not that we say one thing and hypocritically do another, but that the rhetoric of "equality" and "opportunity" confounds what we really mean by equal opportunity. The rhetoric suggests that equal opportunity is a single and ideal state of affairs—difficult to attain, perhaps, but definitely to be desired.[2] The truth is quite the opposite. Equal opportunity is neither a single state of affairs nor ideal—neither difficult to attain nor inherently desirable.

To see why this is so, we must pierce the rhetoric of "equal opportunity" and examine its constituent elements. Equal opportunity is a "verbal formula"[3] consisting of four simple and recurring elements. The formula is treacherous[4] because three of the elements are covert and the fourth term is derivative. Once the four elements are identified, equal opportunity loses much of its mystery and most of its rhetorical appeal.

THE MEANING OF OPPORTUNITY: THREE COVERT ELEMENTS

The difficulties of equal opportunity do not begin with the word "equal." "Equal" presents challenges, as we shall see; but the puzzle of equal

*I am indebted to my colleagues, Richard Brandt, Allan Gibbard, and Nicholas White, for their thoughtful comments on an earlier draft of this essay.

1. R. H. Tawney, *Equality,* 4th ed. (London: G. Allen & Unwin, 1964), p. 103.

2. See, e.g., William Galston, *Justice and the Human Good* (Chicago: University of Chicago Press, 1980), p. 17, referring to equality of opportunity as "one . . . principl[e]"; Michael Levin, "Equality of Opportunity," *Philosophical Quarterly* 31 (1983): 110–25, p. 110: "Everyone agrees that opportunities should be equal"; John Schaar, "Equality of Opportunity," in *Nomos IX: Equality,* ed. James Pennock and John Chapman (New York: Atherton Press, 1967): 228–49, p. 228: "The one [conception of equality] that today enjoys the most popularity is equality of opportunity. The formula has few enemies—politicians, businessmen, social theorists, and freedom marchers all approve of it."

3. Charles Frankel, "Equality of Opportunity," *Ethics* 81 (1971): 191–211, p. 192.

4. John Lucas, *The Principles of Politics* (Oxford: Clarendon Press, 1966), p. 249.

Ethics 95 (July 1985): 837–850

opportunity inheres less in the meaning of "equal" than in the meaning of "opportunity."[5]

Every statement of opportunity consists of three covert elements. The first covert element is the agent, or class of agents, to whom the opportunities belong. Opportunities do not float freely about, unattached to persons. Opportunities, by definition, are of people—whether the people be black people alone, or blacks and whites together, or women, or rich people, or poor people, or rich and poor together, or people from one region alone, or older people, or children, or whatever.[6] The particular agent or class of agents will differ from one opportunity to another, but every opportunity entails an agent or a class of agents. As T. D. Campbell has said, "We may therefore always intelligibly ask about an opportunity—as we may always ask about any liberty or freedom—to whom it belongs."[7]

The second covert element in all statements of opportunity is the goals or set of goals toward which the opportunities are directed. An opportunity is a relationship of some agent to some desired thing. "All opportunities are opportunities to do or enjoy some benefit or activity."[8] The goal of an opportunity may be a job, or an education, or medical care, or a political office, or land to settle, or housing, or a financial investment, or a military promotion, or a life of "culture," or the development of one's natural abilities, or whatever.[9] The particular goal or

5. See Levin: "Before attempting to say what equality of opportunity, or opportunity rights, are, one must say something about what an opportunity is" (p. 110).

6. See, e.g., Weber v. United States Steel Workers, 443 U.S. 193 (1979), an affirmative action program for blacks alone; Women's Educational Equity Act of 1978, 20 U.S.C. Sec. 3341 (b) (1) (1983 Supp.), providing opportunities for women; Naval Petroleum Reserves Production Act of 1976, 10 U.S.C. Sec. 77430 (d) (1980 Supp.), providing an opportunity to major and independent oil producers and refiners to acquire petroleum from naval petroleum reserves; 42 U.S.C. Sec. 5318 (b) (1) (1977 Supp.), providing housing and employment opportunities for low and moderate income persons; 43 U.S.C. Sec. 151 (1976), providing homestead opportunity to persons to settle federal lands; Kentucky Equal Opportunities Act, Ky. Rev. Stat. Sec. 207.130 (1971), providing opportunities to persons "within the state"; 42 U.S.C. Sec. 30001 (1976), providing opportunities to "older people."

7. T. D. Campbell, "Equality of Opportunity," *Proceedings of the Aristotelian Society* 75 (1975): 51–68, pp. 51–52.

8. Onora O'Neill, "How Do We Know When Opportunities Are Equal?" in *Feminism and Philosophy*, ed. Mary Vetterling-Braggin, Frederick Elliston, and Jane English (Totowa, N.J.: Rowman & Littlefield, 1977), pp. 177–89, p. 178. See also Alistair Macleod, "Equality of Opportunity: Some Ambiguities in the Ideal," in *Equality and Freedom*, ed. Gray Dorsey (New York: Oceana Publications, 1975), vol. 3, pp. 1077–84, p. 1077: "Opportunity is always opportunity *to x*—that is, to do, be, become, or receive something or other. Consequently, statements about the existence or nonexistence of equality of opportunity are fully intelligible only if an answer can be supplied to the question, 'opportunity to *what?*'" Compare Frankel, p. 207; John Stanley, "Equality of Opportunity as Philosophy and Ideology," *Political Theory* 5 (1977): 61–74, p. 63.

9. See, e.g., 42 U.S.C. Sec. 2003 (1976), prescribing opportunities for employment; 20 U.S.C. Sec. 1221e (a) (1) (1983 Supp.), prescribing opportunities for education; 10 U.S.C. Sec. 1076 (1976), prescribing opportunities for medical care; 47 U.S.C. Sec. 315

set of goals will differ from one opportunity to another, but every opportunity is a relationship of a specific agent or class of agents (whether explicit or implicit) to a specific goal or set of goals (whether explicit or implicit).

The last and most elusive of the covert elements is the relationship that connects the agent of an opportunity, say, *X*, to the goal of the opportunity, say *Y*. An opportunity of *X* to attain *Y* is not a guarantee that *X* will succeed in attaining *Y* if he so chooses. An opportunity is not a guarantee because agents are rarely (if ever) guaranteed that they will attain their desired goals. Every child born in America has an "opportunity" to become president of the United States, but he has no guarantee of becoming president because he has no assurance that he will overcome the many possible obstacles that stand in the way. Every president-elect in America has an even better opportunity to become president, but he has no guarantee either because he has no assurance that he will overcome the obstacles—illness, natural catastrophe, the discovery of vote fraud, constitutional crisis, death, and so forth—that may stand in the way of his taking the oath of office on January 20. It seems, therefore, that, when we talk about opportunities, we are not necessarily talking about the absence of all possible obstacles between a given agent and a given goal.[10]

Conversely, while an opportunity is something less than a guarantee, it is something more than a mere possibility—more, that is, than merely the possible absence of all obstacles between a given agent and a given goal. An opportunity is more than mere possibility because nearly everything is possible. We might say every foreign-born citizen of the United States has a "possibility" of becoming president because it is possible to remove the constitutional requirement that candidates for president be native-born. Yet we would not say foreign-born citizens have an opportunity to become president because an opportunity requires something more than the possible absence of all obstacles between a given agent and a given goal.

It thus appears that an opportunity falls somewhere between a guarantee and a mere possibility—that is, somewhere between the absence of all possible obstacles between *X* and *Y*, on the one hand, and the

(a) (1976), prescribing opportunities for candidates for public office; 43 U.S.C. Sec. 151 (1976), prescribing opportunities for homesteading public lands; 42 U.S.C. Sec. 5318 (1977 Supp.), prescribing opportunities for housing; 31 U.S.C. Sec. 744 (1983), prescribing opportunities for the purchase of public bonds; 14 U.S.C. Sec. 276 (1976), prescribing opportunities for military promotion; John Rawls, *A Theory of Justice* (Cambridge, Mass.: Harvard University Press, 1971), p. 73, advocating opportunities for persons for lives of "culture"; Tawney, pp. 103–4, advocating opportunities for members of the community to use to the full their "natural endowments of physique, of character, and of intelligence."

10. But see D. A. Lloyd Thomas, "Competitive Equality of Opportunity," *Mind* 86 (1976): 388–404, p. 388: "One has an opportunity to do something or have something provided one can do it or have it if one chooses."

possible absence of all obstacles between *X* and *Y*, on the other hand. But that is not so because one can remove a specified obstacle in the way of *X*'s attaining *Y* without also granting *X* an opportunity to do *Y*. Assume, for example, that having formerly required candidates for governor to be native-born males, a state removes the sex requirement, thus leaving the governorship open to all native-born men and women. We might then say that, by removing one insurmountable obstacle in the way of foreign-born women, the state has given all foreign-born women an increased possibility of becoming governor. But we would not say that the state has also given all foreign-born women an opportunity of becoming governor because, by retaining the requirement that candidates be native-born, the state has explicitly left an insurmountable obstacle directly in their way—an obstacle which, unless removed, permanently precludes foreign-born women from attaining their goal. It thus appears that an opportunity requires at a minimum that an agent have a chance to attain his goal, that is, that no insurmountable obstacles explicitly stand in the way of his attaining his goal.[11]

This may suggest that an opportunity, being no less than a chance, is also no more than a chance, that is, no more than the absence of insurmountable obtacles in the way of *X*'s attaining *Y*. But that is not so either, for opportunities are not confined to the absence of insurmountable obstacles. Suppose, for example, that, having formerly required airline stewardesses to be unmarried women, an airline company removes the marital obstacle, thus opening stewardess jobs to all women. The marital obstacle differs from insurmountable obstacles like race, color, and sex because marriage in America is a legal status that a person himself may change. Yet we would surely use the language of opportunity to describe the airline company's action. We would surely say that, by removing the marital obstacle, the airline company had given married women an opportunity to become airline stewardesses.

What, then, is the relationship between the agent of an opportunity and the goal of an opportunity? The answer should now be clear. An opportunity is not solely the absence of a specified obstacle or solely the absence of insurmountable obstacles. It is a combination of both. "Op-

11. See Levin, p. 111: "An opportunity for a job is a *chance* at it"; Onora O'Neill, "Opportunities, Equalities, and Education," *Theory and Decision* 7 (1976): 275–95, p. 276; "If A has no chance of doing X whenever he chooses, then he has no opportunity to do X." By "insurmountable" obstacles, I mean obstacles that, unless removed, will inevitably preclude an agent from attaining his specified goal. By "explicit" obstacles, I mean specifically enumerated conditions on an agent's attaining a goal. Thus, when we say, Any child born in the United States has an opportunity to become president, we do not mean that there are no insurmountable obstacles of any kind in the way of any child's becoming president, for there are inevitably children who are so ill that they will never overcome the obstacle of poor health that stands in the way of their becoming president. We mean, rather, that there are no insurmountable obstacles of the kind we mean implicitly to be enumerating when we say, Every child born in America has a chance to become president, e.g., obstacles of race, religion, and sex.

portunity" is the word we use to refer to the absence of a specified obstacle or set of obstacles, the absence of which leaves no insurmountable obstacles explicitly in the way of *X*'s attaining *Y*. We say someone has an opportunity when we have in mind a particular obstacle or set of obstacles that is not there, an obstacle the absence of which gives *X* a chance he did not previously possess to attain *Y* if he so chooses. The particular obstacle or set of obstacles the opportunity removes may be insurmountable (e.g., race, color, sex, ancestry, or place of birth), or surmountable (e.g., religious belief, wealth, social class, marital status, minimum age, high school diploma, or residency), or a combination of surmountable and insurmountable obstacles; it may be the obstacle of being excluded from a competitive race altogether, or the obstacle of being admitted to the competition but having to run further than other runners, or the obstacle of running the same distance but having to overcome the "social" disadvantage of poorer training and motivation than other runners,[12] or the obstacle of having the same social advantages but having to overcome the "natural" disadvantage of lesser natural ability than other runners;[13] or the obstacle may be the totality of all the features that distinguish one person from another,[14] or whatever. The particular obstacle will differ from one opportunity to another, but every opportunity is a chance of a specified agent or class of agents, *X*, to choose to attain a specified goal or set of goals, *Y*, without the hindrance of a specified obstacle or set of obstacles, *Z*.

As an illustration, consider the Illinois Human Rights Act of 1979.[15] The act provides statutory opportunities for people in Illinois by making it unlawful for labor organizations to limit the employment of persons in Illinois by discriminating against them on the basis of "race, color, religion, sex, national origin, ancestry, age, marital status, [or] physical or mental handicap." Like all opportunities, the Illinois statutory opportunity consists of a specified agent, *X*, a specified goal, *Y*, and a specified obstacle, *Z*, the absence of which gives *X* a chance he did not previously possess to attain *Y* if he so chooses. The agents, *X*, are the class of persons in Illinois; the goal, *Y*, is employment; and the obstacle, *Z*, is discrimination by labor organizations on the basis of race, color, religion, sex, national origin, ancestry, marital status, or physical or mental handicap. Moreover,

12. See, e.g., Alan Goldman, "The Principle of Equal Opportunity," *Southern Journal of Philosophy* 15 (1977): 473–85, p. 475, advocating a standard of opportunity by which people are given handicaps to correct for "socially relative initial disadvantages."

13. Compare Frankel (p. 204), advocating a standard of opportunity in which people are judged on the basis of their "abilities," with Goldman (p. 474), discussing a "sense" of opportunity by which persons are allowed to compete without the obstacles of their "natural" disadvantages.

14. See Edwina Dorn, *Rules and Racial Equality* (New Haven, Conn.: Yale University Press, 1979), p. 112, advocating a standard of opportunity by which people have a chance to attain their goals without any hindrance other than a pure lottery.

15. Illinois Human Rights Act, Ill. Ann. Stat., ch. 68, Sec. 1–101 (1979) (Smith-Hurd, 1982 Supp.).

like all opportunities, the statutory opportunity consists of a certain relationship between agents, goals, and obstacles. The relationship is less than a guarantee and more than a possibility: less than a guarantee because the act provides for less than the removal of all possible obstacles to employment; more than a possibility because the act provides for more than the possible removal of all obstacles. The relationship does more than simply remove an obstacle because it does so in such a way as to leave no insurmountable obstacles explicitly in the way of *X*'s attaining *Y*; yet the relationship also does more than simply remove insurmountable obstacles because it removes some obstacles (e.g., marital status and religion) that are not insurmountable. In a word (and, indeed, in the words of the act), the relationship constitutes an "opportunit[y]." It gives people in Illinois a chance they did not previously possess to attain employment by removing specified obstacles that would otherwise stand in their way.

To be sure, legislatures do not always specify as explicitly as Illinois did the precise terms of the opportunities they mean to prescribe. Indeed, the obfuscations of opportunity result from expressing opportunities without specifying their three constituent terms. Consider, for example, the statement, Every child in America should have an opportunity to graduate from high school. Like all statements of opportunity, the latter states a relationship between agent *X*, a goal, *Y*, and an obstacle, *Z*. *X* and *Y*, being both explicit, are obvious. *X* is the class of all children in America, and *Y* is the goal of graduating from high school. *Z*, being implicit, is ambiguous. *Z* may be something educationally uncontroversial, like the obstacles of indigency or race, or something educationally controversial, like the obstacle of passing a competency examination. By leaving the content of *Z* unspecified, the statement masks a wide range of possible prescriptions—from the most acceptable prescription of opportunity to the most controversial.

The foregoing analysis supports several conclusions. First, an opportunity is not a particular state of affairs. It is a formal relationship—a relationship of an agent, *X*, to a goal, *Y*, with respect to an obstacle, *Z*. It is a chance on the part of *X*, if he so chooses, to attain *Y* without the hindrance of *Z*. The concept of opportunity can be particularized into specific conceptions of opportunity by replacing the variables *X*, *Y*, and *Z* with specific agents, specific goals, and specific obstacles. One conception of opportunity will differ from another. Some conceptions of opportunity may be more just, or unjust, than others. But every conception of opportunity, qua opportunity, is as much an opportunity as every other.

Second, while opportunities may be divided into subclasses, the subclasses themselves remain relationships of agents, goals, and obstacles. Thus, "descriptive" opportunities are relationships that actually obtain, that is, agents who actually possess chances to attain specified goals without the hindrance of specified obstacles if they so choose. (Descriptively, Americans have an opportunity to engage in national political debate without the hindrance of widespread illiteracy or polyglot because, descriptively, Americans are highly literate in a single common language.) "Prescriptive" opportunities are relationships that ought to obtain, that is, agents who ought to have a chance to attain specified goals without the hindrance of specified obstacles if they so choose. The two subclasses are conceptually distinct. One cannot infer that given agents ought to have an opportunity from the description that they actually have an opportunity or that given agents actually have an opportunity from the prescription that they ought to have an opportunity. Yet as subclasses of opportunity, descriptive and prescriptive opportunities are both chances of agents, *X*, to attain goals, *Y*, without the hindrance of obstacles, *Z*.

Third, one cannot grant prescriptive opportunities to some people without denying prescriptive opportunities to other people. In stating that *X* ought to have a chance to attain *Y* without the obstacle of *Z*, one necessarily states that *W* ought not to have a chance to prevent *X* from attaining *Y* by means of *Z*. This does not mean that one should refrain from creating prescriptive opportunities. It means, rather, that the significant question is, not whether opportunities should be prescribed, but which opportunities should be prescribed and which opportunities proscribed.

THE MEANING OF EQUALITY: A DERIVATIVE ELEMENT

The rhetoric of opportunity accounts for much of the confusion surrounding equal opportunity, but not all. Some confusion derives from the other half of the phrase—from the rhetoric of equality. The confusions of equality, however, are the converse of those of opportunity. Opportunity tends to confuse because it tends to leave essential elements unsaid. Equality confuses because it repeats what opportunity already fully says. The rhetoric of opportunity suffers from being incomplete. The rhetoric of equality suffers from being derivative.

The derivative nature of the word "equal" in "equal opportunity" inheres in the dictionary meaning of "equal." "Equal" means the same thing in the phrase "equal opportunity" as it does everywhere else. To say two persons or things are equal does not mean that they are identical by every possible descriptive or prescriptive measure because no two persons or things can be identical by every possible measure. Nor does it mean that they are identical by any one possible descriptive or prescriptive measure because all persons and things are identical by some measures. It means, rather, that they are identical by the relevant descriptive or prescriptive measure—the relevant measure being the particular measure one stipulates as applicable.[16] To say two persons are equal means that (1) they have each been measured by a stipulated standard of measure, (2) their respective measures have been compared with one another, and (3) the comparison shows their measures to be identical to one another.

16. See Lucas, pp. 244–45; Peter Westen, "The Meaning of Equality in Law, Math, Morals and Science: A Reply," *Michigan Law Review* 81 (1983): 604–63, pp. 607–12.

The same applies to "equal" in "equal opportunity." To say two persons possess (or ought to possess) opportunities means that they possess (or ought to possess) chances to attain given goals without the hindrance of given obstacles. To say further that their opportunities are equal means that (1) their respective opportunities have each been measured by a stipulated standard for measuring opportunities, (2) the opportunities so measured have been compared with one another, and (3) the comparison shows the opportunities to be identical by that standard of measure. Now, what is the standard of "measur[e],"[17] or "predicate,"[18] by which equality of opportunity is ascertained? The standard can be any stipulated opportunity—any "specification"[19] of the three variable terms of which opportunities consist. The specification may be descriptive, that is, it may identify agents who actually possess a chance to attain a specified goal without the hindrance of a specified obstacle. Or the specification may be prescriptive, that is, it may identify the class of agents who ought to possess a chance to attain a specified goal without the hindrance of a specified obstacle. In either event, equality of opportunity is simply the identity that obtains among two or more persons by virtue of their all falling within a class of agents who all possess (or ought to possess) a chance to attain a specified goal or goals without the hindrance of a specified obstacle or obstacles.

It follows, therefore, that two persons can have an equal opportunity to attain a specified goal even though each faces different obstacles of his own, provided that they are both free from the same specified obstacles. Assume, for example, that two runners with different training and talent are both given a chance to win a race by both being allowed to start at the same time and place and run the same distance; assume, too, that the measure of opportunity is the chance to win the race without the obstacles of starting at different times and running different distances; the two runners in that event can truly be said to have an equal opportunity to win the race even though they face different obstacles regarding talent and the training that make it unlikely that both will actually win because they are identical in both being free from the same specified obstacles to attain the same specified goal. Similarly, two persons can lack an equal opportunity to attain a specified goal even though they are identical in both being free from the same obstacles, provided that the obstacles from which they are free are not specified as relevant obstacles. Thus, if the measure of opportunity for the two runners is the chance to win the race without the obstacles of lack of training, the two runners do not

17. Levin, p. 110.

18. O'Neill, "How Do We Know When Opportunities Are Equal?" pp. 177–78.

19. Macleod, p. 1083: one cannot talk about "equality of opportunity" without making a "careful specification of the type of opportunity to be equalized." For the use of an alternative measure of opportunity to determine equality of opportunity, see Frankel, p. 200, discussing the possibility of measuring opportunity by the mathematical probability that agents X and X' will attain their respective goals, Z and Z'.

have an equal opportunity to win the race even though they both start from the same time and place because they are not identical in both being free from the specified obstacles regarding training. People who have equal opportunity by one measure of opportunity will have unequal opportunities by other measures. No two people can have an equal opportunity to attain a specified goal by every measure of opportunity unless they are both guaranteed the result of attaining the goal if they so wish.

It also follows from this that the word "equal" in "equal opportunity" occurs derivatively. It tells us nothing we do not already know about who possesses (or ought to possess) a chance to attain what goal without the hindrance of which obstacle. It refers us, derivatively, to the consequences of measuring agents by stipulated descriptive or prescriptive standards of opportunity. In the absence of stipulated standards of opportunity, one has no way of identifying agents who are equal in respect of possessing such opportunities. Yet in the presence of stipulated standards of opportunity, one has no need to identify equality among agents because the standards themselves tell us everything we need to know: the standards themselves tell us who has (or ought to have) such opportunities and who lacks (or ought to lack) them. If two or more people fall within the class of agents who possess (or ought to possess) a chance to attain a relevant goal without the hindrance of a relevant obstacle, it follows that they are (or ought to be) equals in respect of possessing that opportunity. If two people do not fall within the class of agents who possess (or ought to possess) a chance to attain a relevant goal without the hindrance of a relevant obstacle, it follows that they are not (or ought not to be) equal in respect of possessing that opportunity. In either event, to say that two people possess (or do not possess) equal opportunities is simply a derivative way of saying that they both fall (or do not both fall) within the class of agents who possess specified opportunities in common.

The Public Telecommunications Act of 1978[20] nicely illustrates the derivative nature of "equal" in "equal opportunity." The operative provision of the act can be divided into two parts: part 1 prescribes "equal opportunity in employment" and part 2 prescribes that "no person" seeking employment with the Public Broadcasting Service (PBS) or National Public Radio (NPR) shall be subjected to "discrimination" on grounds of "race, color, religion, national origin, or sex." The prescription of equal opportunity in part 1 is meaningless without part 2 because without part 2 one has no prescriptive standard of opportunity for determining equality among agents—no stipulation of specified goals and specified obstacles by which one can identify the class of agents who are identical in respect of the opportunity they possess. Yet with part 2, one has no need for part 1 because part 2 contains everything one needs to know. Part 2 specifies the three essential terms of the opportunity by which equality of opportunity

20. Public Telecommunications Act, 47 U.S.C. Sec. 398 (1983 Supp.) (1978).

obtains. Part 2 specifies that a specified class of agents (i.e., "all persons" desiring employment) shall have a chance to attain a specified goal (i.e., "employment" at PBS or NPR) without being hindered by a specified obstacle (i.e., discrimination "on grounds of race, color, religion, national origin, or sex"). Given the prescription in part 2, it follows that all members of the agent class are equal in respect of the opportunity the act proscribes. (It also follows that, as far as the Public Telecommunications Act is concerned, they are not prescriptively equal in respect of opportunities the act does not proscribe, for example, the opportunity to work for PBS without discrimination on grounds of sexual preference.) The equality of opportunity that thus obtains among members of the agent class does not add anything to the opportunity that exists without it. The act would continue to mean everything it now means if "equal" were simply deleted from the text and the statute stated without it: "No person shall be subjected to discrimination in employment by [PBS or NPR] on grounds of race, color, religion, national origin, or sex."

To be sure, like statements of opportunity, statutes that prescribe equal opportunity sometimes omit one or more of the essential terms of which the constitutent opportunity consists. Some equal opportunity statutes fail to specify essential terms altogether, while other statutes use "equal opportunity" as a surrogate for implied terms. Yet even where "equal opportunity" is used as a surrogate for implied terms, "equal" still occurs derivatively because it still refers, derivatively, to the identity that obtains among agents by virtue of their falling within a common class of agents for whom the statute implicitly prescribes a particular opportunity.

Now it might be asked, What difference does it make that "equal" works derivatively? Why does it matter whether one specifies the three essential terms of an opportunity directly or uses "equal" to imply them derivatively? It makes a difference because statements of equal opportunity are less perspicuous—and, hence, more "ambiguous"[21]—than are direct specifications of the opportunities for which they stand.

The ambiguities of equal opportunity take two forms: (1) cases in which "equal opportunity" obscures the precise obstacles the prescribed opportunity eliminates and (2) cases in which "equal opportunity" also obscures the precise class of agents who possess the prescribed opportunity. The first confusion occurs with statutes of the form, *A* shall have an equal opportunity with *B* to attain goal *Y*. Such statutes tend to be ambiguous because, while they specify the agents of the prescribed opportunity (i.e., the combined class of *A* and *B*) and the goal of the opportunity (i.e., *Y*), they do not specify the precise obstacle the prescribed opportunity removes. By using "equal opportunity" as a surrogate for the unstated obstacle, they imply that the obstacle consists of one of three things—(1) the obstacle of being classified as *A* (as opposed to *B*), (2) the obstacle

21. Macleod. See also Peter Westen, "To Lure the Tarantula from Its Hole," *Columbia Law Review* (1983), pp. 1186–1208, p. 1204.

of being classified as *B* (as opposed to *A*), or (3) the obstacle of being classified either as *A* (as opposed to *B*) or as *B* (as opposed to *A*). Yet by not specifying which of the three obstacles they have in mind, the statutes obscure their own content.

To illlustrate, assume a statute takes the following form:

> American women shall have an equal opportunity with American men to serve as astronauts in the space-shuttle program.

The statute specifies two terms of the prescribed opportunity but uses "equal opportunity" as an ambiguous surrogate for the third term. It specifies the agents who possess the prescribed opportunity and the goal of the prescribed opportunity; and it uses "equal opportunity" in conjunction with "men" and "women" to imply that the obstacle being removed consists of sex discrimination of one form or another. Unfortunately, because there are at least three different kinds of sex discrimination, there are also at least three kinds of prescribed opportunity:

> O_1: American men and women shall have whatever chances are otherwise prescribed for men to serve as astronauts in the space shuttle program without the hindrance of any special preference for men.
>
> O_2: American men and women shall have whatever chances are otherwise prescribed for women to serve as astronauts in the space shuttle program without the hindrance of any special preference for women.
>
> O_3: American men and women shall have a chance to serve as astronauts in the space shuttle program without the hindrance of any special preference for persons of a particular sex.

The foregoing prescriptions of opportunity each differ significantly from one another. O_1 protects women, but not men, from discrimination; O_2 protects men, but not women, from discrimination; O_3 protects men and women both from discrimination. Yet each prescription creates equal opportunity between men and women because each of them defines men and women as members of an agent class possessing a common opportunity. Each prescription squares with what the statute says. To know which prescription squares with what the statute means, one would have to pierce the "vague"[22] language of equal opportunity for the unspecified prescription that underlies it.

The second major confusion occurs with statutes of the form, *A* and *B* shall have an equal opportunity to attain *Y*. Such statutes tend to be ambiguous because they fail to specify either the agents who possess the prescribed opportunity or the obstacle the opportunity removes. The statutes mask three distinct prescriptive standards of opportunity by which *A* and *B* may be rendered equal: (1) *A* and *B* may be equal to one another

22. Lucas, p. 247.

in their chances to attain *Y* without being disfavored vis-à-vis one another; (2) *A* and *B* may be equal to an unnamed but implied third-party agent, *C*, in their chances to attain *Y* without being disfavored vis-à-vis *C*; or (3) *A* and *B* may be equal to *C* in their chances to attain *Y* without being disfavored vis-à-vis one another. Instead of specifying the particular prescriptive opportunity *A* and *B* share in common, the statutes obscure the pertinent opportunity by referring instead to the equality that obtains between *A* and *B* by virtue of their possessing the unspecified opportunity in common.

The Export Expansion Finance Act of 1971 is a good example.[23] The Act requires the Export-Import Bank (Eximbank) to accord "equal opportunity to . . . independent export firms [and] small commercial banks in the formulation and implementation of its programs." Like all prescriptive statements of equal opportunity, the act necessarily presupposes a prescriptive standard of opportunity by which the equality obtains. Unfortunately, instead of specifying the three essential terms of the prescribed opportunity, the act specifies only one of them, leaving two of them to be inferred. The act specifies the goal of the prescribed opportunity (i.e., the formulation and implementation of Eximbank programs): and it uses "equal opportunity" in conjunction with "independent export firms" and "small commercial banks" in such a way as to raise implications about both the agents who possess the prescribed opportunity and the obstacles the prescribed opportunity removes; but its implications are sufficiently ambiguous to encompass at least three distinct prescriptions of opportunity:

O_1: Independent export agents and small commercial banks shall have a chance, if they so choose, to participate in the formulation and implementation of Eximbank programs without being disfavored vis-à-vis one another.

O_2: Independent export agents, small commercial banks, and large commercial banks shall have a chance, if they so choose, to participate in the formulation and implementation of Eximbank programs on whatever terms are otherwise open to large commercial banks without being disfavored vis-à-vis large commercial banks.

O_3: Independent export agents, small commercial banks, and large commercial banks shall have a chance, if they so choose, to participate in the formulation and implementation of Eximbank policy programs without being disfavored vis-à-vis one another.

Each of the three prescribed opportunities creates a class of agents who are identical—and, hence, equal—in their chances to attain the specified goals. Each of the prescribed opportunities includes independent export agents and small commercial banks within its respective class of agents.

23. Export Expansion Finance Act, 12 U.S.C. Sec. 635 (b) (1) (B) (1983 Supp.) (1971).

Each of the three prescriptions thus accords equal opportunity to independent export agents and small commercial banks. Yet the three prescriptions differ significantly. Legislative history in fact suggests that the act was enacted to codify O_2 rather than O_1 or O_3.[24] Without independent information of that sort, however, one has no way of knowing from the language of equal opportunity which of the three opportunities the act actually prescribes.

CONCLUSION

Popular wisdom on equal opportunity turns it upside down. People commonly think of equal opportunity as a desideratum—a single state of affairs that is highly desirable in theory, though perhaps unattainable in practice. The truth is rather the contrary. Equal opportunity is not a single state of affairs, not unattainable, and not necessarily desirable.

"'Equality of Opportunity' is no more the name of a single ideal than is 'Liberty.'"[25] Rather it is a way of talking about countless states of affairs—about the countless opportunities, descriptive and prescriptive, that groups of people everywhere possess in common. An opportunity is a chance of an agent, *X*, to choose to attain a goal, *Y*, without the hindrance of an obstacle, *Z*. Equal opportunity is the identity of opportunity that obtains among any two persons who fall within a common class of agents. One cannot conceive of particular opportunities without first substituting specified agents, specified goals, and specified obstacles in the place of the variable terms *X*, *Y*, and *Z*. Once one specifies the variables, one can talk about the resulting opportunity in one of two ways: one can talk about the opportunity directly, by speaking directly to the agents' chances to attain their specified goal without being hindered by the pertinent obstacle, or one can talk about the opportunity indirectly, by speaking of the identity—the equality—that obtains among its agents by virtue of their possesssing a common chance to attain the goal without being hindered by the pertinent obstacle. The reference to equality does not add to the content of the opportunity. It is simply a way of talking about the opportunity by reference to the identity that obtains among the agents who possess it in common.

It thus follows, then, that, although particular opportunities may indeed be unattainable,[26] equal opportunity itself is unavoidable. Equal opportunity exists everywhere two or more people have a chance to attain a specified goal without being hindered by a specified obstacle. Descriptively,

24. See H. Rpt. No. 92-303, accompanying H.R. 8181, 92d Congress, 1st session, in *1971 U.S. Congressional and Administrative News* (St. Paul, Minn.: West Publishing Co., 1972), vol. 2, pp. 1414, 1427; Hearings on H.R. 8181, House Subcommittee on International Trade, 92d Congress, 1st session, pp. 513, 514, 517, 522–23, 524.

25. Macleod, p. 1077.

26. See, e.g., James Fishkin, *Justice, Equal Opportunity, and the Family* (New Haven, Conn.: Yale University Press, 1983), pp. 51, 106–7, 132, 145, arguing that "family autonomy" is irreconcilable with a certain specification of equal opportunity.

equal opportunity exists wherever two or more people fall within a class of agents who are all free from the same obstacle to attain the same goal. Prescriptively, equal opportunity exists wherever two or more people fall within a class of agents who we believe ought to be all free from the same obstacle to attain the same goal. Equal opportunity exists prescriptively wherever we wish to prescribe it.

It also follows from the meanings of "opportunity" and "equal" that equal opportunity itself (in contrast to particular equal opportunities) is neither desirable nor undesirable. Equality of opportunity is the identity of opportunity that obtains between two persons by virtue of their both being free from a specified obstacle to attain a specified goal. The desirability of their equality of opportunity thus depends entirely on the desirability of their both being free from the pertinent obstacle to pursue the pertinent goal. Just as opportunities can be good or bad, equal opportunities can be good or bad. The equal descriptive opportunity Americans possess to commit homicide without the hindrance of a national handgun shortage seems to many people to be bad; the equal descriptive opportunity they possess to live a long life without the hindrance of smallpox seems to be good. The equal prescriptive opportunity American property holders possessed under *Dred Scott* to take chattel property into the free territories today seems bad; the equal prescriptive opportunity they still possess to receive just compensation for takings of their property today seems good. The equalities and inequalities that obtain among people under given descriptions and prescriptions of opportunity are as good (and as bad) as the descriptions and prescriptions from which they derive.

The American paradox of equal opportunity—namely, that we profess to believe in equal opportunity and, yet, allow unequal opportunity to prevail in many spheres of life—rests on false premises. We do not really believe in equal opportunity as such. We believe in particular equal opportunities, just as we believe in particular unequal opportunities. We believe in prescribing particular opportunities, and, hence, we believe in the respective equalities and inequalities that obtain among those who do and do not possess such opportunities in common. We do not contradict our professed values by prescribing unequal opportunity. We vindicate them. Logically, we cannot prescribe equal opportunity for some persons without prescribing unequal opportunity for other persons.[27] Logically, we cannot prescribe equal opportunity in respect of some goals and obstacles without withholding equal opportunity in respect of other goals and obstacles. Ultimately, we prescribe equal opportunities for the same reasons we prescribe unequal opportunities—because equal and unequal opportunities obtain as a logical consequence of the opportunities we wish to prescribe.

27. See Felix Oppenheim, *Political Concepts: A Reconstruction* (Chicago: University of Chicago Press, 1981), p. 119; Peter Westen, "The Empty Idea of Equality," *Harvard Law Review* 95 (1982): 537–96, pp. 572–73, n. 124.

[12]

EQUAL OPPORTUNITY AND GENETIC INTERVENTION*

By Allen Buchanan

I. Introduction

What does the prospect of being able to alter a human being's "natural assets" by genetic engineering imply for our understanding of the requirements of justice, and of equal opportunity in particular? Although their proponents are reluctant to admit it, some of the most prominent contemporary theories of justice yield a quite radical conclusion: If safe and effective intervention in the genetic "natural lottery" becomes feasible, there will be at least a strong prima facie case for doing so in the name of equality of opportunity (or of some other egalitarian principle of justice, such as the principle that persons are entitled to equal concern and respect), if this is the most effective way to meet the demands of justice.

The most obvious instance of such a theory is that of John Rawls. His view is that equal opportunity requires that individuals be compensated for the fact that they have *lower life-prospects* as a result of the influence of not only social but also natural contingencies, because these factors are "arbitrary from a moral point of view."[1] Being born into a family that values education or born with a natural talent for developing complex cognitive skills is "morally arbitrary," Rawls thinks, because in both cases it is an *undeserved* natural asset or resource that gives rise to special advantages, to higher life-prospects. The fundamental idea here is that simply to allow such undeserved differences to lead to lower life-prospects for those who had bad luck in the social or natural lotteries is somehow unjust, because it is incompatible with a proper recognition of the basic moral equality of persons.

According to Rawls, the proper response is first to implement a principle requiring that persons of equal talent and motivation have equal prospects of attaining social offices and positions. But this, he thinks, is not sufficient. In addition, in order to compensate those who will still suffer disadvantages in the distribution of natural assets, social and economic inequalities must be so "arranged" as to maximize the life-prospects of the worst off. Since talent and even motivation are presumably in part genet-

* I am indebted to David Benatar and Dan Brock for their astute comments on an earlier draft of this essay.

[1] John Rawls, *A Theory of Justice* (Cambridge, MA: Harvard University Press, 1971), p. 72.

ically based (and hence result from "morally arbitrary" natural contingencies), merely ensuring that those with the same talent and motivation have the same prospects of attaining offices and positions will not be enough. What is required is that social primary goods be redistributed in such a way as to compensate for the unequal distribution of natural primary goods.[2]

[2] In this essay I will be concerned mainly with Rawls's views on equal opportunity and natural inequalities as expressed in *A Theory of Justice*. One reason for doing this is that my aim is to explore a type of position, namely resource egalitarianism, which, while held by others in addition to Rawls (such as G. A. Cohen and Ronald Dworkin), seems to rely upon Rawls's discussion of these matters in *A Theory of Justice*. Resource egalitarians tend to take as relatively uncontroversial the very Rawlsian views I intend to subject to critical examination in this essay. In particular, resource egalitarians in the so-called "Equality of What?" literature (with the possible exception of Amartya Sen) appear to take it as a given that justice requires some sort of compensation for those whose life-prospects are lower as a result of undeserved lesser shares of social or natural resources. (See, e.g., Ronald Dworkin, "What is Equality? Part 1: Equality of Welfare," *Philosophy and Public Affairs*, vol. 10, no. 3 [Summer 1981]; Ronald Dworkin, "What is Equality? Part 2: Equality of Resources," *Philosophy and Public Affairs*, vol. 10, no. 4 [Fall 1981]; John Roemer, "Equality of Talent," *Economics and Philosophy*, vol. 1, no. 2 [October 1985]; G. A. Cohen, "On the Currency of Egalitarian Justice," *Ethics*, vol. 99 [July 1989]; Norman Daniels, "Equality of What: Welfare, Resources, or Capabilities," *Philosophy and Phenomenological Research*, vol. L Supplement [Fall 1990]; and Richard Arneson, "A Defense of Equal Opportunity for Welfare," *Philosophical Studies*, vol. 62 [1991].) It is this claim about a fundamental connection between natural inequalities and the requirements of justice, what I have called the Resource Redress Principle, which seems to play a prominent role in at least some of Rawls's arguments for the Difference Principle and for the Principle of Fair Equality of Opportunity in *A Theory of Justice*.

In Rawls's later work *Political Liberalism* (New York: Columbia University Press, 1993), there may be less reliance on the Resource Redress Principle as a relatively free-standing principle that is to be supported by appeal to "moral intuitions" or "considered judgments." Instead, *Political Liberalism* can be interpreted as basing the Principle of Fair Equality of Opportunity and the Difference Principle (as well as Rawls's First Principle of Justice) ultimately upon the complex idea (or ideal) of what free and equal persons who regard themselves as such would agree to as fair terms of social cooperation. Although I will attempt no systematic critique of the later Rawlsian position here, I will remark that attempting to base everything ultimately on the idea or ideal in question seems to place Rawls on the horns of a dilemma. Either he must admit that the concept of what free and equal persons who regard themselves as such (and seek fair terms of social cooperation) would agree to is too vague to yield anything as determinate and controversial as his principles; or he must lean heavily on what I have elsewhere called his Traditionist mode of justification, attempting to ground his principles in the ideals of freedom and equality (and of fair cooperation) that he claims are distinctive of "our" political culture, that of modern (post 1776) political liberalism. The difficulty with the latter response is that it is dubious indeed to assume, as Rawls does, that there is a clearly dominant political culture in actual modern liberal-democratic societies that endorses his principles or in whose commitments his principles can be grounded. And, remarkably, Rawls does nothing to substantiate his claim that the materials for developing a consensus on his principles are actually present in the political consciousness or major documents and institutions of, say, American society.

If I understand it properly, the Traditionist mode of justification has a substantial empirical component insofar as it requires establishing that the principles of justice that are to be justified are based on ideals that are actually present in a given political culture or tradition. A political culture or tradition is embodied in the political consciousness of a group of people and in the major political institutions of that society, and is expressed in the major documents (such as the constitution) of the society. Yet any attempt to connect the actual features of American (or other modern liberal-democratic) political consciousness and of our society's major documents or institutions with Rawls's interpretation of the ideal of a free

Several other prominent theorists, who differ from Rawls and each other in a number of respects, share with him the assumption that justice requires distributing (and redistributing) social resources in order to compensate for inequalities in natural resources, that is, the genetic assets with which individuals are born.[3] These theorists see the chief task for achieving a just society to be that of devising a resource-allocation mechanism to achieve an overall equal distribution of resources by compensating for inequalities in the distribution of natural resources by special redistributions of social resources.[4]

There are a number of bold if not extravagant assumptions at work in this approach. First of all, to talk of compensating for a lack of natural resources by special allocations of social resources is to assume that natural and social resources can be compared and aggregated. This is not the assumption I wish to question, however. Instead, I want to focus on the fundamental moral assumption that gets this whole approach going in the first place, namely:

> Equality of opportunity (or, more simply, the fundamental equality of persons) requires that individuals be compensated for having lower life-prospects as a result of their (less fortunate and undeserved) natural *or* social endowments.

For reasons that will become apparent as we proceed, I shall call this assumption the "Resource Redress Principle."

Notice that the Resource Redress Principle appears to be a *second-best principle*. It seems to rest on a more basic principle to the effect that, at least from the standpoint of the most basic principles of justice, the ideal situation is one in which all resources, natural and social, are distributed equally, at least so far as they are undeserved. Let us call this the Resource Equality Principle:

---

and equal person is conspicuously absent from Rawls's work. For example, he does not examine the U.S. Constitution and attempt to show that it expresses an ideal of free and equal persons that is determinate enough to ground a construction of the original position capable of yielding his principles.

[3] It seems that all of the theorists who offer a "resourcist" answer to the question "Equality of What?" fall into this category. In other words, to the extent that one is committed to the thesis that defines resourcist versions of egalitarian theories of justice—namely, that justice requires an equal distribution of resources—it seems that "resources" here must be understood broadly enough to include natural resources (genetic endowments) as well as social resources. Included in this class of theorists are Ronald Dworkin and G. A. Cohen. John Roemer regards himself as a resource theorist but apparently restricts the resources to be distributed equally to a type of social resource, alienable productive assets. (See Dworkin, "What Is Equality? Part 2"; Cohen, "On the Currency of Egalitarian Justice"; and Roemer, "Equality of Talent.")

Whether there is a principled way an egalitarian resource theorist can restrict resources to social as opposed to natural resources is quite unclear. Dworkin, at least, is quite explicit that the term "resources" is to include natural assets as well as social ones.

[4] Roemer, "Equality of Talent," p. 154.

All resources, natural and social, ought to be distributed equally (at least so far as they are undeserved).

So long as we assume that we cannot intervene directly in the distribution of natural assets so as to achieve an equal distribution of them, the proper response to such natural "inequities" is to compensate for them by redistributing social resources. Thus, in the face of the unfeasibility of redistributing natural assets, we retreat from the first-best Resource Equality Principle to the second-best Resource Redress Principle.

However, given the prospect of genetic engineering, the assumption that redistributing natural assets is unfeasible is no longer plausible. We are now left wondering why theorists such as Rawls, G. A. Cohen, and Ronald Dworkin have not considered the possibility of responding directly to genetic "inequities." If equal opportunity (or a more fundamental principle of the equality of persons) requires us to redress or compensate for inequalities influenced by genetic differences, why not simply avoid them by eliminating the genetic differences in the first place?[5]

How we answer this question will depend upon how literally the Resource Equality Principle is to be interpreted. There are two relevant interpretations. One takes equality in an *aggregative* sense, the other in a *distributive* sense. According to the former, the Resource Equality Principle requires only that each person's *overall* bundle of resources be equal to every other person's overall bundle. Understood in this way, the Resource Equality Principle can be satisfied even if there are inequalities among some of the resources composing the overall bundle, so long as the totals are equal. In particular, on the aggregative interpretation of the Resource Equality Principle, two persons could have equal resources even if one had inferior genetic endowments, so long as her share of social resources compensated for this deficiency in such a way as to yield an overall resource bundle equal to the other person's overall resource bundle. In contrast, the distributive interpretation would require that resources be equal among all persons at least for every major category of resources, including natural and social resources as major categories. Thus, according to the distributive interpretation of the Resource Equality Principle, natural resources (genetic endowments) are to be equal, unless there is no way to render them equal, in which case we retreat to the Resource Redress Principle, which allows us to compensate for uncorrectable natural inequalities by redistributing social resources.

[5] On p. 100 of *A Theory of Justice*, Rawls appears to equate the principle of equal opportunity with what he calls the principle of redress, and to say that equal opportunity thus understood is required by the principle of equal respect for persons:

> This [the principle of redress] is the principle that undeserved inequalities call for redress; and since inequalities of birth and natural endowment are undeserved, these inequalities are to be somehow compensated for. Thus the principle holds that in order to treat all persons equally, to provide genuine equality of opportunity, society must give more attention to those with fewer native assets and those born into the less favorable social positions.

For our purposes, the chief difference between the two interpretations of the Resource Equality Principle is this: the distributive interpretation *requires* genetic equality, where it is possible to achieve it; the aggregative interpretation allows us to forgo efforts to achieve genetic equality *if* overall resource equality can be achieved by compensating for inequalities in natural endowments by redistributing social resources, but also *permits* efforts to achieve genetic equality.

Those who are disturbed by the prospect that justice, and in particular equal opportunity, might require massive genetic intervention may therefore find the aggregative interpretation of the Resource Equality Principle more congenial. However, it is worth emphasizing that although the aggregative interpretation does not by itself require genetic intervention to equalize resources, it does *permit* it. On the aggregative interpretation, the Resource Equality Principle simply says that overall resources are to be equal. It neither requires equalization of natural endowments nor rules it out as impermissible. However, if it turned out that in certain circumstances the *only* way to achieve overall resource equality were to intervene genetically, then the Resource Equality Principle would require this. And presumably, if genetic intervention were the *most efficient* way to achieve overall resource equality, then the Resource Equality Principle (again on the aggregative interpretation) would also require it, barring some powerful countervailing moral consideration. In sum, opting for the aggregative as opposed to the distributive interpretation of the Resource Equality Principle does not eliminate the possibility that this principle would allow genetic intervention. Indeed, it does not even rule out the possibility that the principle would require it under certain circumstances.[6]

At this point it might be well to address a note of skepticism about the

[6] As compared with the distributive interpretation, the aggregative version of the Resource Equality Principle also has a serious drawback. It requires very strong (cardinal) comparability among all types of resources, including natural and social. To compare different individuals' bundles of total resources of all types, presumably requires knowing what contribution each item in the bundle makes toward an individual's well-being.

However, at this point the aggregative version of the Resource Equality Principle threatens to undermine the whole rationale for a *resourcist* as opposed to a welfarist approach to matters of justice. That rationale consists of two elements. On the one hand, those advocating a resourcist approach (and rejecting welfarist approaches) typically deny the possibility of making such strong welfare comparisons. On the other hand, some resourcists, including Rawls and Ronald Dworkin, contend that due respect for the autonomy of persons makes it impermissible to impose any public standard of comparability (which the aggregative version of the Resource Equality Principle would require), because doing so would violate the right of each person to determine his or her own good. The appeal of a resourcist approach to questions of justice is supposed to consist chiefly in the fact that it avoids both the problem of devising a strong metric for comparing welfare across persons and the violation of respect for individual autonomy that would occur if there were an attempt to impose any such measure as a public standard of justice. Thus, while it is true that opting for the aggregative rather than the distributive interpretation of the Resource Equality Principle avoids the implication that resource equality *requires* genetic intervention in the "natural lottery" in all circumstances, it does so at the cost of adopting a version of the Resource Equality Principle that threatens to undermine the whole resourcist approach.

real-world consequences of the fact that a particular moral theory would permit or under certain circumstances require genetic intervention to equalize resources. Given that our society (or at least the most politically powerful segment of it) tolerates if not encourages quite substantial social and economic inequalities, how serious a risk is there that those who influence public policy would ever contemplate genetic intervention in the name of equal opportunity or justice?

Neither I, nor anyone else, so far as I can tell, is able to predict reliably whether there will ever be a serious effort to justify genetic intervention by appeal to equal opportunity or justice. However, there is some evidence that the probability of this occurring is not negligible. After all, there is, unfortunately, the recurrent tendency, at times rather pronounced in American history in the last hundred years, to fix upon rather facile biological-determinist explanations of deep social ills, and to propose morally dubious or even despicable social programs in the name of high moral principles. From the early eugenic notion that poverty and sexual immorality are the result of "defective germ plasm," to the more recent attempt to explain certain forms of socially deviant behavior as the phenotypical expression of a "criminal gene" (the so-called super-male or XYY genotype), significant segments of the public, as well as some public officials and some leaders of the scientific community, have leaped to biological-determinist conclusions in a disturbingly wide range of cases, and shown a facility for rationalizing repressive programs on moral grounds.[7]

As the technical possibilities for direct and dramatic genetic manipulation rapidly increase, the temptation to try to solve major social problems by genetic intervention – conclusively and in one generation – will no doubt increase. Moreover, if the history of biological determinism is any guide to its future manifestations, proposals for such genetic "quick fixes" or "magic bullets" will surely be publicly justified by their proponents by appeal to moral principles. For these reasons, it is by no means fanciful to consider the implications of ethical theory for genetic intervention.

Accordingly, in the remainder of this essay I will examine two arguments that purport to support the Resource Redress Principle and the Resource Equality Principle. The first is a very basic argument that takes as its first premise the so-called "Formal Principle of Justice," according to which equals are to be treated equally (and unequals unequally). The second focuses on the notion of equal opportunity, arguing that a proper understanding of it requires adherence to the Resource Equality Principle or, at the very least, the Resource Redress Principle. In the end I will conclude that neither of these two arguments is sound. The second of the

[7] For excellent analyses of various aspects of the American eugenics movements in international comparative perspective, see *The Well-Born Science*, ed. Mark Adams (Oxford: Oxford University Press, 1990); especially illuminating is Adams's introductory essay. See also Daniel Kevles, *In the Name of Eugenics* (New York: Alfred A. Knopf, 1985).

two, the argument from equal opportunity, will be dealt with in considerable detail, mainly because of the importance of distinguishing it from quite distinct arguments appealing to different notions of equal opportunity that do have some implications for genetic intervention.

II. The Argument from the Formal Principle of Justice

I will begin by focusing upon Rawls, because I believe his work is the *locus classicus* of the ideas to be explored.[8] (Indeed, it seems as though others, including Dworkin and Cohen, who subscribe to the Resource Redress Principle tend to take Rawls's view as an unquestioned point of departure.) At least so far as they can be gleaned from Rawls's discussions in *A Theory of Justice*, the two arguments to be subjected to critical scrutiny in this essay have this in common: both appeal to the idea that justice requires redressing or compensating undeserved (and hence "morally arbitrary") inequalities in life-prospects.

On one interpretation, at least, the first argument, the argument from the Formal Principle of Justice, simply *assumes*, apparently on the basis of an appeal to widely and confidently held considered moral judgments, that any inequalities in life-prospects due (at least in part) to "undeserved" differences in natural or social endowments are "morally arbitrary" in the particular sense relevant to the Formal Principle itself: undeserved differences in natural or social endowments are *not* the sorts of inequalities that may serve as bases for *unequal treatment* of persons as such. As a first approximation, the argument from the Formal Principle of Justice is as follows:

(1) Equals are to be treated equally (and unequals unequally) (The Formal Principle of Justice).
(2) All persons, as such (that is, as persons), are equal (The Principle of the Equal Worth of Persons).
(3) To allow some persons to have lower life-prospects than others, as a result of differences in natural or social resources that are undeserved, would be to treat persons, as such, unequally. Therefore,
(4) (So far as it is possible to do so,) social and natural resources are to be distributed equally, so far as they are undeserved (The Resource Equality Principle).

The weakness of this argument lies not in the Formal Principle of Justice, nor in the Principle of the Equal Worth of Persons, but in the third premise. The Formal Principle of Justice, taken in its bare abstractness, is unexceptionable. In order to apply it, however, it is necessary to specify

[8] See Rawls, *A Theory of Justice*, pp. 63–80.

what counts as a *relevant* equality, since any two (or more) persons, simply because they are different individuals, will be unequal in *some* respects. Those who appeal to the Principle of Formal Justice to try to justify the Principle of Resource Equality focus upon personhood itself as the relevant equality, since they are attempting to base a principle requiring the equal distribution of resources ultimately on the notion of *the fundamental equality of persons*.

Now it is true that respect for the fundamental equality of persons requires treating persons equally whenever one is in a situation in which one can be said to be treating them in one way or another. But of course, what counts as treating them equally will vary with the context. For example, if the context is that of considering job applicants, then treating them equally usually will mean being objective in the application of whatever the relevant criteria for job qualifications happen to be. In another context, treating persons equally may mean paying everyone the basic respect of listening to their grievances or concerns, or of taking everyone's interests, or certain of their interests, into account.

It is quite a jump, however, from these relatively uncontroversial statements to the thesis that treating persons as such equally requires eliminating or compensating for all differences in people's resources, social or natural, so far as these are undeserved. In one obvious sense, in simply allowing these differences to persist by not intervening in the natural lottery, one is not *treating* people at all, either equally or unequally.

If the so-called natural lottery were in fact a lottery run by society (or by the state, assumed to be the agent of society), then it would make straightforward sense to talk about the use of this lottery as a way of treating people. However, the "natural lottery" is not a lottery at all, or at least not one run by society or by any other human agent or agency. As it stands, then, the argument from the Formal Principle of Justice is unsound. The third premise is false, or at least dubious and in need of convincing support.

For the argument to have a chance of working, it must be revised to include the following premise:

(2′) If a process is technically within our collective control, such that we can intervene to change its effects, then in *allowing* that process to affect people in certain ways (i.e., in *not preventing* it from affecting them in those ways), we are thereby treating people in those ways (The Extended Responsibility Principle).

Also needed is another premise:

(2″) We can intervene to redress or compensate for differences in natural or social endowments that are undeserved.

Premises (2′) and (2″)—along with (1) and (2)—do yield (3), and the conclusion (4), the Resource Equality Principle, does follow.

For the sake of this essay, I will simply accept premise (2″) as given, without exercising a healthy skepticism about the probable limits of genetic intervention. Instead, I shall concentrate on an obvious but nonetheless serious difficulty with premise (2′). It says, in effect, that anything that *can* be brought within our control is *already*, morally speaking, *an exercise of our agency*, at least in the sense that it is something for which we are morally responsible. Not preventing persons from being affected unequally, where they do not deserve the factors that lead to these inequalities, is said to be *treating those persons* with less than equal respect, and is in that sense morally identical to voluntary acts that violate the fundamental respect owed to all persons as such. Instead of the familiar "ought implies can," we have a principle of "can implies ought," or rather: "If we can prevent X from occurring, but do not prevent X from occurring, then we are responsible for X."

Unless something like this is accepted, it is hard to see how one can bridge the gap between the Formal Principle of Justice and the Principle of the Equal Worth of Persons in such a way as to derive the desired conclusion, namely, the Principle of Resource Equality. However, as a general principle, (2′) is not only not self-evident, but implausible. A commitment to (2′) would expand the domain of moral responsibility to tyrannical dimensions. To attribute to a person's moral account everything that he or she does not prevent but could prevent seems incompatible with recognizing and respecting the fact that we each have one life to live, that we are each separate persons who are entitled to give our own projects a certain preference, and who accordingly require a substantial moral space within which to pursue our own conceptions of the good.

To appreciate just how demanding the Extended Responsibility Principle is, it is necessary to distinguish it clearly from a much more reasonable principle of moral responsibility, the Harm Prevention Principle.

> HPP: If one is able to prevent a *harm* to an innocent person, knows one can do so, and can do so without excessive costs, but allows the harm to occur, then one is responsible for the harm befalling that person (barring any countervailing moral obligation that precludes one from acting so as to prevent the harm).

HPP implies that we can, under certain circumstances, be morally responsible (blameworthy) for harms that we do not literally cause. It emphasizes the importance for moral responsibility of *the ability to control* potentially harmful outcomes, rather than limiting responsibility to cases in which an individual causes a harmful outcome. For example, HPP implies that I am morally responsible for the death of an innocent hit-and-run victim who lies bleeding on the roadside, who will die if I do not act, and whom I could save without excessive costs (merely by calling an ambulance), at least so far as no moral duty forbids my doing so.

It is not my aim here to argue in favor of HPP (or for some further qualified formulation of it), though I believe some such principle is in fact defensible. Instead, my purpose in articulating HPP is to avoid the error of confusing it with (2′) and thereby failing to appreciate how dubious (2′) really is. The latter principle, the Extended Responsibility Principle, is a much broader principle of moral responsibility than HPP, the Harm Prevention Principle. HPP only ascribes moral responsibility for *harms* that we can prevent (and then only subject to important qualifications). The Extended Responsibility Principle goes much farther, by ascribing responsibility to us for *all* outcomes that we could prevent, whether they constitute harms or not. Such a principle of responsibility places a very great burden on us all—indeed, one might say, an indefinitely large burden.

Of course, it is possible to render the Extended Responsibility Principle *equivalent* to the Harm Prevention Principle when applied to cases of lower life-prospects due to bad luck in the genetic lottery. We could achieve this equivalence by simply *stipulating* that just by having life-prospects lower than others', as a result of one's genetic endowment, one is *thereby harmed*.

However, to indulge in this stipulation would be to beg the question, either by assuming a very unpersuasive conception of harm or by assuming the truth of the very principle to be established, namely the Resource Equality Principle. For it would make sense to say that simply having lower life-prospects than others due to my genetic endowment is a harm *only* if it were granted that equality of life-prospects is the normative baseline, as it were, such that any shortfall from equality *constitutes* a harm. But that assumption, so far as I can tell, is not only undefended by those who advocate the Resource Principles (Resource Equality or Resource Redress), but indefensible as well. At any rate, some argument in support of it is needed; and the argument in question could not, on pain of a vitiating circularity, rely on a premise to the effect that lower life-prospects due to genetic luck constitute a harm because everyone is entitled to equal resources or to equal life-prospects.

Merely having lower life-prospects would perhaps be a harm (or a harm in the sense of a wrong, a violation of a right) *if* it were assumed that justice or equal opportunity requires equality of life-prospects. However, that assumption, as I also observed earlier, begs the question in favor of the resource-egalitarian view. Thus, in the absence of an argument to show that equal opportunity or some other principle of justice requires equal life-prospects, or an argument to show that having lower life-prospects is itself a harm, recasting the argument from the Formal Principle of Justice by utilizing the principle that one is as responsible for omissions (failures to prevent) as for commissions appears to fail.

Now one might make the mistake of assuming that having lower life-prospects constituted a harm if one erroneously assumed that when we talk of "bad luck" in the genetic lottery in this context, we mean such rotten luck that a person's prospects are so low as to constitute a *deprivation*. It is one thing to say that being born blind or paralyzed is a harm; quite another to say that being born with a somewhat lower compliment of talents than someone else is a harm (no matter how small the difference in talents is). The former case of bad luck in the genetic lottery could, without any hyperbole, be called a deprivation; the latter could not. Later, in examining the second main argument for the Resource Principles, I will attempt to show that that argument too is unpersuasive, once we distinguish carefully between our responsibility to prevent deprivations and our alleged responsibility to prevent any inequalities whatsoever.

One last attempt to rescue the argument from the Formal Principle of Justice (to the Resource Redress Principle) may be worth considering. Suppose the claim is not that any lower life-prospect is itself a harm (or deprivation), but instead that a situation in which some persons, through no fault of their own, have lower life-prospects than others is *suboptimal*—less good than it would be if none had lower prospects because of undeserved natural differences. I am not even sure that this claim is correct, but suppose for a moment that it is. Its truth would still not be sufficient to show that treating persons equally as persons requires avoiding such a "suboptimal" situation, *unless* another premise were added:

> To not prevent a suboptimal state of affairs (when one could do so) is to fail to treat persons as persons equally.

Or, perhaps the same basic idea might be put a bit less ponderously, and with more emphasis on the idea that it is outcomes that are suboptimal *for persons* that are of concern, as follows:

> (i) We ought to prevent persons' lives from being less good than they would otherwise be, and
> (ii) to fail to do so when we can is to treat them as less than equal, as persons, at least if others enjoy the benefits in question and there is no justification for this inequality on grounds of desert.

The difficulty with this proposition is that both components of it are quite questionable and in need of argumentative support.

With regard to (i), it is one thing to say that one ought to try to prevent undeserved deprivations or to prevent harms, quite another to say that one ought to prevent any condition in which persons' lives are less good than they could be, no matter how good in fact they are. (It might be more plausible to say that if one *did* prevent such "suboptimal" outcomes it would be a good, or a commendable thing, but "ought" here seems dubious.)

Further, even if (i) were granted, there is no obvious link between the obligation or moral imperative it states and the notion of treating persons (as persons) equally that is indicated in (ii). For remember, the whole point is supposed to be that, as a matter of justice, persons are to be treated equally, so far as they are persons. Yet there seems to be a lack of connection here between the idea that one ought to avoid suboptimal outcomes for persons and the idea that if one failed to do so, one would somehow violate a principle of justice that persons as such are not to be treated unequally or indeed treated unjustly in any other way. To put the same point differently, it does not seem that a person could justifiably charge that he had been treated as less than an equal, qua person, and thereby *wronged*, simply because others could have prevented his having lower life-prospects as a result of less-fortunate natural endowments but did not do so, regardless of how good his life-prospects were.

Of course, we might imagine a society in which persons' self-understandings were ingrained with a thoroughgoing egalitarianism according to which preventable lower prospects *would* be seen as a lack of equal regard—as a failure to treat persons equally, as persons, as a matter of justice. However, the question before us is whether we should adopt such a point of view. My suggestion is that, at least for many of us, this is not in fact our point of view, and that so far no good reasons have been given to persuade us to adopt it.

My tentative conclusion, then, is that attempts to base the Resource Equality Principle on the Formal Principle of Justice do not work. Accordingly, I propose to examine now the second argument to support the Resource Principles, without dogmatically assuming that no other argument might be given in favor of them.[9]

[9] Some of Rawls's later writings suggest a different argument in favor of the Resource Equality Principle, one which does not rely (at least not explicitly) upon the Formal Principle of Justice. The central idea of this line of argument is that free and equal persons, who regard themselves as such, and as members of a society of such persons understood as a cooperative venture for mutual advantage, would not simply accept a situation in which they had lower life-prospects due to undeserved differences in natural or social resources. See, for example, Rawls, *Political Liberalism*, pp. 15–34.

Perhaps Rawls's point can be expressed this way: In a situation in which undeserved differences in natural or social resources led to some having lower life-prospects, those with lower life-prospects, so far as they viewed themselves as free and equal persons in a cooperative venture for mutual benefit, would *take* themselves to be treated as less than equals by their fellows. That is to say, *from* the perspective of persons conceived as Rawls conceives them for purposes of a liberal theory of justice, allowing undeserved differences in natural or social resources *would count* as unequal treatment of equals, thus violating the Formal Principle of Justice. In other words, if the members of society allowed such factors to result in lower life-prospects for some persons then those persons, so far as they conceived of themselves as free and equal and so far as they conceived of society as a cooperative venture for mutual advantage among free and equal persons, would take this state of affairs to

III. The Argument from Equal Opportunity

If, as I have suggested, arguments from the Formal Principle of Justice are to no avail, then a more promising approach for those who endorse the Resource Principles might be to try to ground them in a moral principle or concept that already enjoys wider support—the concept of equal opportunity. Of course, for this strategy to work, the concept of equal opportunity with which the argument begins must be one that is not only recognizable as a concept of equal opportunity but also one which enjoys wide acceptance. The most obvious way to achieve this would be to appeal to the actual history of the development of the concept of equal opportunity, beginning at an early stage in its development, and showing that the underlying principle that drives its expansion leads ultimately to the desired destination: the Resource Principles. On one interpretation of his discussion of equal opportunity in *A Theory of Justice*, this is precisely what Rawls attempts to do. For convenient reference, let us call this the Expansive Instability Argument.[10]

As Rawls states the Expansive Instability Argument, the moderate, relatively uncontroversial notion of equal opportunity with which he begins is that of Careers Open to Talents. This conception is then shown to be *unstable*: once we acknowledge that equal opportunity requires the elimination of legal barriers to opportunity and once we see the principle upon which the requirement of the elimination of legal barriers rests, we must ultimately—on pain of inconsistency—admit that justice requires redressing or compensating *all* inequalities in life-prospects due to "morally arbitrary factors," whether the latter be natural or social. Rawls's strategy is to begin with this very limited conception of equal opportunity and show how the principle which underlies it, when properly understood, leads to a succession of *more expansive* conceptions of equal opportunity, and ultimately to the conclusion that a proper and consistent

be a public *expression* of a collective judgment that they were not in fact moral equals to those with higher life-prospects.

The problem with this argument for the Resource Equality Principle is that the crucial statement about the beliefs of individuals who view themselves as free and equal cooperators for mutual advantage with others seems to beg the question. As a descriptive, empirical generalization about the way people, or most people in a society like ours, in fact see themselves, it remains wholly unsupported by anything Rawls says. Moreover, in a pluralistic society such as ours in which there are significant differences of judgment about what recognition of the fundamental equality of persons requires (not to mention about whether justice requires intervening in or compensating for bad luck in the natural lottery), it is unlikely that the needed empirical support would be forthcoming. On the other hand, if the statement is taken as a normative one that is somehow implicit in the normative concept of a free and equal person, then it seems equally dubious. (There may be disputes, of course, over which is the correct normative concept of a person.) Precisely what is at issue is whether a proper recognition of the equality of persons as free beings requires implementation of something like the Resource Equality Principle.

[10] See Rawls, *A Theory of Justice*, pp. 72–73.

appreciation of equal opportunity requires the acceptance of his Second Principle of Justice.

The underlying principle—the one which is supposed to ground the initial, very limited conception of opportunity (Careers Open to Talents), but pull us inevitably beyond it to the more expansive conceptions—is simply what I have called the Resource Redress Principle. Upon reflection we are bound to admit, Rawls thinks, that it is not the legal character of obstacles to social positions and offices that makes them unacceptable, but that they are instances of something much more general: undeserved (and hence "morally arbitrary") inequalities in the distribution of primary goods (resources) resulting in lower life-prospects for some.

In other words, according to the Expansive Instability Argument, it is allegiance to the Resource Equality Principle that underlies our commitment to careers being open to talents, and once we recognize this, we see that we must, in all consistency, go beyond Careers Open to Talents to more expansive conceptions of equal opportunity. As we shall see, by using the eighteenth-century phrase "careers open to talents," Rawls strongly suggests that a historical view of the development of the concept will help illuminate the principle that underlies its development.

There are two reasons to believe that the most promising interpretation of the Expansive Instability Argument anchors it in the actual history of the development of the concept of equal opportunity. The first is that doing so increases the chances that the basic strategy of the argument—that of beginning with what is widely accepted and relatively uncontroversial and moving toward a conclusion that is, at the outset, highly controversial—will actually be implemented. By showing how the "inner logic" of the concept led us in the past to accept one expansion of the notion of equal opportunity, Rawls may smooth the way for a new expansion. If we accept the initial expansion and see that it was required by a certain principle we accept, then we should accept the next expansion that same principle requires.

The second reason for appealing to the historical development of the concept is distinct but related: In his later works Rawls has increasingly relied upon what I have elsewhere labeled a Traditionist mode of justification.[11] He now sees his task as that of showing how his principles of justice are rooted in certain ideals that are embedded in the political culture and institutional traditions of modern liberal-democratic society. Therefore, focusing on the historical origins and actual development of the concept of equal opportunity in the political doctrines, major legal documents, and political institutions of that sort of society is not only appropriate, but necessary if this mode of justification is to be employed.

[11] Allen Buchanan, *Secession: The Morality of Political Divorce from Fort Sumter to Lithuania and Quebec* (Boulder: Westview Press, 1991), pp. 82–83.

It is the Expansive Instability Argument that I shall show to be unsound. Four important conclusions will emerge from the analysis. First, there *is* a historically rooted concept of equal opportunity that has undergone a process of expansion; but it does *not* require intervening in the natural lottery in order to reduce natural inequalities, nor does it require compensating for all undeserved natural inequalities by redistributing social resources. Second, equal opportunity, under certain conditions, may require constraints on the uses of technologies for genetic enhancement. Third, given a somewhat controversial extension of the concept of equal opportunity – yet one far less extreme than that employed in the two Resource Principles – justice may, under certain conditions, require genetic interventions, not to avoid natural inequalities, but to prevent *deprivations*. Fourth, at least in its most familiar and uncontroversial applications, equal opportunity is not a *primary* principle of justice; rather, it is a *remedial* principle, one that comes into play as a response to violations of other, more basic principles of justice. And none of these more basic principles includes anything so extreme as the two Resource Principles.

IV. The Concept of Equal Opportunity: Its Historical Development

In criticizing the Expansive Instability Argument, my strategy will be to provide a characterization of the principles that underlie the successively more expansive conceptions of equal opportunity without relying upon anything so extreme as the Resource Principles. In doing so, I will show that Rawls simply begs the question in favor of the Resource Redress Principle by invoking it gratuitously to explain the first stage of the development of the concept of equal opportunity and then arguing that consistency requires that we accept its more radical implications.

We can begin where Rawls does, with a sketch of the development of the concept of equality of opportunity. Rawls first notes what we would now regard as a rather conservative and uncontroversial conception of equal opportunity: the requirement that careers be open to talents. One of the slogans of the French Revolution and of liberalism's struggle against hereditary privilege, this is simply the idea that there should be no *legal* barriers to preclude *qualified* persons from seeking and obtaining valued social positions and offices.

In the historical setting in which it was first voiced, this principle of equal opportunity had a dual thrust. First, it could be seen simply as the demand that a preexisting *meritocratic principle* be consistently applied – that the best-qualified candidates, regardless of their class background, be awarded positions in fair competition. Second, the slogan "careers open to talents" was a rejection of the presumption that the hereditary aristocracy was in fact an aristocracy, that noble lineage conferred superior qualifications for office. Understood in this way, the slogan only required a consistent application of a preexisting commitment to a meritocratic principle, but was revolutionary nonetheless because it rejected a theory of natural superiority that created a presumption of superior merit in a certain class of individuals.

This interpretation of the concept of equality of opportunity as Careers Open to Talents is compatible with the recognition that there may have been, at the same time, a shift in the way merit was understood, or, more precisely, a change in the conception of the bases of merit. As the idea took hold that the social good was best secured through increases in productivity spurred by the motive of private gain, a different conception of the qualifications needed for various social positions developed along with it. Thus, the "talents" required for various positions were those needed to enhance national wealth, and the earlier aristocratic (and largely military) virtues became less relevant. Nevertheless, even if the standard of merit was being transformed by the same socioeconomic forces that produced the challenge to aristocratic privilege, the fact remains that the moral appeal of equal opportunity as Careers Open to Talents is easily explained simply by recourse to the idea that, at least in certain areas of social life, justice requires a consistent application of the meritocratic principle. That principle states that, at least for positions where ability to perform the tasks that define the role is the proper criterion for filling the position, individuals should be allowed to compete fairly, simply on the basis of their qualifications, without any legal barriers to doing so.

There is no need whatsoever to invoke any grand principle about the role of social (or natural) contingencies as such in shaping individuals' life-prospects. It is entirely gratuitous to assume, as Rawls apparently does, that the commitment to equal opportunity as Careers Open to Talents rests on the conviction that justice requires redressing or compensating for differences in life-prospects due to morally arbitrary social contingencies. All that is needed is an appeal to what is obvious: the idea that, in areas of life in which the meritocratic principle is appropriate, it is to be applied consistently.

Similarly, a second stage in the evolution of the concept of equal opportunity can also be explained without recourse to anything so extreme as the Resource Principles. Rawls is right in saying that once we see that there should be no legal barriers to qualified persons competing for valued social positions and offices, we must, in all consistency, be concerned about less formal barriers. The same moral concern which led us to embrace the conception of equal opportunity as Careers Open to Talents requires that we attempt to eliminate barriers to opportunity posed by racial, gender, or ethnic prejudice, even if these manifest themselves in extralegal ways, through the operation of less formal social practices and individual acts of discrimination. Let us call this second conception of equal opportunity, which does seem to be a natural and indeed inevitable expansion of Careers Open to Talents, equal opportunity as Comprehensive Nondiscrimination.

Again, notice how far this expansion of the concept of equality of opportunity (beyond the narrow confines of Careers Open to Talents to that of Comprehensive Nondiscrimination) is from the radical notion that justice requires redressing or compensating for any and every lowering of life-prospects due to any morally arbitrary social or natural contingency.

This second conception of equality of opportunity is indeed an expansion of the first conception: it recognizes that qualified persons may be precluded from positions and offices not only by laws that literally bar them from competing, but also by pervasive social practices that embody deeply ingrained racial, ethnic, or gender prejudices. Especially when it is applied to cases of racial discrimination in filling positions, this second conception of equality of opportunity also invokes, perhaps more explicitly than the conception of Careers Open to Talents, a principle of the equality of persons. When a qualified African American is excluded from being seriously considered for an apprenticeship in a union, he has a double grievance: not only has he been treated unfairly because his qualifications were disregarded, but also he has been insulted and demeaned so far as his equal status as a person is concerned. Racial discrimination in hiring of this sort, whether it achieves exclusion by laws (as in apartheid South Africa or the Jim Crow South) or by informal racist practices, violates two principles of justice: the meritocratic principle and the principle of equal respect for persons. The principle forbidding racial discrimination, then, is simply one application, focused on the case of race, of a more basic principle of equal respect for persons that provided part of the impetus for rejecting the presumption that the hereditary nobility were natural superiors. Nothing more need be said to explain the moral appeal of this second conception of equal opportunity, and nothing more need be added to explain how it developed out of the earlier concept of Careers Open to Talents. The second conception simply applies the meritocratic principle more consistently, by recognizing that qualified persons can be excluded from meritocratic competition by informal as well as legal barriers, and also rejects the idea of a class of natural superiors as the latter is applied to the case of race, gender, or ethnicity.

The only concept of respect for the fundamental equality of persons at work here, however, is a purely negative one: persons are not to be victimized by discrimination, legal or otherwise. No notion that all "morally arbitrary" social (or natural) inequalities are a matter of concern from the standpoint of justice is present or needed.

It is essential to emphasize that what Rawls identifies as the successor to the concept of equal opportunity as Careers Open to Talents is something quite different from what I have described as the first natural expansion of the concept: Comprehensive Nondiscrimination. In the following passage he states that the principle underlying Careers Open to Talents requires us next to embrace what he calls Liberal Equality.

> [Where the requirement of careers open to talents is satisfied there is formal equality] . . . in that all have at least the same legal rights of access to all advantaged social positions. But since there is no effort to preserve an equality, or similarity, of social conditions, except insofar as this is necessary to preserve the requisite background institutions, the initial distribution of assets for any period of time is strongly influenced by natural and social contingencies. . . . Intuitively, the most obvious injustice of the system of natural liberty [in which careers are open to talents] is that it permits distributive shares to be improperly influenced by these factors so arbitrary from a moral point of view.[12]

What is needed, Rawls then concludes, is an expansion of the concept of equal opportunity from Careers Open to Talents to what he calls Fair Equality of Opportunity. The idea is that "positions are to be not only open in a formal sense, but that all should have a fair chance to attain them."[13] By a "fair chance" Rawls means something quite specific: "[T]hose who are at the same level of talent and ability, and have the same willingness to use them, should have the same [i.e., equal] prospects of success. . . ."[14]

At the very least, something of a leap has occurred here. As indicated earlier, the most obvious candidate for a successor to the concept of equal opportunity as Careers Open to Talents is the concept of equal opportunity as Comprehensive Nondiscrimination. The latter simply extends the notion of impermissible barriers to opportunity from legal discrimination to extralegal, informal discrimination. And in fact, not only in this country, but in other countries as well, at least in the West, when legal systems and social practices have extended the notion of equal opportunity beyond Careers Open to Talents, they have done so by recognizing that informal modes of discrimination, such as racist or sexist hiring practices, are unjust restrictions on opportunity and must be eliminated where possible.

It is surprising, given the strategy of the Expansive Instability Argument as I have characterized it, that Rawls simply passes over this stage in the development of the concept of equal opportunity. It is even more surprising that he moves instead to the much more radical notion that equal opportunity requires that those with equal talent and motivation

[12] Rawls, *A Theory of Justice*, p. 72.

[13] *Ibid.*, p. 73.

[14] *Ibid.*

have equal chances of attaining valued positions. It would be very difficult—in fact, I think, impossible—to find evidence of the wide acceptance of this concept of equal opportunity among individuals in modern liberal societies. At any rate, Rawls fails to provide any such evidence. It would also be difficult to show that such a concept is somehow embodied in the political doctrines or legal institutions of such societies. Again, Rawls offers nothing to indicate that it is.

More importantly, if my analysis of the principles that underlie Careers Open to Talents and its successor, Comprehensive Nondiscrimination, is even approximately accurate, Rawls's concept of Fair Equality of Opportunity has no apparent logical or conceptual connection with either. It certainly cannot be seen as an expansion of the concepts of Careers Open to Talents or Comprehensive Nondiscrimination if, as I have argued, these concepts rely upon the consistent application of the meritocratic principle and a negative concept of the equality of persons that only rejects the spurious natural inegalitarianism of views that hold aristocrats or whites or women, etc., to be natural superiors.

The concept of equal opportunity as Comprehensive Nondiscrimination, then, is a much better candidate for a successor to Careers Open to Talents than is Rawls's concept of Fair Equality of Opportunity; and the principles that underlie the former two concepts, and that account for the expansion of the first into the second, do *not* require the adoption of Fair Equality of Opportunity.

Although Rawls does not mention it, there is a third stage in the historical development of the concept of equal opportunity, at least in our (U.S.) society, that is somewhat more controversial. This conception of equal opportunity is sometimes labeled Strong Affirmative Action. It differs from the second (Comprehensive Nondiscrimination) and from the first (Careers Open to Talents), requiring that in some cases equal opportunity mandates departures from the strict meritocratic principle. A policy requiring that African Americans, Hispanics, Native Americans, or women be given preference in hiring for faculty positions even when there are *better*-qualified male or nonminority candidates would fall under this third conception.

It is not my aim here to defend or to attack this third conception of equal opportunity. Instead, I only wish to show that *if* this conception is justifiable, its justification does *not* commit us to anything resembling the Resource Principles. To my knowledge, the best, indeed the only, plausible justification for Strong Affirmative Action, is the idea that under certain historical circumstances a strict adherence to the meritocratic principle would simply perpetuate the negative effects of past injustices toward members of certain groups. For example, since blatantly racist legal and social practices in the past (if not in some cases in the present) have led to African Americans' having, on average, lower educational attainments than whites, requiring them to meet the same educational qualifications as other candidates will perpetuate the tendency for African Americans to not attain higher-paying, more rewarding jobs. In order to compensate for the continuing disadvantages of lower educational attainment—disadvantages due to violations of meritocratic and nondiscrimination principles of justice—it may be necessary to depart from the meritocratic principle. In its most powerful version, this argument includes the idea that by thus departing from the meritocratic principle we will give unjustly disadvantaged persons the opportunity to develop their qualifications, to become better qualified through more challenging work than they would otherwise be able to obtain.

Whether or not this justification for Strong Affirmative Action is compelling is disputable. What is clear is that nothing in this justification relies on or entails the thesis that *every* lowering of life-prospects due to *any* "morally arbitrary" social contingency requires redress or compensation. Instead, the third conception of equal opportunity, Strong Affirmative Action, focuses only on one kind of social contingency—that which is the result of, and tends to perpetuate, the negative effects of injustices.

To summarize: Rawls's Expansive Instability Argument contends that once we acknowledge the more limited conceptions of equal opportunity, beginning with Careers Open to Talents, we are led, as a matter of consistency, to the idea that not only "morally arbitrary" social contingencies but also natural contingencies require redress or compensation, so far as they lead to some having lower life-prospects. As we have seen, however, the earlier stages in the development of the concept of equal opportunity lead to the Resource Redress Principle *only* if one erroneously reads into them the idea that equal opportunity requires redressing or compensating for any lowering of life-prospects due to social contingencies. The slide from redressing or compensating for social contingencies to redressing or compensating for natural contingencies occurs only if one begins with the assumption that equal opportunity requires that all "morally arbitrary" social contingencies be redressed or compensated. But nothing in the development of the concept of equal opportunity so much as hints at anything so radical as this latter assumption.

In each of the three stages of the development of the concept of equal opportunity—from Careers Open to Talents, to Comprehensive Nondiscrimination, to Strong Affirmative Action—"social contingencies" are a focus of concern. However, what makes them so is not that they are "morally arbitrary" (in the sense of being things individuals cannot be said to deserve). Rather, it is the fact that these "social contingencies" that act as barriers to opportunity are *injustices* or (in the case of the third conception) that they help perpetuate the negative impact of injustices. And what counts as an injustice in all of these cases is characterized by appeal to relatively uncontroversial principles of justice *other* than the principle of equality of opportunity itself: namely, the meritocratic principle and a fundamental prohibition against racial, ethnic, or gender discrimination, in the case of the first two conceptions, and a principle of rectificatory justice in the case of the third. (What is controversial about Strong Affirmative Action is not the idea that we should try to eliminate the effects of past injustices, but rather the assumption that this obligation overrides the meritocratic principle.)

As I indicated earlier, my assumption is that if there is an argument from the concept of equal opportunity to the Resource Principles, then it is Rawls's Expansive Instability Argument. That argument has proven to be not only unsound, but irretrievably so. The problem with that argument is not that it lacks an easily supplied premise or that it grounds only an approximation of the conclusion it seeks to establish. On the contrary, nothing in the history of the development of the concept of equal opportunity seems to provide material that could be used to reach Rawls's radical conclusion.

Consequently, in the absence of some other argument to support them, the Resource Principles seem dubious indeed. Neither of these principles is self-evident, and, as we have seen, we cannot hope to justify either of them by showing that it is needed to explain our more widely and confidently held considered judgments about what equality of opportunity requires. Nor, given my own revised account of Rawls's appeal to the historical development of the concept of equal opportunity, does it appear that the Resource Principles are good candidates for what I have referred to as Rawls's Traditionist mode of justification for principles of justice. If my sketch of the development of the concept of equal opportunity is even roughly correct, Rawls will have to do much more to show that the Resource Principles are somehow latent in "our" political traditions or culture.

In the next section, I will consider some of the consequences of accepting the more basic (first-best) Resource Equality Principle. Once these are appreciated, the principle will be seen not only to be unsupported, but quite dubious on its face.

V. Resource Equality and the Disappearance of Persons

Perhaps the first thing to notice is that none of those who subscribe to the Resource Redress Principle have advocated the redistribution of natural assets (genes or complexes of genes). Instead, they have only called for the redistribution of social assets to compensate for some having lesser natural assets than others. However, to my knowledge none of them have given a clear argument to show why, contrary to appearances, allegiance to the Resource Redress Principle, and to the Resource Equality Principle upon which it rests, does *not* create a presumption in favor of redistributing genes in the name of justice, once the technology for doing so is at hand.

It might be replied that Rawls, at least, has an answer: Intervening in the genetic lottery is forbidden by the lexically prior principle of equal liberty. The latter principle includes a right to the freedom and integrity of the person that such intervention would violate.

The inadequacy of this reply becomes apparent once one realizes how radical the notion of intervening in the genetic lottery to equalize natural resources really is. We should not assume that this would merely be a matter of *redistributing* genes among antecedently existing persons (something like the analog of redistributing organs among a group of individuals). Instead, it seems likely that true genetic equality (if we can even make sense of that idea) would best be approximated by engineering individuals from scratch. If this is the case, then it is not at all clear that the right to freedom and integrity of the person of those who were thus engineered would be violated, since those individuals *would not exist prior to the intervention*.

Of course, Rawls might argue that it would not be the freedom or integrity of the person of the recipients of redistribution, but of their parents (or, rather, of those who would not be permitted to be parents of their "natural offspring") that would be violated. The difficulty is that once the possibility of profound genetic intervention is acknowledged and the presumption in favor of engaging in it to equalize natural assets is in place, it is hard to see why, on Rawls's own account, the right of parents to engage in genetic roulette should take absolute priority over the right of persons to be spared the (alleged) inequity of having lower life-prospects as a result of their inferior native endowments. For remember, according to Rawls, to allow this inequity is somehow to fail to honor the fundamental moral equality of persons and thus to do them a profound injustice.

For resource theorists who do not endorse a lexically prior principle of liberty that could provide a possible restraint on the commitment to achieving an equal distribution of all resources, natural and social, the problem is more severe. They seem to be committed to a view of distributive justice that, in effect, shatters what one would have thought was a framework assumption of any theory of distributive justice: namely, that we can distinguish between (1) persons, as the subjects among whom assets are to be distributed, and (2) the assets that are to be distributed among them.

In their defense it might be said that if the egalitarian resource theorists are willing to make a highly problematic metaphysical move, this framework assumption is salvageable. The trick would be to distinguish between those features of a person that are person-constituting, that are elements of his or her identity as a subject of justice, and those nonessential characteristics that are resources, and thus subject to the principle that resources are to be distributed equally or that inequalities in resources are to be eliminated.

Unfortunately, even if some principled way of distinguishing these two sorts of properties of the individual could be devised, it is not at all clear that this would solve the problem. Unless an additional argument were given to show that a person has a *right* to his or her identity-constituting properties, the mere fact that radical genetic manipulation might violate the conditions of identity would seem to carry no conclusive *moral* weight against it. For if the fundamental commitment is to ensuring that, so far as this is possible, persons have equal resources, then why should it bother one that in order to achieve the desired equality one must act so as to determine which persons, having which characteristics, will exist?

It is important to understand that an appeal to the idea that all valid moral principles are "person-affecting" principles provides no bar against the presumption in favor of achieving equality of resources by radical genetic manipulation, that is, by intervention on gametes or embryos. On any reasonable interpretation of the idea that moral principles must be "person-affecting," the scope of moral principles cannot be limited to actual persons—otherwise there would be no moral constraints on any of our behavior that we know will affect future generations. In other words, the thrust of the injunction that all moral principles must be "person-affecting" is simply that, so far as there are any valid moral principles that constrain our behavior that affects future persons, those principles ultimately must be based upon regard for the well-being and freedom of those persons who will come to exist, rather than upon some impersonal standard or value such as maximizing the total of utility per se or creating a better world, where what makes the world better is something other than the benefits for persons who exist in it. The principle that radical genetic intervention is to be employed, where necessary, to achieve equal resources is not, therefore, incompatible with the requirement that moral principles be person-affecting, on a reasonable interpretation of the latter requirement. Thus, appeal to the idea that valid moral principles are person-affecting cannot itself rule out radical genetic manipulation for the sake of achieving resource equality.

VI. A Fourth Concept of Equal Opportunity: The Absence of Undeserved Deprivation

Earlier I argued for a more conservative interpretation of the stages of development of the concept of equal opportunity. There I appealed, in effect, to what are sometimes called our "moral intuitions" (or considered moral judgments) as well as to the actual development of political doctrines concerning equal opportunity and of the legal institutions embodying them, at least in those Western countries that have institutionalized conceptions of equal opportunity. I argued that in the various cases falling under the three concepts of equal opportunity previously discussed, our intuitions, as well as the development of the institutional embodiment of these concepts, can be explained by less radical moral principles than the Resource Principles. (The three concepts of equal opportunity were Careers Open to Talents, Comprehensive Nondiscrimination, and Strong Affirmative Action.)

Perhaps, however, there are some widely and confidently held considered judgments about equal opportunity that have not yet been examined. Perhaps accounting for *them* will drive us in a more radical direction, toward the Resource Principles—and into the arms of the genetic engineers.

What I have in mind is the following sort of case. Suppose that a child is born blind or with some other major disability. Do we not believe that some special treatment is required, as a matter of justice? In the clearest case, assume that a relatively simple (though not costless) operation will restore the child's sight, that neither the child nor her family can afford it, and that no private philanthropic organization steps forward to pay for the operation. Unless there are public resources available for the child, she will go through life with a profound disability. Alternately, consider a case in which sight cannot be restored, but a special allocation of resources can enable the child to learn braille, to live independently (perhaps with the help of a guide dog, a specially equipped apartment, etc.). Again, suppose that private resources are not forthcoming. Some might say not only that a special allocation of the needed resources is required as a matter of justice, but also that the relevant principle of justice is that of equal opportunity.

It should be clear that none of the three conceptions of equal opportunity sketched earlier can encompass such cases. It need not be the case that the blind individual is being excluded, either legally (contrary to the concept of Careers Open to Talent) or by informal barriers of prejudice (contrary to the concept of Comprehensive Nondiscrimination), from competing successfully for positions for which she is qualified. Nor need the obstacle be the lingering effects of former discriminatory practices (to which the concept of Strong Affirmative Action is addressed).

Instead, her blindness—in the absence of special resources—may simply prevent her from being qualified for many if not all desirable positions. So we cannot make a case for allocating special resources to her or to anyone with a major natural disability on the ground that the meritocratic principle is not being applied consistently in her case or on the ground that she is being excluded as a result of prejudice which demeans and devalues her by classifying her as an inferior individual (as in the case of racist, ethnic, or gender discrimination). Nor will these sorts of cases fit under the third (Strong Affirmative Action) conception of equal opportunity: she lacks the qualifications for various positions, not because she is the victim of past injustices, but because she is naturally blind.

Some, of course, might argue that this is not a matter of equal opportunity because it is not a matter of *justice* at all, but rather of charity.[15] My strategy will not be to try to settle the issue of how we are to distinguish the domain of justice from that of charity. Instead, I will simply argue that the most plausible interpretation of these sorts of cases as falling under the notion of equal opportunity does not require anything so extreme and far-reaching as either of the Resource Principles.

I would like to suggest that what bothers us about the situation of severely disabled persons, and what leads us toward a commitment to improving their opportunities through special allocations, is not the conviction that every lowering of life-prospects due to misfortune in the natural lottery requires social redress or compensation. Instead, what is at work here, what fully accounts for our intuitions, is a much more modest principle, namely that, other things being equal, no person should be barred from the chance to have a minimally decent life as a result of undeserved natural (or social) deficits. In other words, the relevant concern here is with *deprivation*, not with inequality as such.[16]

To believe that concern for equal opportunity requires doing something to compensate for this severe disability, we need not believe that there is something inequitable about every case in which an individual's life-prospects are lower than they would be if he had the natural assets which those with superior natural assets have. All that is necessary is that we believe that we have an obligation, perhaps a very limited one, to prevent an undeserved natural *deficit* or *disability* from resulting in a *deprivation*, a lack of some important constituent of a minimally decent or adequate human life.[17]

No doubt many will find the notion of deprivation, dependent as it is on the notion of a minimally decent or adequate life, to be unsatisfyingly vague. It is vague. However, when properly contextualized by reference

[15] For a discussion of the difficulty of drawing a coherent line between justice and charity, see Allen Buchanan, "Justice and Charity," *Ethics*, vol. 97, no. 3 (1987), pp. 558–75.

[16] Those familiar with the extremely illuminating work of Amartya Sen will no doubt find familiar this notion of deprivation and of justice requiring a distribution of resources to prevent deprivation rather than to achieve equality as such. (See, e.g., Amartya Sen, "Equality of What?" in Sterling M. McMurrin, ed., *Liberty, Equality, and Law* [Salt Lake City: University of Utah Press, 1987]; and Sen, *Commodities and Capabilities* [New York: Elsevier Science Publishing Co., 1985].) To my way of thinking, though in some respects Sen's approach seems to be in the resourcist camp, he succeeds in avoiding the mistakes I attribute in this essay to other resourcists such as Cohen, Dworkin, and Rawls.

[17] It is important to add the qualifier "other things being equal" here. For one thing, a serious effort to bring the most severely disabled up to the level of a "decent" life would be prohibitively expensive—would entail an unacceptable restriction on the opportunities of the better off. What I have in mind is what is sometimes referred to in bioethics circles as the "medical black holes problem." Given an indefinitely expanding technology, efforts to bring every severely disabled person up even to some rather modest level of functioning could absorb vast amounts of resources. My own view is that justice does not require this sort of sacrifice.

to the specific forms of human flourishing available in particular societies and by reference to the constraints which the resource base available in particular societies places on reasonable redistribution, it is not vacuous. (For most people in modern, industrial societies, for example, having a chance at a decent life requires the ability to read and access to certain basic health care, at least during childhood, among other things.) In spite of its inherent vagueness, I find the idea that justice requires a response to deprivation much more reasonable than the sweeping thesis that justice requires redressing or compensating for any lowering of life-prospects due to having a lesser share of natural assets.[18]

VII. The Remedial, Derivative Nature of Equal Opportunity

Our analysis of the principles that account for the development of the concept of equal opportunity yields an important conclusion about the role of equal opportunity in a theory of justice. That analysis revealed three distinct conceptions of equal opportunity: (1) Careers Open to Talents, requiring the removal of legal barriers to advancement by qualified individuals; (2) Comprehensive Nondiscrimination, requiring removal

[18] A resource egalitarian might make one last attempt to persuade me otherwise. Suppose that we know that unless something is done to prevent it, a child will be born, not with a deprivation, but with somewhat lower life-prospects due to some genetic condition. Are we not morally obliged to prevent this from occurring, if we could do so without excessive costs and without violating any important moral principles (such as that requiring respect for the rights of prospective parents, etc.)?

It is hard to know what to make of this example, even if one shares the intuition of those who advance it—namely, that we should intervene. Part of the problem, of course, is the vague inclusiveness of the qualifying phrase "without excessive costs and without violating any important moral principles." But even apart from that, there is another problem with such an appeal to intuitions. Even if one admits that it would be *a good thing*, morally speaking, to undertake the intervention (subject to the qualifying phrase), and even if one goes further to say that we *ought* to undertake it, we are still quite some distance from the desired conclusion, namely, that equal opportunity or some other principle of *justice* (such as equal respect for persons as such) requires us to do so.

For one thing, it does not seem to be the case that justice demands that we bring about everything that is morally good or everything that would be an improvement for some person. For another, we usually think of what is a matter of justice as being that which a person is entitled to *as a matter of right*, such that if she is not provided with it, she may correctly charge that she has been *wronged*. However, from the fact (if it be such) that in some sense it would be better (even better from a moral point of view) if persons did not have lower life-prospects due to undeserved natural differences, or that it would be a good thing if we prevented this, it does not follow that by not intervening to bring about this better state of affairs we wrong those with lower life-prospects. Moreover, I think it is fair to say that many of us simply do not share the resource egalitarian's intuition that if we (society? the state?) do not intervene in such a case, then the individual in question can reasonably say that she has been wronged, that some right of hers has been violated.

If, on the other hand, the point of this appeal to intuitions is not that to not intervene is to wrong the person, but rather simply that it would be a good thing to intervene and that, other things being equal, we ought to bring about good things, then one can hardly quarrel with it. Considerably more will be needed, however, to show that not intervening is incompatible with a proper recognition of the fundamental equality of persons, with equal opportunity, or with some other principle of justice.

not only of legal barriers but also of informal barriers of racial, gender, or ethnic prejudice; and (3) Strong Affirmative Action, requiring special preference for candidates who may be less than optimally qualified, in order to combat the lingering effects of discrimination. In each of these three conceptions, equal opportunity was seen to be a derivative principle, to be explained entirely by reference to other, more basic principles of justice. In (1) and (2), the relevant principles were the meritocratic principle and a principle of nondiscrimination against persons (on the basis of race, ethnicity, gender, etc.); in (3), the relevant principle was a principle of rectificatory justice requiring special efforts to negate the harmful effects of past injustices.

When one characterizes the moral significance of any instances falling under any of these three conceptions, the phrase "equal opportunity" is eliminable without loss. In that respect, at least with regard to (1), (2), and (3), equal opportunity is at best a *derivative* (or secondary) principle of justice, and one that is remedial in the sense that it requires a response to violations of other, more basic principles. Contrary to what Rawls and others have assumed, it is not even the *sort* of principle that could be basic in a theory of justice.

It will be recalled, however, that we also considered another area in which some might think a conception of equal opportunity applies: cases where we believe that special allocations of resources are required to prevent persons from suffering deprivation, due to misfortune in the distribution of natural endowments. Here two questions arise. First, is this properly described as a matter of equal opportunity? Second, if it is, is the notion of equal opportunity here playing a merely secondary (and in principle eliminable) role, as in the other cases of equal opportunity analyzed above, or is this a case where equal opportunity is a basic principle of justice?

I see nothing that forces us to describe the fourth situation (providing special resources to the naturally severely disabled) as a case of equal opportunity. It may be more natural, in fact, to describe it as a case in which what best explains our intuition that some special allocation of resources is called for is a principle of basic welfare (or basic welfare rights), rather than of opportunity. Such a principle would state, roughly, that every person has a right to the resources necessary for a minimally decent life.[19]

[19] If it is to be plausible, such a principle would have to be qualified by a clause to the effect that supplying a person with resources necessary for a minimally decent life does not involve excessive costs. Given the apparently indefinitely expanding horizon of medical technology, enormous amounts of resources could be expended in making marginal improvements in the lot of the most severely disabled individuals (the "medical black holes problem" I mentioned in note 17). A proper recognition that persons have only one life to lead and that they should be allowed to give special preference to their own projects and to those with whom they are closely affiliated surely places significant limits on what each of us owes to others as a matter of general obligations of justice.

Suppose, however, that one were to insist that the principle underlying our intuitions in such cases is in fact a principle requiring that every person have *an opportunity* to lead a minimally decent life, and that it is this principle that underlies the notion that we should prevent deprivations, where the latter are thought of as harmful conditions or deficits so serious as to pose obstacles to achieving a minimally decent life. This way of understanding our response to the case of naturally severely disabled persons apparently would see the response as resting on a principle of equal opportunity, though only in this attenuated sense: it would require a kind of minimal egalitarianism with regard to opportunity – everyone is to have an opportunity for a minimally decent life. This principle of equal opportunity, however, is presumably to be understood as a basic, rather than a derivative, principle of justice. It does not appear to rest upon other principles of justice such as the meritocratic principle, a principle of nondiscrimination, or a principle of rectification.

If this is the correct characterization of the principle that underlies our intuitions in the fourth type of case, then the conception of equal opportunity at work in it is quite distinct from, and indeed sharply discontinuous with, that implicit in the three stages of the development of the concept of equal opportunity sketched above. As I have noted, there is a clear continuity among the first three stages – the same principles, of merit and nondiscrimination, are appealed to in all three (although in the case of Strong Affirmative Action the principle of rectification is said to override the current application of the meritocratic principle whose past violation calls for rectification).

In contrast, none of these principles is appealed to in the case of special allocations for the naturally severely disabled, if we describe this as an instance falling under the principle that every person is to have an opportunity to lead a minimally decent life. This suggests either that it is misleading to characterize all four types of cases as matters of equal opportunity or that there are two quite distinct concepts of equal opportunity at work: a remedial, secondary, and eliminable conception in the first three, and a quite distinct, primary conception in the fourth. If this is correct, then there is no univocal answer to the question of whether equal opportunity is a basic or a derivative principle in the theory of justice. Some principles of equal opportunity (the first three) are derivative, and at least one other (the fourth) is basic.

VIII. The Role of Equal Opportunity in the Uses of Genetic Intervention: Preventing Genetically Based Deprivations and Placing Constraints on Genetic Enhancement

So far I have argued that the Resource Redress and Resource Equality Principles are not supported either by the Formal Principle of Justice or by a principle of equal opportunity (whether the latter is understood as a remedial principle of justice based on the meritocratic principle and a negative principle of respect for persons as nondiscrimination, or as a primary principle of justice requiring that we prevent undeserved deprivations). Thus, unless some *other* justification for it succeeds, the Resource Equality Principle provides no reason to undertake redistributions of genes in the name of equal opportunity, and the Resource Redress Principle, understood as a second-best principle, lacks support because the first-best principle upon which it is supposed to rest is itself groundless. I now wish to suggest, however, that there are two *other* roles that a plausible conception of equal opportunity can play in determining the moral uses of genetic intervention technology.

The first role is that of placing *constraints* on the uses of genetic enhancement technology, under conditions in which differences in access to this technology would be likely to exacerbate other, independently characterizable injustices. The second is that of imposing obligations to use genetic interventions to prevent genetic harms that would constitute deprivations. I will now consider each of these roles briefly, in turn.

Suppose that it becomes possible to identify, synthesize, and implant in embryos complexes of genes that will greatly increase the probability of an individual possessing certain desirable characteristics to a significantly higher degree than the average person in a given population. Such characteristics might include superior memory, ability to concentrate for long periods of time, and general resistance to disease and to psychiatric maladies (such as depression). If access to this technology were solely dependent on ability to pay, then its use might exacerbate and perpetuate disadvantages that certain groups (such as African Americans, Native Americans, and Hispanics) suffer as the result of past injustices of discrimination.

For example, employers would be more likely to select a person for a desirable career track involving much investment in the individual's training over a long period of time if that candidate could produce a "genetic enhancement certificate" credibly stating that his resistance to disease and to psychiatric maladies was much greater than average, his memory was superior, etc. If members of certain disadvantaged minorities were unable to afford this extra credential, they would face yet one more barrier to attaining desirable positions. (Of course, if health insurance continues to be employer-based, as it currently is in the U.S., and if employers are required to offer health insurance to their employees, then employers would have an added incentive to hire those who had benefited from genetic enhancement of their resistance to disease and to shun those who lacked it.)

It is important to see that this effect might occur even if the actual results of genetic enhancement were in fact markedly less significant than was generally supposed. One can easily imagine situations in which job candidates who could produce "genetic enhancement certificates" would have an advantage in competition for positions, even if the benefits of the genetic interventions in question were less pronounced than advertisements portrayed them as being. (Similarly, today, possession of a college diploma serves as a *necessary* qualification for entering competition for most white-collar jobs, even though in many cases the educational experience the individual received in college is of dubious relevance to the actual skills needed for the job, and even though individuals who are excluded because they lack a diploma are in many cases better qualified.)

What are the implications of such a scenario for equality of opportunity? The answer, of course, depends upon which conception of equal opportunity is assumed. *If* equal opportunity requires Strong Affirmative Action, then this principle of justice might, under certain circumstances, require public subsidies for genetic enhancement for groups that have suffered from past discrimination in order to avoid the perpetuation of their injustice-based disadvantages. Or, if providing access to enhancement technology for *everyone* proved too expensive, equal opportunity might conceivably require that *no one* is to have access to certain extremely significant enhancements if this could be demonstrated to perpetuate the effects of past discrimination.

A second, and in some respects more radical role for equal opportunity (yet one still much more conservative than that envisioned by the Resource Equality Principle) would be to require genetic intervention to avoid safely and easily correctable genetically based deprivations that would not respond to treatment after birth. As I observed earlier, whether efforts to avoid deprivations properly fall under the heading of equal opportunity is far from obvious. Supposing that they do, however, there is obviously a strong prima facie case for intervening with gene surgery on the embryo (or perhaps even the gamete) to prevent blindness, if there is a case for allocating special resources to correct or compensate for blindness after the child is born. At least in cases in which postnatal remedies are not available, genetic intervention to prevent deprivations has a great deal of moral plausibility.

The most obvious countervailing consideration, of course, would be respect for the privacy and autonomy of the prospective parents. Complex and daunting issues arise here, not just concerning the permissibility of requiring genetic surgery against the will of one or both of the prospective parents, but also regarding the impact on individual and familial privacy and on confidentiality in the physician-patient relationship that would result from any serious effort to screen embryos or gametes for deprivation-producing defects. My purpose here, however, is not to settle—or even systematically to delineate—these problems. Instead, I only wish to note that there is a strong prima facie case for using genetic interventions to prevent deprivations, while at the same time distinguishing *this* justification for intervening in the "natural lottery" from appeals to the Resource Principles advocated by Rawls and other radical egalitarian resource theorists of justice.

Business Ethics, Philosophy, and Medical Ethics,
University of Wisconsin–Madison

[13]

Published by Blackwell Publishers Ltd. 1997, 108 Cowley Road, Oxford OX4 1JF, UK
and 350 Main Street, Malden, MA 02148, USA.
Ratio (new series) X 3 December 1997 0034–0006

EQUALITY OF OPPORTUNITY[1]

Janet Radcliffe Richards

1. The problem

Some people are against equality of opportunity. Peter Singer, for instance, says that it rewards the lucky and penalises the unlucky;[2] John Schaar,[3] who regards the whole idea as a cruel deception, claims that it entrenches the inequalities of the status quo, makes people feel their failure is their own, and cheats them into thinking they have a chance to achieve what they could never possibly achieve. Most people, however, seem to take it for granted that equality of opportunity is a Good Thing we should be aiming for, and that our only problem lies in achieving it. So institutions and governments declare their commitment to it, employ experts to tell them how to get it, and study statistics to find out how far they are falling short.

All this being the case, you might not unnaturally presume that everyone knew what *it* was, and (as a good many people seem to) that its persistent elusiveness, despite all the efforts put into its achievement, was yet another indication of ordinary selfishness and lack of political will. But philosophers, whose invaluable technique of armchair thought experimenting allows them to race ahead of the laborious empirical variety, do not have to look far before they see that whatever the truth about selfishness and politics, a deeper root of the elusiveness lies in the concept itself.

Consider, for instance, Christopher Jencks's Ms Higgins,[4] a school teacher committed to equality of opportunity. She is anxious to share her time and attention among her pupils accordingly, but quickly finds herself baffled. Should she make herself equally available to all, or give equal time to all, or give more to

[1] I am grateful to Andrew Mason, Derek Parfit and Cass Sunstein for very helpful comments on drafts of this paper.

[2] Peter Singer, *Practical Ethics*, 2nd edn. (Cambridge: Cambridge University Press, 1993), p. 39.

[3] John H. Schaar, 'Equality of opportunity, and beyond', in J.R. Pennock and J.W. Chapman (eds) *Nomos IX: Equality* (New York: Atherton Books, 1967).

[4] Christopher Jencks, 'What must be equal for opportunity to be equal?', in N. Bowie (ed) *Equal Opportunity* (Boulder, CO: Westview Press, 1988).

children from deprived backgrounds, or try to make all the children equally proficient in everything by the end of the year, or even think beyond the classroom to the effects her children will eventually have on the community or the world at large? Dozens of incompatible policies seem plausible candidates for equality of opportunity, and Jencks thinks this is why the idea is so popular. People mean different things by the term, but because they think of it as implying whatever version they are in favour of, they are under the illusion that there is a single ideal about which they agree.

No doubt this catches part of the truth, but a more fundamental and familiar difficulty is one implied in an earlier article by Bernard Williams,[5] and in many others since. Williams considers difficulties rather like the one faced by Ms Higgins, but brings out a different aspect of the problem, which (in an extremely free amalgamation of his examples) goes something like this. Imagine a society with a traditionally privileged warrior caste, which now decides to commit itself to equality of opportunity. It will start by removing caste restrictions and selecting the warrior elite by merit; but if the previously excluded lower castes are poor and undernourished, they will still not be able to reach the necessary standards, and so still cannot be said to have equality of opportunity. On the other hand, even if there were arrangements to feed everyone equally well, that would still leave the naturally puny with unequal chances of success. True equality of opportunity must therefore involve special remedial treatments, or perhaps even, eventually, genetic engineering. But that seems to make equality of opportunity identical with equality of outcome.

What emerges from this kind of argument, in other words, is not so much the existence of plausible competing conceptions of equal opportunity, as a difficulty in finding *any* conception adequate to our intuitions. At any stage it looks as though genuine equality of opportunity requires more, until, eventually, it seems that nothing will do short of the equality of outcome with which it is usually contrasted. To set out in pursuit of equality of opportunity is rather like landing on a snakes-and-ladders snake, which you find yourself slithering down, with no holds in sight, until you land on a square disconcertingly like one you left some time before. And this is a difficulty that any institution committed to

[5] Bernard Williams, 'The idea of equality', in Williams, B., *Problems of the Self* (Cambridge: Cambridge University Press, 1973).

equality of opportunity will instantly recognize, even though it will usually misdiagnose the problem as a failure to reach the target, rather than of there being no clear target at all.

If equality of opportunity is to be a political ideal, it must be brought out of its well-intentioned blur. There is no point in setting out in pursuit of something you would not recognize if you caught it, and it is cheating to complain that other people are falling short of it if you cannot tell them what would count as success. Before any possibility arises of implementing equality of opportunity, we need a description of what would count as having achieved it. That is the first problem.

1.1. How to find a description

Still, what *kind* of problem is it? How can Ms Higgins, for instance, tell when she has succeeded in finding a suitable description? There is no point in her pondering the meanings of words – asking herself whether this can really count as equality, or that be plausibly described as opportunity – since she already knows that language cannot solve the problem; and anyway, even if it could it would be beside the point. Her starting commitment to equality of opportunity is the expression of a vague *moral* concern, and that is how she should interpret her problem. If this concern is to be got into a state where it is of any practical use, she needs to review her candidate policies and make a *moral* judgment between them. Her problem is ultimately one of ethics, not of language; and when she has reached her own conclusion, her arguments with people who are recommending rival conceptions, or opposing equality of opportunity altogether, will also be about ethics.

Anyone who has reached that stage, however, and is poised to begin the process of moral assessment of candidate equal opportunity standards, is going to have to wait some time for the rest of us to catch up. What the snakes-and-ladders snake suggests is that most of us have failed not only to make a moral judgment between rival conceptions of equality of opportunity, but even to draw the candidates out of the blur and array them for inspection. How, then, can we set about doing that?

Consider again what the problem is. What is needed is a description of what would count as *success* in providing equality of opportunity: a clear, positive account, that would show what was to be aimed at, and provide a measure of how far there was still to go. When you look at the snake with this in mind, it becomes clear that although it *seems* to lure you along by offering a distant prospect of equal opportunity which you never quite reach – at least, until you arrive at the paradoxical equality of outcome – no positive account ever seems to be given. All the work is done by rhetorical negatives: 'How can you say there is equality of opportunity, when . . . ?', or 'Nobody could call this equality of opportunity'. Such negatives need to be replaced with positives.

Rather than asking whether each stage on the descent looks like equality of opportunity, therefore, (which it will not, as long as we have conflicting ideals that generate conflicting intuitions,) we can try to work out what *positive standards* of equal opportunity would, if accepted, *justify* the move from each stage to the next. This should at least produce some candidate criteria whose relationship and moral merits can be investigated.

1.2. A typical descent

For this purpose an illustration is needed; and here it should be said that although I have referred to a snakes-and-ladders snake, you do not need to dabble for long in its natural history before you find you are dealing with a whole nest of snakes, most of which turn out to be reverse hydras, sprouting new tails as quickly as you find ways to avoid the original one. There are endless philosophical and real-life variations on the Williams theme, which between them slither over every aspect of the broad problem of equality, taking in their path equal consideration of interests, equal outcomes, equal rights, group and individual equality, and equality of treatment. One case study will obviously not cover all the ground. Still, one will have to do; and indeed it should be enough to illustrate the main problems.

Consider, then, another free adaptation of one of Williams's illustrations, this time the case of a sought-after school. It is, let us say, old and well-endowed, renowned for its high teaching standards and traditional culture, and with a headmaster strongly committed to both. Tradition has always limited its intake to the Sons of Gentlemen; but now the serpent, in the guise of a well-meaning critic, persuades the headmaster that he should be adopting equal opportunity policies, and in doing so lures him to the start of the downward slide.[6] This we can think of in four stages, thus:

[6] I know serpents do not lure you into sliding down themselves, but this metaphor seems (appropriately) to have a life of its own. A mixed metaphor will, anyway, turn out to be well matched to the confusions it is supposed to represent.

1. Obviously the school cannot be said to provide equality of opportunity as long as it rules out girls (upper class girls are culturally perfect), and even working class children who have managed to acquire the right manners. So the headmaster drops those barriers, and agrees to let in any reasonably intelligent applicant who will keep up the cultural tone of the school.
2. But, the critic argues, this still cannot count as equality of opportunity, because educational opportunities should be available to everybody, not just to children with a particular cultural background. So the headmaster agrees to admit pupils strictly on the basis of academic ability.
3. The critic is still not satisfied. By the time candidates apply they have already had unequal backgrounds, and many bright but unprivileged children are far behind others of lesser natural ability.[7] So the headmaster, although now rather puzzled, wonders about offering remedial classes, and starts to wrestle with counterfactuals about what the children would have been like if they had had each other's backgrounds.
4. He will not be left with this particular puzzle for long, however, since the critic will already have moved on to matters still more perplexing. Even equality of background could not give genuine equality of opportunity, since the children's different genetic endowments would still leave them with unequal chances of success. This seems to imply that genuine equality of opportunity requires the admission of everybody – the equality of outcome to which the headmaster always thought equality of opportunity was opposed – or, since this is impossible, either closing down the school or admitting pupils by lot. Neither of these is anything like what he had in mind when he started off in pursuit of equal opportunities, but he can now see no escape.

The problem now is to see what happens to stories like this if, instead of asking at each stage whether equality of opportunity has been achieved, and feeling obliged to press on with the efforts because it apparently has not, we ask what *positive* standard

[7] And they are now so far behind that they will never catch up in an equal environment. If they had had the potential to catch up in spite of being behind at the time of selection, they would have been admitted by a sufficiently sophisticated selection process at the second stage.

of equal opportunity might justify the move from each stage to the next.[8]

I shall argue that such an analysis shows the apparently seamless descent to be nothing of the kind. What really happens is that at each stage on the slide the equal opportunity standard changes; and the reason why this is not obvious, and leads to paradox, is that these different standards, unlike the rival candidates Ms Higgins contemplates as she tries to make her moral decision, are not competitors on the same ground. Ms Higgins's problems run, as it were, at right angles to the main line of descent: her decision has to be made *within* one of its steps. The ideas that propel the main descent, in contrast, represent irreducibly different *kinds* of standard, *compatible* with each other, which introduce equality in different ways and imply different interpretations of opportunity, and whose confusion turns all practical discussions of equal opportunity into convoluted fallacies of equivocation. The whole appearance of an ideal of equal opportunity is produced by a series of intellectual slips into which the serpent has, appropriately, beguiled us.

2. *First slip: impartiality to positive rights*

2.1. *Justifying the first step*

Consider first the headmaster's opening move, of doing away with sex and class barriers, and applying a single selection criterion to everybody. This certainly looks like a step in the direction of equal opportunity. If so, what *standard* of equal opportunity was he falling short of in the original situation, but is now meeting, or perhaps moving towards, in this change of policy?

The most obvious standard to which he is actually conforming, after his change, is that of considering only (above a basic level of educability) the contribution of prospective pupils to the school's traditional culture. And indeed, the school probably now does better from that point of view than it did before, since there is now a larger range of children to choose from, and the cultural threshold can be set higher. But keeping the school's culture intact does not sound like an equal opportunity principle. In itself, it implies

[8] 'Justify' contains a tiresome ambiguity which causes endless trouble in this area and many others. Here I mean only what might be called *formal* justification: the action's actually being supported by the standard in question, whatever that is. Substantive justification would require the acceptability of the standard as well.

nothing about either opportunities or equality. Why, then, does the change of policy look like an equal opportunity move?

Consider first whether there is anything resembling a standard of equal opportunity that the headmaster has actually *met* (as opposed to progressed towards) in this first move. Opportunities are now equal in the sense that sexes and classes are allowed to compete on the same terms, but what is the *standard* involved? Is it that everyone should be allowed to try? But the culturally inappropriate are still as decisively ruled out from the start as girls and the lower classes used to be. Is it that the previous exclusion was on irrelevant grounds? That seems closer, but the exclusion was relevant to the school's former policy of providing for the sons of gentlemen.

Consider, then, this possibility (which I should say in advance needs at least two more papers to spell out and defend in full). The original policy was to admit only sons of gentlemen, which gave that group an advantage over everyone else. That in itself is not significant: virtually any policy that involves selection will give somebody an advantage over somebody. But many such advantages can be justified as incidental effects of other, general, aims: if you are looking for a winning football team, your choosing the best players available may be an advantage to the individuals chosen, but that advantage is only incidental to your general aim of winning. But what general reason could the headmaster have given for his policy of preferring the sons of gentlemen? The only possibility in the story as we have it would have been the importance of culture.[9] That, however, does not work as a justification. When he eventually does let in the previous outcasts he actually raises his cultural standards. So unless he can show some other general principle to explain it, his policy of privileging of the sons of gentlemen over working class boys and everyone's daughters is in effect (though not necessarily recognized as) *an end in itself.* One group is *simply* preferred to the others.

Now that must seem, to most people's intuitions, a clear case of *inequality* of opportunity, and it suggests by contrast one kind of equal opportunity principle being met in the headmaster's abandoning of his original sons-of-gentlemen policy. Its detailed spelling out would be complicated, but it is something on the lines of equal consideration of interests, or positional indifference, or

[9] We could tell a different story in which the headmaster did have an explanation to offer, and in that case a different analysis would have to be given. That is an illustration of the hydra point.

ground-level impartiality.[10] It amounts to the claim that any action or policy must ultimately be justifiable in terms of principles of perfect generality – which is of course a version of the meta-ethical principle of universalizability – without any *not further-explainable* preference of some groups over others.

If this is an equal opportunity principle (about which more later) it is a very limited one, since it has no positive, normative content of its own. It works only as a constraint, specifying that you must *not* have such arbitrary preferences – where 'arbitrary' means, roughly, not justifiable in terms of general principles held by whoever has those preferences – but not what positive general principles you should be following. What follows in practice from the removal of arbitrary preferences is determined by the default of whatever general principles were previously in the background. This is why the the commitment to a certain cultural atmosphere – the only principle to which the headmaster seems to be conforming after his first equal opportunity move – does not look like an equal opportunity principle. It is not; it is just what happens to remain when the arbitrary privilege has been abandoned. The relevant principle here is a *negative* one that demands the removal of an arbitrary obstruction – a constraint, rather than a positive specification of what should be done – which therefore drops out of sight once it has been acted on.

This may not be the only candidate for an equal opportunity standard met by the headmaster's first move, though I think pinning down any other would prove more difficult than might appear. Perhaps, however, the first step should be seen not as *achieving* any kind of equality of opportunity, but as a move in its direction. This possibility will be reviewed later, after the justification of the second step has been considered.

2.2. *Justifying the second step*

Since the critic is not satisfied with the first move, of abandoning the sons-of-gentlemen policy, the headmaster now drops the culture requirement, and changes to a policy of choosing the children who are academically best. This again looks like an equal opportunity move, but if so, what is the standard of equal opportunity to which it conforms?

[10] See, e.g. Singer, *op cit.*, pp 12–13, 21 ff.; Thomas Nagel, *Equality and Partiality* (Oxford: Oxford University Press, 1991), Ch 2, *passim.*

The waters are deep here, but the most obvious idea of an equal opportunity kind seems to be something in the area of the liberal principle of careers open to talents: the idea that people should have the right to move and rise in society according to their abilities and inclinations, and that other people have corresponding duties to refrain from placing various obstacles in their paths. It is a plausible aspect of this idea that since the purpose of the school is to give a good education, access to it should be open to competition on purely educational grounds. It is also clear that ideas of liberal rights of these kinds are a strong component of familiar ideas of equality of opportunity. Some standard of this kind, therefore, seems to be underpinning the second move.

It is important to stress 'of this kind'. Although the idea of careers open to talents is distinguishable from other kinds of political ideal, it is not in itself a single principle, but a marker for a range. Within that range there are endless incompatible variations, depending on which opportunities you want to equalize over which people. And this, of course, is the Ms Higgins problem. Should she regard 'opportunity' as time actually spent with each child, or just her availability to them all, or what? Should the equalizing be confined to the children in her class, or extended further, and if so how far? She needs to answer these questions *before* she can adapt her conduct to the moral requirements of equal opportunity, because the answer she gives *is* the standard at this stage. She may incline towards the idea of careers open to talents in general, but that means only that her intuitions lie in that range, *not* that there is a higher-level equal opportunity principle that will tell her which part of the range to go for. The choice within the range must be made and justified in more fundamental terms, just as the choice of the range itself must be.

I shall say more about the significance of this shortly. For now all that needs to be noted is that this second equal opportunity move seems to be underpinned by some principle in the general range of careers open to talents, and that this is both recognizable as equal opportunity territory, and widely accepted as right.

No matter how recognizable or acceptable some version of this may be as an equal opportunity principle, however, it is *totally different* from the impartiality principle implied by the headmaster's first step. The first is a negative principle, demanding only the removal of arbitrary preference and the achievement of ground-level impartiality against any generalizable background. The second kind is positive and normative, specifying what admissions policy the headmaster should have, and saying something to the effect that he must not take into consideration anything except academic merit. The two are quite independent, and accepting the principle one that would support the first step provides not the smallest justification for moving on to the second. If the first two steps are both equal opportunity moves, this account makes them so in irreducibly different senses.

2.3. Justifying both steps at once

Perhaps, however, we can avoid this problem, and keep a single ideal of equal opportunity, by looking at the matter from the other direction. Perhaps we should see *both* moves as justified by the *second* principle. That would be in keeping with the appearance given by the snake, which works by implying that there is an equal opportunity principle pulling from the tail end, and making all the changes seem like progression towards it. If so, it could still be maintained that there was a progression from the first to the second stage of the slide.

But if this seems plausible, consider the details. How is the second principle to be specified? In its simplest form it demands selection by academic ability, rather than by culture. But the headmaster's first step makes no move at all in that direction. Getting rid of sons-of-gentlemen requirements has in itself nothing whatever to do with giving places to the children who are academically best. If we are to say the first move is a step in the direction of the second principle, that principle must somehow be respecified to make it so.

The most obvious way to do this is simply to add the first, and say there should be ground-level impartiality *as well as* a change to selection by academic ability. But that is not very helpful: all it means is that you now have a composite equal opportunity specification with two separate elements, which is like saying your ideal house must be both beautiful and energy efficient. You may well want both these things, but you cannot turn beautifying into a step towards energy efficiency by saying they are both aspects of your ideal.

But, it may be said, it is *obvious* that you want ground-level impartiality; that is in effect *part* of the academic ability requirement, not something separate. You would not be selecting by academic ability if you kept to the sons of gentlemen, and since that is so, you can see the first move as on the way to fulfilling the second principle. The first move deals with one aspect of the principle, the second with another.

However, there are two problems with that idea. First, it seems simply false. You *could* change from cultural to academic selection while limiting the candidate pool to the sons of gentlemen. If the headmaster could override his original culture requirement by a sons-of-gentlemen constraint, he can override the new academic requirement in the same way. And this is not just a theoretical possibility; it is a perfectly familiar phenomenon. The sweeping changes of political principle brought about by the French and American Revolutions managed to go ahead without many people's noticing that the arbitrary disadvantaging of women and blacks was carrying on regardless. So even if the two elements could somehow be regarded as logically one, they would still need separate specification in practice, and moving towards one would not entail moving towards the other.

The second problem with trying to see ground-level impartiality as an implied part of the principle of careers open to talents, rather than a separate element, is that if ground-level impartiality can be taken for granted in this case, that must be because it can be regarded as a requirement of *any* general principle. Certainly no reason has been given for taking it to be specific to this case but not others. This means that the first move is no more in the direction of equal opportunity (specified as selection by academic ability) than it is in the direction of all possible general policies *incompatible* with that one. You cannot count the impartiality move as an equal opportunity move in virtue of its being a move equally towards equal opportunity *and* towards everything incompatible with it.

So it looks as though there is no way round this problem. If the first step on its own is to be seen as an equal opportunity move at all, it cannot be explained in terms of meeting the requirements of the principle underpinning the second step. To move from one to the next in the name of equality of opportunity is not progression in the direction of a single ideal, but a slip arising from the conflation of radically different ideas.

2.4. *Differences and dangers*

To say that the first two moves are underpinned by different principles, and cannot be seen as progress towards a single ideal of equal opportunity, is not to deny that there could be *moral* progress from the first to the second, or that a moral ideal of equal opportunity might be composite in form. Although the standards implied by the two moves seem to be radically different, they are not like the different standards contemplated by Ms Higgins. Her options were competitors on the same ground, such that accepting the next standard meant abandoning the first, but the different standards so far identified in the headmaster's slide are so different in kind as to be *compatible*. You can accept both at once (or rather, the first and one particular version of the second); and if you do, you will obviously think it a moral improvement to comply with the second standard as well as the first.

On the other hand, to decide whether each move does constitute moral improvement, you must assess individually the standards to which they conform. It is not a mistake to want them both, but it is a mistake to *slip* from the acceptance of one to some version of (let alone a particular version of) the other, by way of some fuzzy generic idea of equal opportunity. Paradoxically, it is probably because the two kinds of standard are so different – operating at different levels and in different ways – that they are compatible and easy to conflate. Their moral assessment, however, requires the recognition not only of their distinctness, but of the nature of the differences between them, which is systematic, striking, and potentially sinister.

The impartiality principle is a ground-level, metaethical principle, with no positive normative content of its own. It acts only as a constraint: practical implications appear only in particular contexts, and are derived from other values and principles in the background. It is simple and clear in nature, and also quite general in scope, specifying that there shall be *no* non-impartial treatment, so that everything recommended or done must be justifiable in general terms. It is also, when properly spelt out, virtually beyond controversy, as appears in the lengths to which people go to claim that they are conforming to it.[11]

Liberal rights principles, on the other hand, of the kind exemplified by the idea that the headmaster should make academic ability the only determinant of admission, are positively normative, providing criteria for the direct assessment of policies and conduct. Since the range of incompatible normative principles, even within the range plausibly counted as involving equality of opportunity, is in principle endless, the various candidates need detailed spelling out (the precise kind of opportunity, and the range over which it is to be equalized) before they can be applied. From this it follows that there must be endless possible

[11] See, e.g. Williams, *op cit.*, p. 233.

controversy about particular recommended versions, as well as about whether to go for liberal ideas of this kind at all. Standards of equal opportunity at this level provide for the assessment of actions and policies, but must themselves be defended in more fundamental terms.

So the equal opportunity standards implied by the headmaster's first two moves are not only different, but different to the extent that (as Ruskin said of men and women) each has what the other has not; and this striking reciprocity provides particularly dangerous potential for confusion if the two are conflated. If the first gives the impression that the idea of equality of opportunity is clear, specific, unlimited in scope and morally certain, and if no distinction is seen between the two kinds of ideal, these impressions may carry over to the territory of the second, where they are totally out of place. The result of the conflation may be – and I suspect typically is – to produce the idea that that there is a true equal opportunity standard *to which normative standards at the second level should conform*, rather than that the normative standards at that level *are themselves* offered as equal opportunity standards to which *actions and policies* should conform, and themselves needing justification in *other* terms. The effect is to give the problem of finding normative standards at this second level a spurious appearance of simplicity and moral clarity.

That claim and its implications, however, would take a great deal of spelling out. All that is possible here is to mark the spot.

3. Second slip: giving to having

Consider now the move from the second stage to the third, when the headmaster is reminded of the candidates' unequal family backgrounds, and starts trying to formulate policies to meet that concern. What standard of equal opportunity is implied in this complaint? Here the answer seems obvious, in outline at least. There is equality of opportunity when people have equally rich formative backgrounds. Of course this idea, like the previous one, is an umbrella covering endless incompatible variations, with the problem of which variant to go for once again running at right angles to the descent of the snake. But again what is important here is to identify the general category, rather than to engage with details.

Since inequality of formative background does sound like inequality of opportunity, it is easy once again to assume that this move is yet another towards some ideal of equality of opportunity, and that in trying to address it the headmaster is still pressing on in a single direction. But it takes only a moment, once you are thinking in such terms, to see that there is yet again a quite independent standard at work. The requirement to choose the children who are academically best, irrespective of culture, has in itself nothing to do with levels of formative environment. Either concern could be acted on while nothing was done about the other; either standard could be accepted while the other was rejected.

It should be added that that is not necessarily the end of the matter in this case. Given that the second kind of equal opportunity standard, whichever variation is recommended, needs justification in more fundamental terms, perhaps some single underlying principle could justify the proposed equal opportunity standards at both levels and link them in that way. If, for instance, it is taken for granted at the second level that the *reason* for the entitlement is that the cleverest children *deserve* the places in the school, it may well seem that an aspect of that same entitlement is their having equally good starting positions. So, depending on what justification you gave for whatever conclusion you reached at the second level, you might find an underlying connection between that and the conclusion you reached at the third.[12]

But even if any any such connection could be established, there is yet another distinction separating the second and third stages of the descent, which is of even more significance. Once again, they are not only different in substance, but also radically different in kind.

Up to this point, the standards of equal opportunity implied by the critic have been ones with direct implications for what the headmaster should *do*. He must not practise exclusions that have no general justification, and he must admit pupils purely on the basis of academic ability. But here there is a major shift, because at this third stage the proposed standard of equal opportunity has, as it stands, *no direct implications at all* for what anyone ought

[12] A more complicated point is that (a) such an underlying principle would probably not itself be a principle of equal opportunity (the desert idea is not), and (b) even if it were, it would be so in yet another quite different sense from the equal opportunity of each of the two standards it was supposed to be justifying. Hence, it would be *coincidence* that all three seemed to be linked by ideas of equality of opportunity: the connection between them would not be that the equal opportunity standards at the second and third levels were such in virtue of being manifestations of an underlying principle of equal opportunity.

to be doing. It is a standard for the assessment not of what people do, but of *states of affairs.* The critic's implied complaint here is about the unequal distribution of something called opportunity through the population as a whole; and this is quite different, because even if you think there should be such equality, and even if you have decided whch version of the ideal to accept, that does not – even for the simplest of act consequentialists – translate directly into any conclusion about what anyone should do. It is strongly relevant to that question, of course, but that is all.

The distinction between the idea of opportunity as something people *give* equally (as with equal opportunity employers), and as a good that people should *have* equal amounts of, is well signalled in ordinary language, but apparently not much noticed. Often it seems to be vaguely assumed that if an employer provides equality of opportunity there must be something of which people end up with equal amounts, and, conversely, that if they have unequal amounts of something called opportunity that must be because someone is failing to act according to equal opportunity principles. But unless you make such claims true by definition, they are obviously false. If 'opportunity' means something like early education and resources, it is clear that no possible headmaster or employer could equalize over a whole population either it or any equivalent.

This is why, in the story as we have it, the headmaster is puzzled about how to proceed at this point. Even if he accepts that there should be equality of this kind of opportunity, and even if he is willing to do his bit towards it, it is not in the least obvious what his bit should be. But if he does not notice the radical change of equal opportunity standard between the second and third moves, he will find himself struggling, with most other equal opportunity institutions, in the trap of presuming that as long as the people he is dealing with *have* unequal opportunity, he, *as* an equal opportunity appointer, must have certain obligations to them.

Once again, (some version of) this new standard of equal opportunity, in being independent of the previous two, may be independently defensible, and if so the headmaster may once again be making *moral* progress in taking some such view of the standards by which states of affairs should be assessed. Furthermore, unless he is an unmodified deontologist, conclusions reached at this states-of-affairs level are bound to form part of the justification he gives for whatever normative conclusion he reaches at the actions-and-policies level, so he may want to adjust his second-level conclusions in the light of his third-level deliberations. But if he is to make such a judgment, rather than slipping into so-called equal opportunity policies by mistake, he must recognize both the separateness of this new kind of equal opportunity ideal, *and* the further separateness of conclusions about his own conduct. Neither can be derived from any general commitment to equality of opportunity. Once again in the descent of the snake there is a slip to another principle altogether, rather than progress along an increasingly radical continuum.

4. Third slip: illusory oppositions

4.1. Means and prospects

In a way, that is as far as the headmaster needs to go. He can now tell the critic that even if he accepts this new standard of equal opportunity – that children should have equal education from birth – it has no direct implications for what he should do. If the critic wants to add to or replace the standard he was originally proposing at the second level – admission by academic ability – let him say what the revised standard is, so that it can be given due moral consideration. But the first problem is his. The critic is not entitled to say that some standard is being fallen short of without saying what that is.

Even though we have now left the direct matter of the headmaster's duties, however, we are not yet at the end of the snake; and as he may well be influenced in his judgment of what he should do by whether he can be lured into making the final descent, he should press on with the analysis. It is, furthermore, also from this point onwards that the downward slide is most familiar to philosophers. This is the stage at which it is claimed that the equalizing of formative environments cannot count as equality of opportunity either, because people's genetic differences will give them unequal opportunities for success. The idea is that genuine equality of opportunity should be seen as people's having equal *probabilities*, or prospects, of success, rather than equal external input: as what Douglas Rae calls *prospect-regarding* as opposed to *means-regarding* equality of opportunity.[13]

[13] Douglas Rae, *Equalities* (Cambridge, MA: Harvard University Press, 1981), Ch 4, p. 65ff.

Once again, at this point on the descent, there arises a lateral problem of which version to go for. There are indefinitely many possible variations on the prospect-regarding theme, just as there are with the means-regarding, depending on what the probabilities are for, when they are to be equalized, and over whom. Things are also complicated at this stage by the fact that the natural end of the snake seems sometimes to be regarded as prospect-regarding equality of opportunity, and sometimes as a shift from equality of opportunity to equality of outcome. Perhaps the most plausible idea is that you should go for equality of outcome as far as you can, and then try to get prospect-regarding equality of opportunity for all the things you cannot equalize; but there are many proposals in the area. Once again, however, the details are less important than the fact that there is here *yet another kind* of candidate standard for equality of opportunity, different from the previous three.

Here, however, there is a change in the pattern (the cunning of this serpent knows no bounds), because at this stage, for the first time in the descent, the new arrival is *not* compatible with its predecessor. This prospect-regarding range of equal opportunity standards is also, like the means-regarding one, offered for the assessment of states of affairs, and it is a *rival* range of standards. Since there are natural differences between people, identical formative environments will preclude equal chances of success. It is therefore impossible to equalize both these varieties of opportunity at once, so it seems that a choice must be made between them.

What I want to argue, however, is that setting up the problem in this way is the serpent's final trick, and that the presentation of the problem in this form is itself the slip at the bottom of the slide. In fact no such choice needs to be made, and the whole equal opportunity issue is distorted by the appearance that it does.

To see this, consider the following thought experiments.

4.2. Thought experiments

These conflicting views of equality of opportunity in states of affairs – means-regarding and prospect-regarding – both have their roots in perfectly familiar contexts. The idea of fairness underlying the idea of means-regarding equality of opportunity is that of a level playing field: in competitions whose aim is to find the best football team, or runner, or sheepdog, the environment must be controlled in such a way as to make success turn on what it is supposed to turn on, and inequality of the relevant controls constitutes unfairness. The idea of fairness underlying the prospect-regarding variant, in contrast, is that of a lottery, in which fairness requires each ticket's having a precisely equal chance of winning.

Obviously, however, both these familiar ideas of fairness are highly context dependent. Fairness of these kinds consists in controlling means, or in giving equal chances, *given* that competitions have been set up in one of these ways, and therefore can provide no justification for their being set up in either way in the first place. So, a fortiori, they can provide no justification for setting up the whole of life in either of these ways. If we are to claim that opportunity of either of these kinds should be equalized between people in general, we need first to justify arranging society as either kind of competition in the first place, and that calls for a separate argument.

This is why the philosophical debate about the problem, which is perhaps the most familiar in the area of equality of opportunity, rapidly becomes enmeshed in problems about free will, ultimate desert and the nature of the self. People who think that ultimately nobody is responsible for anything tend to take the prospect-regarding route, opposed by people who think that if we do this we can no longer make any sense of the self, and that we cannot avoid regarding ourselves as ultimately responsible for our own choices.

However, both these ways of looking at the matter presuppose that the point of equalizing opportunity in states of affairs is to provide a fair way of determining the distribution of some *further* good, and the argument between them is about which way is fair. This presupposition, then, is the first thing to investigate, because if it is inadequately founded – if the idea of fair competition for some further good is not what underlies our intuitions that opportunity, somehow interpreted, should be equally distributed – there may be no need to grapple with the problem of which of the two kinds of competition to go for.

Consider directly, then, the fuzzy intuitions we have about equality of opportunity in states of affairs, and compare them with the context-dependent attitudes to fairness that seem to underpin the ideas of means-regarding and prospect-regarding equalities of opportunity. This is not easy, because in real situations it is necessary to take into account the instrumental value of any equalizing of opportunity,[14] as well as its intrinsic value from

[14] Compare Derek Parfit, 'Equality and Priority', this volume, p. 206.

the point of view of fairness to the individuals involved. But since it is impossible to assess the merits of a complex whole without a view of the weight of its elements, the abstraction must be attempted. The following thought experiments are to be understood in that way.

1. Suppose first that you are inclined to favour some *means-regarding* version of equality of opportunity, and you feel that people should have equal amounts of education and early resources. What is the basis of this intuition? Is it that of a fairly controlled competition, ensuring that the prizes go to the people who deserve them? If so, you should regard *equality* of the relevant means as of overriding importance; and, as Derek Parfit implies,[15] you can test your attitude to the importance of equality as such by a consideration of your willingness to *level down.*

Consider, for instance, two different kinds of situation in which some people have shoes and others have not: first a cold winter, and second a race. In the first case, most people would be against levelling down. They would not be willing to take the shoes away from the people who had them just to achieve equality, since that would be worse for some people and better for none. They would probably find their intuitions about the importance of equality more plausibly expressed in trying to get shoes to the shoeless before the already shod got less important things. But if half the competitors had their shoes stolen before a race, matters would be different. It would probably seem better that everyone should run without shoes, to make sure the outcome was not influenced by matters irrelevant to the purpose of identifying the best runner. Since the shoes in this context are a control, rather than (primarily) a good in their own right, *equality* matters in a way that it does not when shoes are wanted for warmth. That is why it may seem appropriate, all things considered, to level down in this case but not the first.

If attitudes to levelling down can be used in this way as a test of whether inclinations towards means-regarding interpretations of equal opportunity really are rooted in ideas of fair competition, they can be used to work out what really underpins impulses towards equal education and resources for children. Suppose, then, there are some good teachers but many more mediocre ones, some towns with playing fields and others with none, some schools with fine musical facilities and others whose instruments

[15] Parfit, *op. cit.*, pp. 205, 211 and *passim.*

are adequate at best. Also assume you have done all the sharing out you can. Still considering only the matter of intrinsic fairness to the children, rather than instrumental effects, would you level down – by sacking the good teachers, closing the playing fields, locking up the Steinways and wheeling in honky-tonk pianos – to make provision equal, or would you work on the inequality only by trying to get better provisions for the less well provided? Most people, presumably, would go for the second,[16] and if so, this is significant. If *equality* is not your overriding concern, to the extent of your being willing to level down, and you care about the *amount* of opportunity (i.e. education and resources) people have as well as how it is distributed, you are not thinking primarily in terms of a fairly run competition with the prizes going to the people who deserve them. Your inclination towards means-regarding equality of opportunity apparently takes the form of regarding these 'opportunities' as *goods in themselves*, rather than as controls to guarantee the fair achevement of some further end.[17]

2. Now consider a similar pattern of analysis for *prospect-regarding* equality of opportunity: the idea that what people should have is equal amounts not of goods like education, but of chances of success. If your intuitions about equality of opportunity seem to take this form, is this because you think this is the only fair way to run a competition for limited prizes? If so, you should regard absolute equality of prospects as a requirement of fairness, because a lottery is unfair to the extent that there is any deviation from equality. (You do not give one competitor a handful of spare tickets to avoid wasting paper.)

Consider, then, various goods that cannot be had by everyone, like being Prime Minister or presenting the nine o'clock news. People have very different prospects of achieving such things, but

[16] That is, they would probably go for the second *all things considered*. The argument as it stands ignores the Parfit refinement of distinguishing between good in one way and good all things considered. This distinction is important in itself, and does apply in this context, but it makes no difference to the outcome here.

[17] This is just as well, since if you are going to control the means you need to have decided what the end is to be. In a fair competition there is a necessary connection between *which* means are equalized and what the selection is *for* – but what *that* should be is an entirely open question. (You do not have to decide competitors should not take drugs, for instance. You might have a competition to see who could run fastest by any means whatever.) When this point is considered in the context of the idea that means-regarding equality of opportunity is a way of making sure that the people who win are the ones who *deserve* to win, it becomes clear that it generates strings of absurdities of its own.

if you wanted to make these prospects more equal there are two quite different ways in which you could set about it. You could improve the education and access to resources of the people with the lowest chances, or you could simply draw lots for the goods in question. Since no possible adjustment of resources could give everyone equal probabilities of success, the second is immeasurably better from the point of view of setting up a fair competition. If your inclinations to prospect-regarding equality of opportunity are rooted in the idea of a fair competition, therefore – and still disregarding instrumental effects – that is what you should be in favour of. If not, and you prefer the route of increasing resources to the people whose chances are low, you seem once more to be regarding these resources as desirable *in themselves,* rather than as providing a fair way to determine the achievement of other desirable things.

3. Finally, for good measure, consider the Rawlsian compromise between these two, of equal prospects for equal abilities.[18] Once again, there are two ways in which you might set about increasing equality of this kind. You could give more resources to the people in a particular range of ability who had fewer, or you could take up the famous Fishkin proposal[19] of swapping children around, by lot, at birth. The second would guarantee absolutely equal prospects for equal abilities, but it seems unlikely that many people would think there was much to be said for it, even when instrumental effects on the parents had been left out of account. Once again, therefore, it looks as though it is the distribution and amount of the resources themselves which are the real subjects of concern, rather than some further end for the fair achieving of which they are the control.

4.3. Implications

If in contexts of this kind – where opportunity is regarded as a good that people can possess in different amounts – such thought experiments show that our intuitions about equality are not primarily about the arranging of fair competitions for further goods, but directly about the fair distribution of something good in itself, the appearance of *this part* of the equal opportunity

[18] John Rawls, *A Theory of Justice* (Cambridge, MA: Harvard University Press, 1971), p. 73.

[19] James S. Fishkin, *Justice, Equal Opportunity and the Family* (New Haven, CT: Yale University Press, 1983), p. 57. See also the discussion by Brian Barry in 'Equal Opportunity and Moral Arbitrariness', in Bowie (ed.), *Equal Opportunity*.

problem changes completely. Essentially, the apparent competition between opportunity and outcome disappears. What is vaguely thought of as opportunity is not opportunity to achieve some *further* outcome, but is itself a kind of outcome. 'Opportunity' becomes a rough term for the ability to make our lives take whatever direction we want them to take, rather than to achieve any particular end; and when it is understood in this way it becomes a candidate for recognition as the most fundamental kind of human good, in competition with such others as happiness and well being and money, and in the same general area as Rawls's primary social goods, Sen's capabilities and functionings, and various accounts of positive liberty. It is in need of much detailed specification, of course, but to see it in this way is to transform the problem of pinning down the nature of opportunity.

And once this shift is made, the equality part of the question takes a different form as well. If opportunity is understood not as a control on the fair achieving of something else, but as itself the good to be distributed fairly, equalizing as such is unlikely to seem of overriding importance. The relevant question becomes the familiar one of how to regard different quantities and arrangements of whatever is regarded as fundamentally good. What needs to be decided is the relative importance of increasing its overall amount and adjusting its distribution, and whether equality as such, rather than priority for the worse off, matters at all.[20]

This conclusion does not, of course, hold for anyone whose thought experiments led in the other direction, to the conclusion that what matters is the fair setting up of society as a competition; but anyone who accepts the account of the matter proposed here, of seeing 'opportunity' as signifying a good in itself, should find it solves or avoids most of the usual difficulties. The means-regarding/prospect-regarding confrontation simply disappears, along with innumerable other problems about both forms.[21] For instance, there is no longer the problem of what these equal opportunities are supposed to be *for*,[22] since on this account 'opportunity' is to be understood in an undirected sense, rather than as for anything. And there is no longer any need to tangle with the metaphysics of ultimate desert or the limits of the self,

[20] See Parfit, *op cit*, pp. 213ff.

[21] These are not discussed here, but see, e.g. Rae, *op. cit.* Ch 4, *passim*.

[22] If you are actually *equalizing* opportunity in either way, there are very few things you can equalize it for. Equal opportunity for one thing entails inequality of opportunity for others.

since the idea that we should arrange fairly between people the power to live the lives they want to lead does not depend on anyone's ultimately deserving anything.

This approach also explains, and shows how to deal with, familiar objections to equality of opportunity such as those expressed by Singer and Schaar. Most of these turn out to be objections to equalizing starting positions and letting things rip thereafter,[23] and the account offered here removes that problem at the roots. The fair distributing of opportunity has nothing to do with controls at the beginning of a knockout competition, and one important implication of this interpretation of opportunity as undirected is that questions of how it should best be distributed arise not only between lives, but within them. The ability to direct and change the course of life is valuable at any time during its span, and may support such ideas as continuing education, job mobility, and pressures to employ older people. The approach is also entirely compatible with the equal consideration of interests that Singer, for instance, says should replace equality of opportunity as an ideal. All it does is offer a particular interpretation of the nature of interests.

If all this is right, what it suggests is that most of the familiar debate about equal opportunity in terms of means, prospects and personal responsibility is a tangle generated by the slide itself. If our intuitions about the distribution of such things as education and other opportunity-giving matters are considered directly, and if the thought experiments come out as I have suggested, the problems simply do not arise. It is only when questions about opportunity as something people *have* are approached through an unnoticed slip from ideas of equality of opportunity as something people *give* – where the opportunity to be equalized typically *is* directed to the achievement of something else, such as admission to the school – that the confusions set in. Contemplation of the acreage of paper devoted to attempts to solve this illusory problem must leave that snake feeling pretty pleased with itself.

5. *Conclusion*

This analysis has, of course, been only of one particular snake, and, since I have already suggested that the issue of equal opportunity

[23] There is no necessary connection between these two anyway; you could equalize starting positions and at the same time set up a framework which restricted the resulting inequality.

involves a whole nest of them, I am certainly not claiming that all are open to exactly this analysis, or pass over this very ground. There can be any number of variations,[24] and even slight differences in this story would have raised different issues. I am not even claiming to have offered the only possible analysis of what is going on in this particular case, though I think something like it is probably needed. But what I do think the argument has shown is that you cannot deal with the complex intuitions that jostle under the heading of equality of opportunity without encountering *at least* this range of issues, and that their confusion is the deepest root of the familiar problems.

Consider again the problem identified at the outset of this paper. If you want to implement equal opportunity policies, it is necessary first to identify a standard, to aim at and to assess your progress by. Ms. Higgins's musings illustrate one of the difficulties of doing this: that people may have incompatible or confused interpretations of what to count as opportunity, and when to equalize it over whom. But what the snake phenomenon shows is that there is more to it than that. Intuitions about equality of opportunity seem to involve at least three irreducibly different *kinds* of standard, each of which generates detailed problems of its own. Separate questions need to be asked first about ground-level impartiality and its status, then about standards for the assessment of the actions and policies of moral agents, and then about standards for the comparison of states of affairs. And then, once the philosophical problems have been resolved and the appropriate standards identified, questions about their application to practice will arise independently for all three kinds.

This is not to deny that there can be connections between the appropriate standards to accept at the different levels. The ground-level impartiality constraint, for instance, applies over the whole range of ethics, so it limits the range of conclusions that can be reached at the second and third levels; and for anyone whose ethical views contain any element of consequentialism, conclusions reached at the third level – about the relative merits of states of affairs – will be strongly relevant to conclusions about standards for actions and policies at the second. Nevertheless, there is no entailment between them, and, within a broad range, standards accepted at one level will be compatible with most possible standards at the others. And even to the extent that there

[24] The group equality issues, for instance, produce striking convolutions of their own.

are connections or incompatibilities, it is impossible to identify them without first recognizing the distinctness of the issues, and the confusion that will result from their conflation into the appearance of a single ideal.

So the problem of identifying a standard for equality of opportunity is not just that people disagree about what it should be, or that they have difficulty as individuals in deciding between different competing versions. It is that familiar claims and arguments about equality of opportunity cannot be reduced to a single issue, and there will be no possibility of sorting out the moral issues as long as the impression persists that there is.

Should we then be thinking in terms of (at least) three separate equal opportunity ideals, rather than just one? That may seem the obvious conclusion, in spite of the obvious pragmatic infelicity of calling by the same name ideas that are habitually run together and that you are trying to disentangle. But in fact the difficulties are more than pragmatic. I want to conclude by suggesting that none of these kinds of standard, considered individually and uncoloured by spurious associations with the others, can plausibly be counted as concerned with equality of opportunity at all.

The principle of ground-level impartiality, even though it does not on its own require any particular kind of equal outcome or treatment, does consist of a strong and quite universal requirement of equal consideration. But on the other hand, it has in itself nothing to do with opportunity. If its implementation goes ahead against a background where opportunity is valued – as when the emancipation of women and blacks began to take place against a general background of liberalism – the result will be the equalizing between groups of some kind of opportunity; but that comes from the combination of the impartiality principle with background ideals, not from the principle alone. A utilitarian who believed that the greatest happiness could be achieved by people's being allocated to fixed positions in life at birth might implement that idea with complete impartiality, but would not be producing anything plausibly described as opportunity. So the first standard implies (a particular kind of) strong equality, but, in itself, nothing about opportunity.

At the bottom of the descent, where the issue is opportunity as a good that people can have in differing amounts, the situation reverses. If, as I have argued, opportunity should in this context be thought of in an undirected way, as a good in itself, the analysis gives fundamental value to *opportunity* so conceived, but correspondingly removes the weight from equality.[25] If the idea that equality matters survives the levelling-down objection at all,[26] it will still have to be balanced against the total quantity of whatever is valued, and is unlikely to seem of overriding importance.

Finally, at the second stage of this descent, where equal opportunity concerns are about the actions and policies of moral agents, they are typically expressed in the form of liberal ideas about the opening of careers to talents. But although such ideals may usually be expressible in terms of equalizing *particular* opportunities over *particular* ranges of people, it would be as odd and misleading to describe these as exemplifying some general ideal of equality of opportunity, *tout court*, as it would be to say that Georgian architects who maintained a single height for the windows of each storey were working with a general ideal of equality of height.

If this is right, what it implies is that the ideal of equality of opportunity is *essentially* a ragbag, taking elements from the various levels and seeming to make out of the confusion something that is not to be found among them either individually or collectively. The composite idea seems to derive the elements of equality, simplicity and strong moral force from the first level, of opportunity as a good that people can have in differing amounts from the third, and of fair competition from the second; and the consequence is that familiar attempts to think about, and implement, equality of opportunity, are in intellectual – and therefore moral – chaos.

The obvious moral of all this is that the term should just be dropped. Everything that concerns us about equality of opportunity – the assortment of worries that pulls us to the end of the snake – could be far better expressed in other terms, and if that were done nothing would be lost except potential for confusion, oversimplification and political sleight of hand. But since any opposition to the term is bound to be regarded as opposition to its supposed referent, and since it is widely taken for granted that equality of opportunity is a good thing, that proposal is probably a political non-starter.

[25] This does not apply for anyone who concludes that equal opportunity is about the setting up of a fair competition. In that case it would indeed be essential to go for equality of the relevant kind of opportunity.

[26] See Parfit, *op cit.*, p. 211 and *passim*.

Still, it should be possible to achieve much of the same effect by a retreat to less obviously provocative ground. If the serpent's rhetorical negatives are systematically met with demands for clarification of the positive standards we are being asked adopt, and with moral rather than linguistic analysis of any that are offered, that will be enough to stop the slide into paradox. If the term 'equality of opportunity' can be prevented from doing any moral work, the way will at least be open for the real moral work to be done.

Department of Philosophy
Open University
Walton Hall
Milton Keynes MK7 6AA
England

Part IV
The Ideal of Equality for Its Own Sake versus Other Ideals with Similar Implications

[14]

R. M. HARE

Corpus Christi College, Oxford

Justice and Equality

The Senses of 'Just'

There are several reasons why a philosopher of my persuasion should wish to write about justice. The first is the general one that ethical theory ought to be applied to practical issues, both for the sake of improving the theory and for any light it may shed on the practical issues, of which many of the most important involve questions of justice. This is shown by the frequency with which appeals are made to justice and fairness and related ideals when people are arguing about political or economic questions (about wages for example, or about schools policy or about relations between races or sexes). If we do not know what 'just' and 'fair' mean (and it looks as if we do not) and therefore do not know what would settle questions involving these concepts, then we are unlikely to be able to sort out these very difficult moral problems. I have also a particular interest in the topic: I hold a view about moral reasoning which has at least strong affinities with utilitarianism; [1] and there is commonly thought to be some kind of antagonism between justice and utility or, as it is sometimes called, expediency. I have therefore a special need to sort these questions out.

We must start by distinguishing between different kinds of justice, or

 An earlier version of this paper appeared in Polish in *Etyka* (Warsaw).

between different senses or uses of the word 'just' (the distinction between these different ways of putting the matter need not now concern us). In distinguishing between different kinds of justice we shall have to make crucial use of a distinction between different levels of moral thinking which I have explained at length in other places.[2] It is perhaps simplest to distinguish three levels of thought, one ethical or meta-ethical and two moral or normative-ethical. At the meta-ethical level we try to establish the meanings of the moral words, and thus the formal properties of the moral concepts, including their logical properties. Without knowing these a theory of normative moral reasoning cannot begin. Then there are two levels of (normative) moral thinking which have often been in various ways distinguished. I have myself in the past called them 'level 2' and 'level 1'; but for ease of remembering I now think it best to give them names, and propose to call level 2 the *critical* level and level 1 the *intuitive* level. At the intuitive level we make use of *prima facie* moral principles of a fairly simple general sort, and do not question them but merely apply them to cases which we encounter. This level of thinking cannot be (as intuitionists commonly suppose) self-sustaining; there is a need for a critical level of thinking by which we select the *prima facie* principles for use at the intuitive level, settle conflicts between them, and give to the whole system of them a justification which intuition by itself can never provide. It will be one of the objects of this paper to distinguish those kinds of justice whose place is at the intuitive level and which are embodied in *prima facie* principles from those kinds which have a role in critical and indeed in meta-ethical thinking.

The principal result of meta-ethical enquiry in this field is to isolate a sense or kind of justice which has come to be known as 'formal justice'. Formal justice is a property of all moral principles (which is why Professor Rawls heads his chapter on this subject not 'Formal constraints of the concept of *just*' but 'Formal constraints of the concept of *right*',[3] and why his disciple David Richards is able to make a good attempt to found the whole of morality, and not merely a theory of justice, on a similar hypothetical-contract basis).[4] Formal justice is simply another name for the formal requirement of universality in moral principles on which, as I have explained in detail elsewhere,[5] golden-rule arguments are based. From the formal, logical properties of the moral words, and in particular from the logical prohibition of individual references in moral principles, it is possible to derive formal canons of moral argument, such as the rule that we are not allowed to discriminate morally between individuals unless there is some qualitative difference between them which is the ground for the discrimination; and the rule that the equal interests of different individuals have equal moral weight. Formal justice consists simply in the observance of these canons in our moral arguments; it is widely thought that this observance by itself is not enough to secure justice in some more substantial sense. As we shall see, one is not offending against the first rule if one says that extra privileges should be given to people just because they have white skins; and one is not offending against either rule if one says that one should take a cent from everybody and give it to the man with the biggest nose, provided that he benefits as much in total as they lose. The question is, How do we get from formal to substantial justice?

This question arises because there are various kinds of material or substantial justice whose content cannot be established directly by appeal to the uses of moral words or the formal properties of moral concepts (we shall see later how much can be done indirectly by appeal to these formal properties *in conjunction with* other premisses or postulates or presuppositions). There is a number of different kinds of substantial justice, and we can hardly do better than begin with Aristotle's classification of them,[6] since it is largely responsible for the different senses which the word 'just' still has in common use. This is a case where it is impossible to appeal to common use, at any rate of the word 'just' (the word 'fair' is better) in order to settle philosophical disputes, because the common use is itself the product of past philosophical theories. The expressions 'distributive' and 'retributive' justice go back to Aristotle,[7] and the word 'just' itself occupies the place (or places) that it does in our language largely because of its place in earlier philosophical discussions.

Aristotle first separated off a generic sense of the Greek word commonly translated 'just', a sense which had been used a lot by Plato: the sense in which justice is the whole of virtue in so far as it concerns our relations with other people.[8] The last qualification reminds us that this is not the most generic sense possible. Theognis had already used it to include the whole of virtue, full stop.[9] These very generic senses of the word, as applied to men and acts, have survived into modern English to confuse philosophers. One of the sources of confusion is that, in the less generic sense of 'just' to be discussed in most of this paper, the judgment that an act would be unjust is sometimes fairly easily overridden by other moral considerations ('unjust', we may say, 'but right as an act of mercy'; or 'unjust, but right because necessary in order to avert an appalling calamity'). It is much more difficult for judgments that an act is required by justice in the generic sense, in which 'unjust' is almost equivalent to 'not right', to be overridden in this way.

Adherents of the '*fiat justitia ruat caelum*'[10] school seldom make clear whether, when they say 'Let justice be done though the heavens fall', they are using a more or less generic sense of 'justice'; and they thus take advantage of its non-overridability in the more generic sense in order to claim unchallengeable sanctity for judgments made using one of the less generic senses. It must be right to do the just thing (whatever that may be) in the sense (if there still is one in English) in which 'just' *means* 'right'. In this sense, if it were right to cause the heavens to fall, and therefore just in the most generic sense, it would of course be right. But we might have to take into account, in deciding whether it would be right, the fact that the heavens would fall (that causing the heavens to fall would be one of the things we were doing if we did the action in question). On the other hand, if it were merely the just act in one of the less generic senses, we might hold that, though just, it was not right, because it would not be right to cause the heavens to fall merely in order to secure justice in this more limited sense; perhaps some concession to mercy, or even to common sense, would be in order.

This is an application of the 'split-level' structure of moral thinking sketched above. One of the theses I wish to maintain is that principles of justice in these less generic senses are all *prima facie* principles and therefore overridable. I shall later be giving a utilitarian account of justice which finds a place, at the intuitive level, for these *prima facie* principles of justice. At this level they have great importance and utility, but it is in accordance with utilitarianism, as indeed with common sense, to claim that they can on unusual occasions be overriden. Having said this, however, it is most important to stress that this does *not* involve conceding the overridability of either the generic kind of justice, which has its place at the critical level, or of formal justice, which operates at the meta-ethical level. These are preserved intact, and therefore defenders of the sanctity of justice ought to be content, since these are the core of justice as of morality. We may call to mind here Aristotle's [11] remarks about the 'better justice' or 'equity' which is required in order to rectify the crudities, giving rise to unacceptable results in particular cases, of a justice whose principles are, as they have to be, couched in general (i.e. simple) terms. The lawgiver who, according to Aristotle, 'would have' given a special prescription if he had been present at this particular case, and to whose prescription we must try to conform if we can, corresponds to the critical moral thinker, who operates under the constraints of formal justice and whose principles are not limited to simple general rules but can be specific enough to cover the peculiarities of unusual cases.

Retributive and Distributive Justice

After speaking briefly of generic justice, Aristotle goes on [12] to distinguish two main kinds of justice in the narrower or more particular sense in which it means 'fairness'. He calls these retributive and distributive justice. They have their place, respectively, in the fixing of penalties and rewards for bad and good actions, and in the distribution of goods and the opposite between the possible recipients. One of the most important questions is whether these two sorts of justice are reducible to a single sort. Rawls, for example, thinks that they are, and so do I. By using the expression 'justice as fairness', he implies that all justice can be reduced to kinds of distributive justice, which itself is founded on procedural justice (i.e. on the adoption of fair procedures) in distribution.[13]

We may (without attempting complete accuracy in exposition) explain how Rawls might effect this reduction as follows. The parties in his 'original position' are prevented by his 'veil of ignorance' from knowing what their own positions are in the world in which they are to live; so they are unable when adopting principles of justice to tailor them to suit their own individual interests. Impartiality (a very important constituent, at least, of justice) is thus secured. Therefore the principles which govern *both* the distribution of wealth and power and other good things *and* the assignment of rewards and penalities (and indeed all other matters which have to be regulated by principles of justice) will be impartial as between individuals, and in this sense just. In this way Rawls in effect reduces the justice of acts of retribution to justice in distributing between the affected parties the good and bad effects of a system of retributions, and reduces this distributive justice in turn to the adoption of a just procedure for selecting the system of retributions to be used.

This can be illustrated by considering the case of a criminal facing a judge (a case which has been thought to give trouble to me too, though I dealt with it adequately, on the lines which I am about to repeat here, in my book *Freedom and Reason*).[14] A Rawlsian judge, when sentencing the criminal, could defend himself against the charge of injustice or unfairness by saying that he was faithfully observing the principles of justice which would be adopted in the original position, whose conditions are procedurally fair. What these principles would be requires, no doubt, a great deal of discussion, in the course of which I might find myself in disagreement with Rawls. But my own view on how the judge should justify his action is, in its formal properties, very like his. On my view likewise, the judge can say that, when he asks himself what universal principles he is prepared to adopt for situations exactly like the one he is in, and considers examples of such logically possible situations in which *he* occupies, successively, the positions of judge, and of criminal, and of all those who are affected by the administration and enforcement of the law under which he is sentencing the criminal, including, of course, potential victims of possible future crimes—he can say that when he asks himself this, he has no hesitation in accepting the principle which bids him impose such and such a sentence in accordance with the law.

I am assuming that the judge is justifying himself at the critical level. If he were content with justifying himself at the intuitive level, his task would be easier, because, we hope, he, like most of us, has intuitions about the proper administration of justice in the courts, embodying *prima facie* principles of a sort whose inculcation in judges and in the rest of us has a high social utility. I say this while recognizing that *some* judges have intuitions about these matters which have a high social *dis*-utility. The question of what intuitions judges ought to have about retributive justice is a matter for *critical* moral thinking.

On both Rawls' view and mine retributive justice has thus been reduced to distributive; on Rawls' view the principles of justice adopted are those which *distribute* fairly between those affected the good and the evil consequences of having or not having certain enforced criminal laws; on my own view likewise it is the impartiality secured by the requirement to universalize one's prescriptions which makes the judge say what he says, and here too it is an impartiality in distributing good and evil consequences between the affected parties. For the judge to let off the rapist would not be *fair* to all those who would be raped if the law were not enforced. I conclude that retributive justice can be reduced to distributive, and that therefore we shall have done what is required of us if we can give an adequate account of the latter.

What is common to Rawls' method and my own is the recognition that to get solutions to particular questions about what is just or unjust, we have to have a way of selecting principles of justice to answer such questions, and that to ask them in default of such principles is senseless. And we both recognize that the method for selecting the principles has to be founded on what he calls 'the formal constraints of the concept of right'. This measure of agreement can extend to the method of selecting principles of distributive justice as well as retributive. Neither Rawls nor I need be put off our stride by an objector who says that we have not addressed ourselves to the question of what acts are just, but have divagated on to the quite different question of how to select principles of justice. The point is that the first question cannot be answered without answering the second. Most of the apparently intractable conflicts about justice and rights that plague the world have been generated by taking certain answers to the first question as obvious and requiring no argument. We shall resolve these conflicts only by asking what arguments are available for the principles by which questions about the justice of individual acts are to be answered. In short, we need to ascend from intuitive to critical thinking; as I have argued in my review of his book, Rawls is to be reproached with not *completing* the ascent.[15]

Nozick, however, seems hardly to have begun it.[16] Neither Rawls nor I have anything to fear from him, so long as we stick to the formal part of our systems which we in effect share. When it comes to the application of this formal method to produce substantial principles of justice, I might find myself in disagreement with Rawls, because he relies much too much on his own intuitions which are open to question. Nozick's intuitions differ from Rawls', and sometimes differ from, sometimes agree with mine. This sort of question is simply not to be settled by appeal to intuitions, and it is time that the whole controversy ascended to a more serious, critical level. At this level, the answer which both Rawls and I should give to Nozick is that whatever sort of principles of justice we are after, whether structural principles, as Rawls thinks, or historical principles, as Nozick maintains, they have to be supported by critical thinking, of which Nozick seems hardly to see the necessity. This point is quite independent of the structural-historical disagreement.

For example, if Nozick thinks that it is just for people to retain whatever property they have acquired by voluntary exchange which benefited all parties, starting from a position of equality but perhaps ending up with a position of gross inequality, and if Rawls, by contrast, thinks that such inequality should be rectified in order to make the position of the least advantaged in society as good as possible, how are we to decide between them? Not by intuition, because there seems to be a deadlock between their intuitions. Rawls has a procedure, which *need* not appeal to intuition, for justifying distributions; this would give him the game, if he were to base the procedure on firm logical grounds, and if he followed it correctly. Actually he does not so base it, and mixes up so many intuitions in the argument that the conclusions he reaches are not such as the procedure really justifies. But Nozick has no procedure at all: only a variety of considerations of different sorts, all in the end based on intuition. Sometimes he seems to be telling us what arrangements in society would be arrived at if bargaining took place in accordance with games-theory between mutually disinterested parties; sometimes what arrangements would maximize the welfare of members of society; and sometimes what arrangements would strike them as fair. He does not often warn us when he is switching from one of these grounds to another; and he does little to convince us by argument that the arrangements so selected would be in accordance with justice. He hopes that we will think what he thinks; but Rawls at least thinks otherwise.

Formal Justice and Substantial Equality

How then do we get from formal to substantial justice? We have had an example of how this is done in the sphere of retributive justice; but how is this method to be extended to cover distributive justice as a whole, and its relation, if any, to equality in distribution? The difficulty of using formal justice in order to establish principles of substantial justice can indeed be illustrated very well by asking whether, and in what sense, justice demands equality in distribution. The complaint is often made that a certain distribution is unfair or unjust because unequal; so it looks, at least, as if the substantial principle that goods ought to be distributed equally in default of reasons to the contrary forms part of some people's conception of justice. Yet, it is argued, this substantial principle cannot be established simply on the basis of the formal notions we have mentioned. The following kind of schematic example is often adduced: consider two possible distributions of a given finite stock of goods, in one of which the goods are distributed equally, and in the other of which a few of the recipients have nearly all the goods, and the rest have what little remains. It is claimed with some plausibility that the second distribution is unfair, and the first fair. But it might also be claimed that impartiality and formal justice alone will not establish that we ought to distribute the goods equally.

There are two reasons which might be given for this second claim, the first of them a bad one, the other more cogent. The bad reason rests on an underestimate of the powers of golden-rule arguments. It is objected, for example, that people with white skins, if they claimed privileges in distribution purely on the ground of skin-colour, would not be offending against the formal principle of impartiality or universalizability, because no individual reference need enter into the principle to which they are appealing. Thus the principle that blacks ought to be subservient to whites is impartial as between *individuals*; any individual whatever who has the bad luck to find himself with a black skin or the good luck to find himself with a white skin is impartially placed by the principle in the appropriate social rank. This move receives a brief answer in my *Freedom and Reason*,[17] and a much fuller one in a forthcoming paper.[18] If the whites are faced with the decision, not merely of whether to frame this principle, but of whether to prescribe its adoption universally in all cases, including hypothetical ones in which their own skins turn black, they will at once reject it.

The other, more cogent-sounding argument is often used as an argument against utilitarians by those who think that justice has a lot to do with equalitv. It could also, at first sight, be used as an argument against the adequacy of formal justice or impartiality as a basis for distributive justice. That the argument could be leveled against both these methods is no accident; as I have tried to show elsewhere,[19] utilitarianism of a certain sort is the embodiment of—the method of moral reasoning which fulfils in practice—the requirement of universalizability or formal justice. Having shown that neither of these methods can produce a direct justification for equal distribution, I shall then show that both can produce indirect justifications, which depend, not on a priori reasoning alone, but on likely assumptions about what the world and the people in it are like.

The argument is this. Formal impartiality only requires us to treat everybody's interest as of equal weight. Imagine, then, a situation in which utilities are equally distributed. (There is a complication here which we can for the moment avoid by choosing a suitable example. Shortly I shall be mentioning the so-called principle of diminishing marginal utility, and shall indeed be making important use of it. But for now let us take a case in which it does not operate, so that we can, for ease of illustration, treat money as a linear measure of utility.) Suppose that we can vary the equal distribution that we started with by taking a dollar each away from everybody in the town, and that the loss of purchasing power is so small that they hardly notice it, and therefore the utility enjoyed by each is not much diminished. However, when we give the resulting large sum to one man, he is able to buy himself a holiday in Acapulco, which gives him so much pleasure that his access of utility is equal to the sum of the small losses suffered by all the others. Many would say that this redistribution was unfair. But we were, in the required sense, being impartial between the equal interests of all the parties; we were treating an equal access or loss of utility to any party as of equal value or disvalue. For, on our suppositions, the taking away of a dollar from one of the unfortunate parties deprived him of just as much utility as the addition of that dollar gave to the fortunate one. But if we are completely impartial, we have to regard *who has* that dollar or that access of utility as irrelevant. So there will be nothing to choose, from an impartial point of view, between our original equal distribution and our later highly unequal one, in which everybody else is deprived of a cent in order to give one person a holiday in Acapulco. And that is why people say that formal impartiality alone is not enough to secure social justice, nor even to secure impartiality itself in some more substantial sense.

What is needed, in the opinion of these people, is some principle which says that it is unjust to give a person more when he already has more than the others—some sort of egalitarian principle. Egalitarian principles are only one possible kind of principles of distributive justice; and it is so far an open question whether they are to be preferred to alternative inegalitarian principles. It is fairly clear as a matter of history that different principles of justice have been accepted in different societies. As Aristotle says, 'everybody agrees that the just distribution is one in accordance with desert of some kind; but they do not call desert the same thing, but the democrats say it is being a free citizen, the oligarchs being rich, others good lineage, and the aristocrats virtue'.[20] It is not difficult to think of some societies in which it would be thought unjust for one man to have privileges not possessed by all men, and of others in which it would be thought unjust for a slave to have privileges which a free man would take for granted, or for a commoner to have the sort of house which a nobleman could aspire to. Even Aristotle's democrats did not think that slaves, but only citizens, had equal rights; and Plato complains of democracy that it 'bestows equality of a sort on equals and unequals alike'.[21] We have to ask, therefore, whether there are any reasons for preferring one of these attitudes to another.

At this point some philosophers will be ready to step in with their intuitions, and tell us that some distributions or ways of achieving distributions are *obviously* more just than others, or that *everyone will agree on reflection* that they are. These philosophers appeal to our intuitions or prejudices in support of the most widely divergent methods or patterns of distribution. But this is a way of arguing which should be abjured by anybody who wishes to have rational grounds for his moral judgments. Intuitions prove nothing; general consensus proves nothing; both have been used to support conclusions which *our* intuitions and our consensus may well find outrageous. We want arguments, and in this field seldom get them.

However, it is too early to despair of finding some. The utilitarian, and the formalist like me, still have some moves to make. I am supposing that we have already made the major move suggested above, and have ruled out discrimination on grounds of skin colour and the like, in so far as such discrimination could not be accepted by all for cases where they were the ones discriminated against. I am supposing that our society has absorbed this move, and contains no racists, sexists or in general discriminators, but does still contain economic men who do not think it wrong, in pursuit of Nozickian economic liberty, to get what they can, even if the resulting distribution is grotesquely unequal. Has the egalitarian any moves to make against them, and are they moves which can be supported by appeal to formal justice, in conjunction with the empirical facts?

Two Arguments for Equal Distribution

He has two. The first is based on that good old prop of egalitarian policies, the diminishing marginal utility, within the ranges that matter, of money and of nearly all goods. Almost always, if money or goods are taken away from someone who has a lot of them already, and given to someone who has little, total utility is increased, other things being equal. As we shall see, they hardly ever are equal; but the principle is all right. Its ground is that the poor man will get more utility out of what he is given than the rich man from whom it is taken would have got. A millionaire minds less about the gain or loss of a dollar than I do, and I than a pauper.

It must be noted that this is not an *a priori* principle. It is an empirical fact (if it is) that people are so disposed. The most important thing I have to say in this paper is that when we are, as we now are, trying to establish *prima facie* principles of distributive justice, it is enough if they can be justified in the world as it actually is, among people as they actually are. It is a wholly illegitimate argument against formalists or utilitarians that states of society or of the people in it could be *conceived of* in which gross inequalities could be justified by formal or utilitarian arguments. We are seeking principles for practical use in the world as it is. The same applies when we ask what qualifications are required to the principles.

Diminishing marginal utility is the firmest support for policies of progressive taxation of the rich and other egalitarian measures. However, as I said above, other things are seldom equal, and there are severe empirical, practical restraints on the equality that can sensibly be imposed by governments. To mention just a few of these hackneyed other things: the removal of incentives to effort may diminish the total stock of goods to be divided up; abrupt confiscation or even very steep progressive taxation may antagonize the victims so much that a whole class turns from a useful element in society to a hostile and dangerous one; or, even if that does not happen, it may merely become demoralized and either lose all enterprise and readiness to take business risks, or else just emigrate if it can. Perhaps one main cause of what is called the English sickness is the alienation of the middle class. It is an empirical question, just when egalitarian measures get to the stage of having these effects; and serious political argument on this subject should concentrate on such empirical questions, instead of indulging in the rhetoric of equal (or for that matter of unequal) rights. Rights are the offspring of *prima facie*, intuitive principles, and I have nothing against them; but the question is, What *prima facie* principles ought we to adopt? What intuitions ought we to have? On these questions the rhetoric of rights sheds no light whatever, any more than do appeals to intuition (i.e. to prejudice, i.e. to the *prima facie* principles, good or bad, which our upbringings happen to have implanted in us). The worth of intuitions is to be known by their fruits; as in the case of the principles to be followed by judges in administering the law, the best principles are those with the highest acceptance-utility, i.e. those whose general acceptance maximizes the furtherance of the interests, in sum, of all the affected parties, treating all those interests as of equal weight, i.e. impartially, i.e. with formal justice.

We have seen that, given the empirical assumption of diminishing marginal utility, such a method provides a justification for moderately egalitarian policies. The justification is strengthened by a second move that the egalitarian can make. This is to point out that inequality itself has a tendency to produce envy, which is a disagreeable state of mind and leads people to do disagreeable things. It makes no difference to the argument whether the envy is a good or a bad quality, nor whether it is justified or unjustified—any more than it makes a difference whether the alienation of the middle class which I mentioned above is to be condemned or excused. These states of mind are facts, and moral judgments have to be made in the light of the facts as they are. We have to take account of the actual state of the world and of the people in it. We can very easily think of societies which are highly unequal, but in which the more fortunate members have contrived to find some real or metaphorical opium or some Platonic noble lie [22] to keep the people quiet, so that the people feel no envy of privileges which we should consider outrageous. Imagine, for example, a society consisting of happy slave-owners and of happy slaves, all of whom know their places and do not have ideas above their station. Since there is *ex hypothesi* no envy, this source of disutility does not exist, and the whole argument from envy collapses.

It is salutary to remember this. It may make us stop looking for purely formal, *a priori* reasons for demanding equality, and look instead at the actual conditions which obtain in particular societies. To make the investigation more concrete, albeit oversimplified, let us ask what would have to be the case before we ought to be ready to push this happy slave-owning society into a revolution—peaceful or violent—which would turn the slaves into free and moderately equal wage-earners. I shall be able only to sketch my answer to this question, without doing nearly enough to justify it.

Arguments for and against Egalitarian Revolutions

First of all, as with all moral questions, we should have to ask what would be the actual consequences of what we were doing—which is the same as to ask what we should be *doing*, so that accusations of 'consequentialism' [23] need not be taken very seriously. Suppose, to simplify matters outrageously, that we can actually predict the consequences of the revolution and what will happen during its course. We can then consider two societies (one actual and one possible) and a possible process of transition from one to the other. And we have to ask whether the transition from one to the other will, all in all, promote the interests of all those affected more than to stay as they are, or rather, to develop as they would develop if the revolution did not occur. The question can be divided into questions about the process of transition and questions about the relative merits of the actual society (including its probable subsequent 'natural' development) and the possible society which would be produced by the revolution.

We have supposed that the slaves in the existing society feel no envy, and that therefore the disutility of envy cannot be used as an argument for change. If there *were* envy, as in actual cases is probable, this argument *could* be employed; but let us see what can be done without it. We have the fact that there is gross inequality in the actual society and much greater equality in the possible one. The principle of diminishing marginal utility will therefore support the change, provided that its effects are not outweighed by a reduction in total utility resulting from the change and the way it comes about. But we have to be sure that this condition is fulfilled. Suppose, for example, that the actual society is a happy bucolic one and is likely to remain so, but that the transition to the possible society initiates the growth of an industrial economy in which everybody has to engage in a rat-race and is far less happy. We might in that case pronounce the actual society better. In general it is not self-evident that the access of what is called wealth makes people happier, although they nearly always think that it will.

Let us suppose, however, that we are satisfied that the people in the possible society will be better off all round than in the actual. There is also the point that there will be more generations to enjoy the new regime than suffer in the transition from the old. At least, this is what revolutionaries often say; and we have set them at liberty to say it by assuming, contrary to what is likely to be the case, that the future state of society is predictable. In actual fact, revolutions usually produce states of society very different from, and in most cases worse than, what their authors expected—which does not always stop them being better than what went before, once things have settled down. However, let us waive these difficulties and suppose that the future state of society can be predicted, and that it is markedly better than the existing state, because a greater equality of distribution has, owing to diminishing marginal utility, resulted in greater total utility.

Let us also suppose that the more enterprising economic structure which results leads to increased production without causing a rat-race. There will then be more wealth to go round and the revolution will have additional justification. Other benefits of the same general kind may also be adduced; and what is perhaps the greatest benefit of all, namely liberty itself. That people like having this is an empirical fact; it may not be a fact universally, but it is at least *likely* that by freeing slaves we shall *pro tanto* promote their interests. Philosophers who ask for *a priori* arguments for liberty or equality often talk as if empirical facts like this were totally irrelevant to the question. Genuine egalitarians and liberals ought to abjure the aid of these philosophers, because they have taken away the main ground for such views, namely the fact that people are as they are.

The arguments so far adduced support the call for a revolution. They will have to be balanced against the disutilities which will probably be caused by the process of transition. If heads roll, that is contrary to the interests of their owners; and no doubt the economy will be disrupted at least temporarily, and the new rulers, whoever they are, may infringe liberty just as much as the old, and possibly in an even more arbitrary manner. Few revolutions are pleasant while they are going on. But if the revolution can be more or less smooth or even peaceful, it may well be that (given the arguments already adduced about the desirability of the future society thereby achieved) revolution can have a utilitarian justification, and therefore a justification on grounds of formal impartiality between people's interests. But it is likely to be better for all if the same changes can be achieved less abruptly by an evolutionary process, and those who try to persuade us that this is not so are often merely giving way to impatience and showing a curious indifference to the interests of those for whom they purport to be concerned.

The argument in favour of change from a slave-owning society to a wage-earning one has been extremely superficial, and has served only to illustrate the lines on which a utilitarian or a formalist might argue. If we considered instead the transition from a capitalist society to a socialist one, the same forms of argument would have to be employed, but might not yield the same result. Even if the introduction of a fully socialist economy would promote greater equality, or more equal liberties (and I can see no reason for supposing this, but rather the reverse; for socialism tends to produce very great inequalities of *power*), it needs to be argued what the consequences would be, and then an assessment has to be made of the relative benefits and harms accruing from leaving matters alone and from having various sorts of bloody or bloodless change. Here again the rhetoric of rights will provide nothing but inflammatory material for agitators on both sides. It is designed to lead to, not to resolve, conflicts.

Remarks About Methods

But we must now leave this argument and attend to a methodological point which has become pressing. We have not, in the last few pages, been arguing about what state of society would be just, but about what state of society would best promote the interests of its members. All the arguments have been utilitarian. Where then does justice come in? It is likely to come into the propaganda of revolutionaries, as I have already hinted. But so far as I can see it has no direct bearing on the question of what would be the better society. It has, however, an important indirect bearing which I shall now try to explain. Our *prima facie* moral principles and intuitions are, as I have already said, the products of our upbringings; and it is a very important question *what* principles and intuitions it is best to bring up people to have. I have been arguing on the assumption that this question is to be decided by looking at the consequences for society, and the effects on the interests of people in society, of inculcating different principles. We are looking for the set of principles with the highest acceptance-utility.

Will these include principles of justice? The answer is obviously 'Yes', if we think that society and the people in it are better off with *some* principles of justice than without any. A 'land without justice' (to use the title of Milovan Djilas' book) [24] is almost bound to be an unhappy one. But what are the principles to be? Are we, for example, to inculcate the principle that it is just for people to perform the duties of their station and not envy those of higher social rank? Or the principle that all inequalities of any sort are unjust and ought to be removed? For my part, I would think that neither of these principles has a very high acceptance-utility. It may be that the principle with the highest acceptance-utility is one which makes just reward vary (but not immoderately) with desert, and assesses desert according to service to the interests of one's fellow-men. It would have to be supplemented by a principle securing equality of opportunity. But it is a partly empirical question what principles would have the highest acceptance-utility, and in any case beyond the scope of this paper. If some such principle is adopted and inculcated, people will *call* breaches of it unjust. Will they *be* unjust? Only in the sense that they will be contrary to a *prima facie* principle of distributive justice which we ought to adopt (not because it is itself a just principle, but because it is the best principle). The only sense that can be given to the question of whether it is a just principle (apart from the purely circular or tautological question of whether the principle obeys itself), is by asking whether the procedure by which we have selected the principle satisfies the logical requirements of critical moral thinking, i.e. is *formally* just. We might add that the adoption of such a formally just procedure and of the principles it selects is just in the *generic* sense mentioned at the beginning of this paper: it is the right thing to do; we morally ought to do it. The reason is that critical thinking, because it follows the requirements of formal justice based on the logical properties of the moral concepts, especially 'ought' and 'right', can therefore not fail, if pursued correctly in the light of the empirical facts, to lead to principles of justice which are in accord with morality. But because the requirements are all formal, they do not by themselves determine the content of the principles of justice. We have to do the thinking.

What principles of justice are best to try to inculcate will depend on the circumstances of particular societies, and especially on psychological facts about their members. One of these facts is their readiness to accept the principles themselves. There might be a principle of justice which it would be highly desirable to inculcate, but which we have no chance of successfully inculcating. The best principles for a society to *have* are, as I said, those with the highest acceptance-utility. But the best principles to *try to inculcate* will not necessarily be these, if these are impossible to inculcate. Imagine that in our happy slave-society both slaves and slave-owners are obstinately conservative and know their places, and that the attempt to get the slaves to have revolutionary or egalitarian thoughts will result only in a very few of them becoming discontented, and probably going to the gallows as a result, and the vast majority merely becoming unsettled and therefore more unhappy. Then we ought not to try to inculcate such an egalitarian principle. On the other hand, if, as is much more likely, the principle stood a good chance of catching on, and the revolution was likely to be as advantageous as we have supposed, then we ought. The difference lies in the dispositions of the inhabitants. I am not saying that the probability of being accepted is the same thing as acceptance-utility; only that the rationality of trying to inculcate a principle (like the rationality of trying to do anything else) varies with the likelihood of success. In this sense the advisability of trying to inculcate principles of justice (though not their merit) is relative to the states of mind of those who, it is hoped, will hold them.

It is important to be clear about the extent to which what I am advocating is a kind of relativism. It is certainly not relativistic in any strong sense. Relativism is the doctrine that the truth of some moral statement depends on whether people accept it. A typical example would be the thesis that if in a certain society people think that they ought to get their male children circumcised, then they ought to get them circumcised, full stop. Needless to say, I am not supporting any such doctrine, which is usually the result of confusion, and against which there are well-known arguments. It is, however, nearly always the case that among the facts relevant to a moral decision are facts about people's thoughts or dispositions. For example, if I am wondering whether I ought to take my wife for a holiday in Acapulco, it is relevant to ask whether she would like it. What I have been saying is to be assimilated to this last example. If we take as given certain dispositions in the members of society (namely dispositions not to accept a certain principle of justice however hard we work at propagating it) then we have to decide whether, in the light of these facts, we ought to propagate it. What principles of justice we ought to propagate will vary with the probable effects of propagating them. The answer to this 'ought'-question is not relative to what we, who are asking it, think about the matter; it is to be arrived at by moral thought on the basis of the facts of the situation. But among these facts are facts about the dispositions of people in the society in question.

The moral I wish to draw from the whole argument is that ethical reasoning *can* provide us with a way of conducting political arguments about justice and rights rationally and with hope of agreement; that such rational arguments have to rest on an understanding of the concepts being used, *and* of the facts of our actual situation. The key question is 'What principles of justice, what attitudes towards the distribution of goods, what ascriptions of rights, are such that their acceptance is in the general interest?' I advocate the asking of this question as a substitute for one which is much more commonly asked, namely 'What rights do I have?' For people who ask this latter question will, being human, nearly always answer that they have just those rights, whatever they are, which will promote a distribution of goods which is in the interest of their own social group. The rhetoric of rights, which is engendered by this question, is a recipe for class war, and civil war. In pursuit of these rights, people will, because they have convinced themselves that justice demands it, inflict almost any harms on the rest of society and on themselves. To live at peace, we need principles such as critical thinking can provide, based on formal justice and on the facts of the actual world in which we have to live. It is possible for all to practise this critical thinking in cooperation, if only they would learn how; for all share the same moral concepts with the same logic, if they could but understand them and follow it.

Notes

1. See my 'Ethical Theory and Utilitarianism' (*ETU*) in *Contemporary British Philosophy 4*, ed. H. D. Lewis (London, 1976).
2. See, e.g., my 'Principles', *Ar. Soc.* 72 (1972/3), 'Rules of War and Moral Reasoning', *Ph. and Pub. Aff.* 1 (1972) and *ETU*.
3. Rawls, J., *A Theory of Justice* (Cambridge, Mass., 1971), p. 130.
4. Richards, D. A. J., *A Theory of Reasons for Action* (Oxford, 1971).
5. See my *Freedom and Reason*, pt. II (Oxford, 1963) and *ETU*.
6. *Nicomachean Ethics*, bk. V.
7. ib. 1130 b 31, 1131 b 25.
8. ib. 1130 a 8.
9. Theognis 147; also attr. to Phocylides by Aristotle, ib. 1129 b 27.
10. The earliest version of this tag is attr. by the *Oxford Dictionary of Quotations* to the Emperor Ferdinand I (1503–64).
11. ib. 1137 b 8.
12. ib. 1130 a 14 ff.
13. *A Theory of Justice*, p. 136.
14. Pp. 115-7, 124.
15. *Ph. Q.* 23 (1973), repr. in *Reading Rawls*, ed. N. Daniels (Oxford, 1975).
16. Nozick, R. D., *Anarchy, State and Utopia* (New York, 1974).
17. Pp. 106f.
18. 'Relevance', in a volume in honour of R. Brandt, W. Frankena and C. Stevenson, eds. A. Goldman and J. Kim (Reidel, forthcoming).
19. See note 2 above.
20. ib. 1131 a 25.
21. *Republic* 558 c.
22. ib. 414 b.
23. See, e.g., Anscombe, G. E. M., 'Modern Moral Philosophy', *Philosophy* 33 (1958) and Williams, B. A. O., in Smart, J. J. C. and Williams, B. A. O., *Utilitarianism: For and Against* (Cambridge, Eng., 1973), p. 82.
24. Djilas, M., *Land without Justice* (London, 1958).

[15]

Equality as a Moral Ideal

Harry Frankfurt

First man: "How are your children?"
Second man: "Compared to what?"

I

Economic egalitarianism is, as I shall construe it, the doctrine that it is desirable for everyone to have the same amounts of income and of wealth (for short, "money").[1] Hardly anyone would deny that there are situations in which it makes sense to tolerate deviations from this standard. It goes without saying, after all, that preventing or correcting such deviations may involve costs which—whether measured in economic terms or in terms of noneconomic considerations—are by any reasonable measure unacceptable. Nonetheless, many people believe that economic equality has considerable moral value in itself. For this reason they often urge that efforts to approach the egalitarian ideal should be accorded—with all due consideration for the possible effects of such efforts in obstructing or in conducing to the achievement of other goods—a significant priority.[2]

In my opinion, this is a mistake. Economic equality is not, as such, of particular moral importance. With respect to the distribution of economic assets, what *is* important from the point of view of morality is not that everyone should have *the same* but that each should have *enough*. If everyone had enough, it would be of no moral consequence whether some had more than others. I shall refer to this alternative to egalitarianism—

1. This version of economic egalitarianism (for short, simply "egalitarianism") might also be formulated as the doctrine that there should be no inequalities in the *distribution* of money. The two formulations are not unambiguously equivalent because the term "distribution" is equivocal. It may refer either to a pattern of possession or to an activity of allocation, and there are significant differences in the criteria for evaluating distributions in the two senses. Thus it is quite possible to maintain consistently both that it is acceptable for people to have unequal amounts of money and that it is objectionable to allocate money unequally.

2. Thus, Thomas Nagel writes: "The defense of economic equality on the ground that it is needed to protect political, legal and social equality . . . [is not] a defense of equality *per se*—equality in the possession of benefits in general. Yet the latter is a further moral idea of great importance. Its validity would provide an independent reason to favor economic equality as a good in its own right" ("Equality," in his *Mortal Questions* [Cambridge: Cambridge University Press, 1979], p. 107).

Ethics 98 (October 1987): 21–43

namely, that what is morally important with respect to money is for everyone to have enough—as "the doctrine of sufficiency."[3]

The fact that economic equality is not in its own right a morally compelling social ideal is in no way, of course, a reason for regarding it as undesirable. My claim that equality in itself lacks moral importance does not entail that equality is to be avoided. Indeed, there may well be good reasons for governments or for individuals to deal with problems of economic distribution in accordance with an egalitarian standard and to be concerned more with attempting to increase the extent to which people are economically equal than with efforts to regulate directly the extent to which the amounts of money people have are enough. Even if equality is not as such morally important, a commitment to an egalitarian social policy may be indispensable to promoting the enjoyment of significant goods besides equality or to avoiding their impairment. Moreover, it might turn out that the most feasible approach to the achievement of sufficiency would be the pursuit of equality.

But despite the fact that an egalitarian distribution would not necessarily be objectionable, the error of believing that there are powerful moral reasons for caring about equality is far from innocuous. In fact, this belief tends to do significant harm. It is often argued as an objection to egalitarianism that there is a dangerous conflict between equality and liberty: if people are left to themselves, inequalities of income and wealth inevitably arise, and therefore an egalitarian distribution of money can be achieved and maintained only at the cost of repression. Whatever may be the merit of this argument concerning the relationship between equality and liberty, economic egalitarianism engenders another conflict which is of even more fundamental moral significance.

To the extent that people are preoccupied with equality for its own sake, their readiness to be satisfied with any particular level of income or wealth is guided not by their own interests and needs but just by the magnitude of the economic benefits that are at the disposal of others. In this way egalitarianism distracts people from measuring the requirements to which their individual natures and their personal circumstances give rise. It encourages them instead to insist upon a level of economic support that is determined by a calculation in which the particular features of their own lives are irrelevant. How sizable the economic assets of others are has nothing much to do, after all, with what kind of person someone

3. I focus attention here on the standard of equality in the distribution of money chiefly in order to facilitate my discussion of the standard of sufficiency. Many egalitarians, of course, consider economic equality to be morally less important than equality in certain other matters: e.g., welfare, opportunity, respect, satisfaction of needs. In fact, some of what I have to say about economic egalitarianism and sufficiency applies as well to these other benefits. But I shall not attempt in this essay to define the scope of its applicability, nor shall I attempt to relate my views to other recent criticism of egalitarianism (e.g., Larry S. Temkin, "Inequality," *Philosophy and Public Affairs* 15 [1986]: 99–121; Robert E. Goodin, "Epiphenomenal Egalitarianism," *Social Research* 52 [1985]: 99–117).

is. A concern for economic equality, construed as desirable in itself, tends to divert a person's attention away from endeavoring to discover—within his experience of himself and of his life—what he himself really cares about and what will actually satisfy him, although this is the most basic and the most decisive task upon which an intelligent selection of economic goals depends. Exaggerating the moral importance of economic equality is harmful, in other words, because it is alienating.[4]

To be sure, the circumstances of others may reveal interesting possibilities and provide data for useful judgments concerning what is normal or typical. Someone who is attempting to reach a confident and realistic appreciation of what to seek for himself may well find this helpful. It is not only in suggestive and preliminary ways like these, moreover, that the situations of other people may be pertinent to someone's efforts to decide what economic demands it is reasonable or important for him to make. The amount of money he needs may depend in a more direct way on the amounts others have. Money may bring power or prestige or other competitive advantages. A determination of how much money would be enough cannot intelligently be made by someone who is concerned with such things except on the basis of an estimate of the resources available to those with whose competition it may be necessary for him to contend. What is important from this point of view, however, is not the comparison of levels of affluence as such. The measurement of inequality is important only as it pertains contingently to other interests.

The mistaken belief that economic equality is important in itself leads people to detach the problem of formulating their economic ambitions from the problem of understanding what is most fundamentally significant to them. It influences them to take too seriously, as though it were a matter of great moral concern, a question that is inherently rather insignificant and not directly to the point, namely, how their economic status compares with the economic status of others. In this way the doctrine of equality contributes to the moral disorientation and shallowness of our time.

The prevalence of egalitarian thought is harmful in another respect as well. It not only tends to divert attention from considerations of greater moral importance than equality. It also diverts attention from the difficult but quite fundamental philosophical problems of understanding just what these considerations are and of elaborating, in appropriately comprehensive and perspicuous detail, a conceptual apparatus which would facilitate their exploration. Calculating the size of an equal share is plainly

4. It might be argued (as some of the editors of *Ethics* have suggested to me) that pursuing equality as an important social ideal would not be so alienating as pursuing it as a personal goal. It is indeed possible that individuals devoted to the former pursuit would be less immediately or less intensely preoccupied with their own economic circumstances than those devoted to the latter. But they would hardly regard the achievement of economic equality as important for the society unless they had the false and alienating conviction that it was important for individuals to enjoy economic equality.

much easier than determining how much a person needs in order to have enough. In addition, the very concept of having an equal share is itself considerably more patent and accessible than the concept of having enough. It is far from self-evident, needless to say, precisely what the doctrine of sufficiency means and what applying it entails. But this is hardly a good reason for neglecting the doctrine or for adopting an incorrect doctrine in preference to it. Among my primary purposes in this essay is to suggest the importance of systematic inquiry into the analytical and theoretical issues raised by the concept of having enough, the importance of which egalitarianism has masked.[5]

II

There are a number of ways of attempting to establish the thesis that economic equality is important. Sometimes it is urged that the prevalence of fraternal relationships among the members of a society is a desirable goal and that equality is indispensable to it.[6] Or it may be maintained that inequalities in the distribution of economic benefits are to be avoided because they lead invariably to undesirable discrepancies of other kinds—for example, in social status, in political influence, or in the abilities of people to make effective use of their various opportunities and entitlements. In both of these arguments, economic equality is endorsed because of its supposed importance in creating or preserving certain noneconomic conditions. Such considerations may well provide convincing reasons for recommending equality as a desirable social good or even for preferring egalitarianism as a policy over the alternatives to it. But both arguments construe equality as valuable derivatively, in virtue of its contingent connections to other things. In neither argument is there an attribution to equality of any unequivocally inherent moral value.

A rather different kind of argument for economic equality, which comes closer to construing the value of equality as independent of contingencies, is based upon the principle of diminishing marginal utility. According to this argument, equality is desirable because an egalitarian

5. I shall address some of these issues in Sec. VII below.

6. In the Sterling Memorial Library at Yale University (which houses 8.5 million volumes), there are 1,159 entries in the card catalog under the subject heading "liberty" and 326 under "equality." Under "fraternity," there are none. This is because the catalog refers to the social ideal in question as "brotherliness." Under that heading there are *four* entries! Why does fraternity (or brotherliness) have so much less salience than liberty and equality? Perhaps the explanation is that, in virtue of our fundamental commitment to individualism, the political ideals to which we are most deeply and actively attracted have to do with what we suppose to be the rights of individuals, and no one claims a right to fraternity. It is also possible that liberty and equality get more attention in certain quarters because, unlike fraternity, they are considered to be susceptible to more or less formal treatment. In any event, the fact is that there has been very little serious investigation into just what fraternity is, what it entails, or why it should be regarded as especially desirable.

distribution of economic assets maximizes their aggregate utility.[7] The argument presupposes: (*a*) for each individual the utility of money invariably diminishes at the margin and (*b*) with respect to money, or with respect to the things money can buy, the utility functions of all individuals are the same.[8] In other words, the utility provided by or derivable from an *n*th dollar is the same for everyone, and it is less than the utility for anyone of dollar ($n - 1$). Unless *b* were true, a rich man might obtain greater utility than a poor man from an extra dollar. In that case an egalitarian distribution of economic goods would not maximize aggregate utility even if *a* were true. But given both *a* and *b*, it follows that a marginal dollar always brings less utility to a rich person than to one who is less rich. And this entails that total utility must increase when inequality is reduced by giving a dollar to someone poorer than the person from whom it is taken.

In fact, however, both *a* and *b* are false. Suppose it is conceded, for the sake of the argument, that the maximization of aggregate utility is in its own right a morally important social goal. Even so, it cannot legitimately be inferred that an egalitarian distribution of money must therefore have similar moral importance. For in virtue of the falsity of *a* and *b*, the argument linking economic equality to the maximization of aggregate utility is unsound.

So far as concerns *b*, it is evident that the utility functions for money of different individuals are not even approximately alike. Some people suffer from physical, mental, or emotional weaknesses or incapacities that limit the satisfactions they are able to obtain. Moreover, even apart from the effects of specific disabilities, some people simply enjoy things more than other people do. Everyone knows that there are, at any given level of expenditure, large differences in the quantities of utility that different spenders derive.

7. Nagel endorses this argument as establishing the moral importance of economic equality. Other formulations and discussions of the argument may be found in: Kenneth Arrow, "A Utilitarian Approach to the Concept of Equality in Public Expenditures," *Quarterly Journal of Economics* 85 (1971): 409–10; Walter Blum and Harry Kalven, *The Uneasy Case for Progressive Taxation* (Chicago: University of Chicago Press, 1966); Abba Lerner, *The Economics of Control* (New York: Macmillan Publishing Co., 1944); Paul Samuelson, *Economics* (New York: McGraw-Hill Book Co., 1973), and "A. P. Lerner at Sixty," in *Collected Scientific Papers of Paul A. Samuelson*, ed. Robert C. Merton, 3 vols. (Cambridge, Mass.: MIT Press, 1972), vol. 3, pp. 643–52.

8. Thus, Arrow says: "In the utilitarian discussion of income distribution, equality of income is derived from the maximization conditions if it is further assumed that individuals have the same utility functions, each with diminishing marginal utility" (p. 409). And Samuelson offers the following formulation: "If each extra dollar brings less and less satisfaction to a man, and if the rich and poor are alike in their capacity to enjoy satisfaction, a dollar taxed away from a millionaire and given to a median-income person is supposed to add more to total utility than it subtracts" (*Economics*, p. 164, n. 1).

So far as concerns *a*, there are good reasons against expecting any consistent diminution in the marginal utility of money. The fact that the marginal utilities of certain goods do indeed tend to diminish is not a principle of reason. It is a psychological generalization, which is accounted for by such considerations as that people often tend after a time to become satiated with what they have been consuming and that the senses characteristically lose their freshness after repetitive stimulation.[9] It is common knowledge that experiences of many kinds become increasingly routine and unrewarding as they are repeated.

It is questionable, however, whether this provides any reason at all for expecting a diminution in the marginal utility of *money*—that is, of anything that functions as a generic instrument of exchange. Even if the utility of everything money can buy were inevitably to diminish at the margin, the utility of money itself might nonetheless exhibit a different pattern. It is quite possible that money would be exempt from the phenomenon of unrelenting marginal decline because of its limitlessly protean versatility. As Blum and Kalven explain: "In . . . analysing the question whether money has a declining utility it is . . . important to put to one side all analogies to the observation that particular commodities have a declining utility to their users. There is no need here to enter into the debate whether it is useful or necessary, in economic theory, to assume that commodities have a declining utility. Money is infinitely versatile. And even if all the things money can buy are subject to a law of diminishing utility, it does not follow that money itself is."[10] From the supposition that a person tends to lose more and more interest in what he is consuming as his consumption of it increases, it plainly cannot be inferred that he must also tend to lose interest in consumption itself or in the money that makes consumption possible. For there may always remain for him, no matter how tired he has become of what he has been doing, untried goods to be bought and fresh new pleasures to be enjoyed.

There are in any event many things of which people do not, from the very outset, immediately begin to tire. From certain goods, they actually derive more utility after sustained consumption than they derive at first. This is the situation whenever appreciating or enjoying or otherwise benefiting from something depends upon repeated trials, which serve as a kind of "warming up" process: for instance, when relatively little significant gratification is obtained from the item or experience in question until the individual has acquired a special taste for it, has become addicted to it, or has begun in some other way to relate or respond to it profitably.

9. "With successive new units of [a] good, your total utility will grow at a slower and slower rate because of a fundamental tendency for your psychological ability to appreciate more of the good to become less keen. This fact, that the increments in total utility fall off, economists describe as follows: as the amount consumed of a good increases, the *marginal utility* of the good (or the extra utility added by its last unit) tends to decrease" (Samuelson, *Economics*, p. 431).

10. Blum and Kalven, pp. 57–58.

The capacity for obtaining gratification is then smaller at earlier points in the sequence of consumption than at later points. In such cases marginal utility does not decline; it increases. Perhaps it is true of everything, without exception, that a person will ultimately lose interest in it. But even if in every utility curve there is a point at which the curve begins a steady and irreversible decline, it cannot be assumed that every segment of the curve has a downward slope.[11]

III

When marginal utility diminishes, it does not do so on account of any deficiency in the marginal unit. It diminishes in virtue of the position of that unit as the latest in a sequence. The same is true when marginal utility increases: the marginal unit provides greater utility than its predecessors in virtue of the effect which the acquisition or consumption of those predecessors has brought about. Now when the sequence consists of units of money, what corresponds to the process of warming up—at least, in one pertinent and important feature—is *saving*. Accumulating money entails, as warming up does, generating a capacity to derive, at some subsequent point in a sequence, gratifications that cannot be derived earlier.

The fact that it may at times be especially worthwhile for a person to save money rather than to spend each dollar as it comes along is due in part to the incidence of what may be thought of as "utility thresholds." Consider an item with the following characteristics: it is nonfungible, it is the source of a fresh and otherwise unobtainable type of satisfaction, and it is too expensive to be acquired except by saving up for it. The utility of the dollar that finally completes a program of saving up for such an item may be greater than the utility of any dollar saved earlier in the program. That will be the case when the utility provided by the item is greater than the sum of the utilities that could be derived if the money saved were either spent as it came in or divided into parts and used to purchase other things. In a situation of this kind, the final dollar saved permits the crossing of a utility threshold.[12]

11. People tend to think that it is generally more important to avoid a certain degree of harm than to acquire a benefit of comparable magnitude. It may be that this is in part because they assume that utility diminishes at the margin, for in that case the additional benefit would have less utility than the corresponding loss. However, it should be noted that the tendency to place a lower value on acquiring benefits than on avoiding harms is sometimes reversed: when people are so miserable that they regard themselves as "having nothing to lose," they may well place a higher value on improving things than on preventing them from becoming (to a comparable extent) even worse. In that case, what is diminishing at the margin is not the utility of benefits but the disutility of harms.

12. In virtue of these thresholds, a marginal or incremental dollar may have conspicuously greater utility than dollars that do not enable a threshold to be crossed. Thus, a person who uses his spare money during a certain period for some inconsequential improvement in his routine pattern of consumption—perhaps a slightly better quality of meat for dinner every night—may derive much less additional utility in this way than by saving up the

It is sometimes argued that, for anyone who is rational in the sense that he seeks to maximize the utility generated by his expenditures, the marginal utility of money must necessarily diminish. Abba Lerner presents this argument as follows:

> The principle of diminishing marginal utility of income can be derived from the assumption that consumers spend their income in the way that maximizes the satisfaction they can derive from the good obtained. With a given income, all the things bought give a greater satisfaction for the money spent on them than any of the other things that could have been bought in their place but were not bought for this very reason. From this it follows that if income were greater the additional things that would be bought with the increment of income would be things that are rejected when income is smaller because they give less satisfaction; and if income were greater still, even less satisfactory things would be bought. The greater the income the less satisfactory are the additional things that can be bought with equal increases of income. That is all that is meant by the principle of the diminishing marginal utility of income.[13]

Lerner invokes here a comparison between the utility of *G*(*n*)—the goods which the rational consumer actually buys with his income of *n* dollars—and "the other things that could have been bought in their place but were not." Given that he prefers to buy *G*(*n*) rather than the other things, which by hypothesis cost no more, the rational consumer must regard *G*(*n*) as offering greater satisfaction than the others can provide. From this Lerner infers that with an additional *n* dollars the consumer would be able to purchase only things with less utility than *G*(*n*); and he concludes that, in general, "the greater the income the less satisfactory are the additional things that can be bought with equal increases of income." This conclusion, he maintains, is tantamount to the principle of the diminishing marginal utility of income.

It seems apparent that Lerner's attempt to derive the principle in this way fails. One reason is that the amount of satisfaction a person can derive from a certain good may vary considerably according to whether or not he also possesses certain other goods. The satisfaction obtainable from a certain expenditure may therefore be greater if some other ex-

---

extra money for a few weeks and going to see some marvelous play or opera. The threshold effect is particularly integral to the experience of collectors, who characteristically derive greater satisfaction from obtaining the item that finally completes a collection—whichever item it happens to be—than from obtaining any of the other items in the collection. Obtaining the final item entails crossing a utility threshold: a complete collection of twenty different items, each of which when considered individually has the same utility, is likely to have greater utility for a collector than an incomplete collection that is of the same size but that includes duplicates. The completeness of the collection itself possesses utility, in addition to the utility provided individually by the items of which the collection is constituted.

13. Lerner, pp. 26–27.

penditure has already been made. Suppose that the cost of a serving of popcorn is the same as the cost of enough butter to make it delectable, and suppose that some rational consumer who adores buttered popcorn gets very little satisfaction from unbuttered popcorn but that he nonetheless prefers it to butter alone. He will buy the popcorn in preference to the butter, accordingly, if he must buy one and cannot buy both. Suppose now that this person's income increases so that he can buy the butter too. Then he can have something he enjoys enormously: his incremental income makes it possible for him not merely to buy butter in addition to popcorn but also to enjoy buttered popcorn. The satisfaction he will derive by combining the popcorn and the butter may well be considerably greater than the sum of the satisfactions he can derive from the two goods taken separately. Here, again, is a threshold effect.

In a case of this sort, what the rational consumer buys with his incremental income is a good—$G(i)$—which, when his income was smaller, he had rejected in favor of $G(n)$ because having it alone would have been less satisfying than having only $G(n)$. Despite this, however, it is not true that the utility of the income he uses to buy $G(i)$ is less than the utility of the income he used to buy $G(n)$. When there is an opportunity to create a combination which is (like buttered popcorn) synergistic in the sense that adding one good to another increases the utility of each, the marginal utility of income may not decline even though the sequence of marginal items—taking each of these items by itself—does exhibit a pattern of declining utilities.

Lerner's argument is flawed in virtue of another consideration as well. Since he speaks of "the *additional* things that can be bought with equal increases of income," he evidently presumes that a rational consumer uses his first n dollars to purchase a certain good and that he uses any incremental income beyond that to buy something else. This leads Lerner to suppose that what the consumer buys when his income is increased by i dollars (where i is equal to or less than n) must be something which he could have bought and which he chose not to buy when his income was only n dollars. But this supposition is unwarranted. With an income of $(n + i)$ dollars, the consumer need not use his money to purchase both $G(n)$ and $G(i)$. He might use it to buy something which cost more than either of these goods—something which was too expensive to be available to him at all before his income increased. The point is that if a rational consumer with an income of n dollars defers purchasing a certain good until his income increases, this does not necessarily mean that he "rejected" purchasing it when his income was smaller. The good in question may have been out of his reach at that time because it cost more than n dollars. His reason for postponing the purchase may have had nothing to do with comparative expectations of satisfaction or with preferences or priorities at all.

There are two possibilities to consider. Suppose on the one hand that, instead of purchasing $G(n)$ when his income is n dollars, the rational consumer saves that money until he can add an additional i dollars to it and then purchases $G(n + i)$. In this case it is quite evident that his deferral of the purchase of $G(n + i)$ does not mean that he values it less than $G(n)$. On the other hand, suppose that the rational consumer declines to save up for $G(n + i)$ and that he spends all the money he has on $G(n)$. In this case too it would be a mistake to construe his behavior as indicating a preference for $G(n)$ over $G(n + i)$. For the explanation of his refusal to save for $G(n + i)$ may be merely that he regards doing so as pointless because he believes that he cannot reasonably expect to save enough to make a timely purchase of it.

The utility of $G(n + i)$ may not only be greater than the utility either of $G(n)$ or of $G(i)$. It may also be greater than the sum of their utilities. That is, in acquiring $G(n + i)$ the consumer may cross a utility threshold. The utility of the increment i to his income is then actually greater than the utility of the n dollars to which it is added, even though i equals or is less than n. In such a case, the income of the rational consumer does not exhibit diminishing marginal utility.

IV

The preceding discussion has established that an egalitarian distribution may fail to maximize aggregate utility. It can also easily be shown that, in virtue of the incidence of utility thresholds, there are conditions under which an egalitarian distribution actually minimizes aggregate utility.[14] Thus, suppose that there is enough of a certain resource (e.g., food or medicine) to enable some but not all members of a population to survive. Let us say that the size of the population is ten, that a person needs at least five units of the resource in question to live, and that forty units are available. If any members of this population are to survive, some must have more than others. An equal distribution, which gives each person four units, leads to the worst possible outcome, namely, everyone dies. Surely in this case it would be morally grotesque to insist upon equality! Nor would it be reasonable to maintain that, under the conditions specified, it is justifiable for some to be better off only when this is in the interests of the worst off. If the available resources are used to save eight people, the justification for doing this is manifestly not that it somehow benefits the two members of the population who are left to die.

An egalitarian distribution will almost certainly produce a net loss of aggregate utility whenever it entails that fewer individuals than otherwise will have, with respect to some necessity, enough to sustain life—in other words, whenever it requires a larger number of individuals to be below the threshold of survival. Of course, a loss of utility may also occur even when the circumstances involve a threshold that does not separate life and death. Allocating resources equally will reduce aggregate

14. Conditions of these kinds are discussed in Nicholas Rescher, *Distributive Justice* (Indianapolis: Bobbs-Merrill Co., 1966), pp. 28–30.

utility whenever it requires a number of individuals to be kept below *any* utility threshold without ensuring a compensating move above some threshold by a suitable number of others.

Under conditions of scarcity, then, an egalitarian distribution may be morally unacceptable. Another response to scarcity is to distribute the available resources in such a way that as many people as possible have enough or, in other words, to maximize the incidence of sufficiency. This alternative is especially compelling when the amount of a scarce resource that constitutes enough coincides with the amount that is indispensable for avoiding some catastrophic harm—as in the example just considered, where falling below the threshold of enough food or enough medicine means death. But now suppose that there are available, in this example, not just forty units of the vital resource but forty-one. Then maximizing the incidence of sufficiency by providing enough for each of eight people leaves one unit unallocated. What should be done with this extra unit?

It has been shown above that it is a mistake to maintain that *where some people have less than enough, no one should have more than anyone else.* When resources are scarce, so that it is impossible for everyone to have enough, an egalitarian distribution may lead to disaster. Now there is another claim that might be made here, which may appear to be quite plausible but which is also mistaken: *where some people have less than enough, no one should have more than enough.* If this claim were correct, then—in the example at hand—the extra unit should go to one of the two people who have nothing. But one additional unit of the resource in question will not improve the condition of a person who has none. By hypothesis, that person will die even with the additional unit. What he needs is not one unit but five.[15] It cannot be taken for granted that a person who has a certain amount of a vital resource is necessarily better off than a person who has a lesser amount, for the larger amount may still be too small to serve any useful purpose. Having the larger amount may even make a person worse off. Thus it is conceivable that while a dose of five units of some medication is therapeutic, a dose of one unit is not better than none but actually toxic. And while a person with one unit of food may live a bit longer than someone with no food whatever, perhaps it is worse to prolong the process of starvation for a short time than to terminate quickly the agony of starving to death.

The claim that no one should have more than enough while anyone has less than enough derives its plausibility, in part, from a presumption that is itself plausible but that is nonetheless false: to wit, giving resources to people who have less of them than enough necessarily means giving

15. It might be correct to say that he does need one unit if there is a chance that he will get four more, since in that case the one unit can be regarded as potentially an integral constituent of the total of five that puts him across the threshold of survival. But if there is no possibility that he will acquire five, then acquiring the one does not contribute to the satisfaction of any need.

resources to people who need them and, therefore, making those people better off. It is indeed reasonable to assign a higher priority to improving the condition of those who are in need than to improving the condition of those who are not in need. But giving additional resources to people who have less than enough of those resources, and who are accordingly in need, may not actually improve the condition of these people at all. Those below a utility threshold are not necessarily benefited by additional resources that move them closer to the threshold. What is crucial for them is to attain the threshold. Merely moving closer to it either may fail to help them or may be disadvantageous.

By no means do I wish to suggest, of course, that it is never or only rarely beneficial for those below a utility threshold to move closer to it. Certainly it may be beneficial, either because it increases the likelihood that the threshold ultimately will be attained or because, quite apart from the significance of the threshold, additional resources provide important increments of utility. After all, a collector may enjoy expanding his collection even if he knows that he has no chance of ever completing it. My point is only that additional resources do not necessarily benefit those who have less than enough. The additions may be too little to make any difference. It may be morally quite acceptable, accordingly, for some to have more than enough of a certain resource even while others have less than enough of it.

V

Quite often, advocacy of egalitarianism is based less upon an argument than upon a purported moral intuition: economic inequality, considered as such, just seems wrong. It strikes many people as unmistakably apparent that, taken simply in itself, the enjoyment by some of greater economic benefits than are enjoyed by others is morally offensive. I suspect, however, that in many cases those who profess to have this intuition concerning manifestations of inequality are actually responding not to the inequality but to another feature of the situations they are confronting. What I believe they find intuitively to be morally objectionable, in the types of situations characteristically cited as instances of economic inequality, is not the fact that some of the individuals in those situations have *less* money than others but the fact that those with less have *too little*.

When we consider people who are substantially worse off than ourselves, we do very commonly find that we are morally disturbed by their circumstances. What directly touches us in cases of this kind, however, is not a quantitative discrepancy but a qualitative condition—not the fact that the economic resources of those who are worse off are *smaller in magnitude* than ours but the different fact that these people are so *poor*. Mere differences in the amounts of money people have are not in themselves distressing. We tend to be quite unmoved, after all, by inequalities between the well-to-do and the rich; our awareness that the former are substantially worse off than the latter does not disturb us morally at all.

And if we believe of some person that his life is richly fulfilling, that he himself is genuinely content with his economic situation, and that he suffers no resentments or sorrows which more money could assuage, we are not ordinarily much interested—from a moral point of view—in the question of how the amount of money he has compares with the amounts possessed by others. Economic discrepancies in cases of these sorts do not impress us in the least as matters of significant moral concern. The fact that some people have much less than others is morally undisturbing when it is clear that they have plenty.

It seems clear that egalitarianism and the doctrine of sufficiency are logically independent: considerations that support the one cannot be presumed to provide support also for the other. Yet proponents of egalitarianism frequently suppose that they have offered grounds for their position when in fact what they have offered is pertinent as support only for the doctrine of sufficiency. Thus they often, in attempting to gain acceptance for egalitarianism, call attention to disparities between the conditions of life characteristic of the rich and those characteristic of the poor. Now it is undeniable that contemplating such disparities does often elicit a conviction that it would be morally desirable to redistribute the available resources so as to improve the circumstances of the poor. And, of course, that would bring about a greater degree of economic equality. But the indisputability of the moral appeal of improving the condition of the poor by allocating to them resources taken from those who are well off does not even tend to show that egalitarianism is, as a moral ideal, similarly indisputable. To show of poverty that it is compellingly undesirable does nothing whatsoever to show the same of inequality. For what makes someone poor in the morally relevant sense—in which poverty is understood as a condition from which we naturally recoil—is not that his economic assets are simply of lesser magnitude than those of others.

A typical example of this confusion is provided by Ronald Dworkin. Dworkin characterizes the ideal of economic equality as requiring that "no citizen has less than an equal share of the community's resources just in order that others may have more of what he lacks."[16] But in support of his claim that the United States now falls short of this ideal, he refers to circumstances that are not primarily evidence of inequality but of poverty: "It is, I think, apparent that the United States falls far short now [of the ideal of equality]. A substantial minority of Americans are chronically unemployed or earn wages below any realistic 'poverty line' or are handicapped in various ways or burdened with special needs; and most of these people would do the work necessary to earn a decent living if they had the opportunity and capacity" (p. 208). What mainly concerns Dworkin—what he actually considers to be morally

16. Ronald Dworkin, "Why Liberals Should Care about Equality," in his *A Matter of Principle* (Cambridge, Mass.: Harvard University Press, 1985), p. 206. Page numbers in parentheses in the text refer to this work.

important—is manifestly not that our society permits a situation in which a substantial minority of Americans have *smaller shares* than others of the resources which he apparently presumes should be available for all. His concern is, rather, that the members of this minority *do not earn decent livings.*

The force of Dworkin's complaint does not derive from the allegation that our society fails to provide some individuals with as much as others but from a quite different allegation, namely, our society fails to provide each individual with "the opportunity to develop and lead a life he can regard as valuable both to himself and to [the community]" (p. 211). Dworkin is dismayed most fundamentally not by evidence that the United States permits economic inequality but by evidence that it fails to ensure that everyone has enough to lead "a life of choice and value" (p. 212)—in other words, that it fails to fulfill for all the ideal of sufficiency. What bothers him most immediately is not that certain quantitative relationships are widespread but that certain qualitative conditions prevail. He cares principally about the value of people's lives, but he mistakenly represents himself as caring principally about the relative magnitudes of their economic assets.

My suggestion that situations involving inequality are morally disturbing only to the extent that they violate the ideal of sufficiency is confirmed, it seems to me, by familiar discrepancies between the principles egalitarians profess and the way in which they commonly conduct their own lives. My point here is not that some egalitarians hypocritically accept high incomes and special opportunities for which, according to the moral theories they profess, there is no justification. It is that many egalitarians (including many academic proponents of the doctrine) are not truly concerned whether they are as well off economically as other people are. They believe that they themselves have roughly enough money for what is important to them, and they are therefore not terribly preoccupied with the fact that some people are considerably richer than they. Indeed, many egalitarians would consider it rather shabby or even reprehensible to care, with respect to their own lives, about economic comparisons of that sort. And, notwithstanding the implications of the doctrines to which they urge adherence, they would be appalled if their children grew up with such preoccupations.

VI

The fundamental error of egalitarianism lies in supposing that it is morally important whether one person has less than another regardless of how much either of them has. This error is due in part to the false assumption that someone who is economically worse off has more important unsatisfied needs than someone who is better off. In fact the morally significant needs of both individuals may be fully satisfied or equally unsatisfied. Whether one person has more money than another is a wholly extrinsic matter. It has to do with a relationship between the respective economic assets of the two people, which is not only independent of the amounts of their assets and of the amounts of satisfaction they can derive from them but also independent of the attitudes of these people toward those levels of assets and of satisfaction. The economic comparison implies nothing concerning whether either of the people compared has any morally important unsatisfied needs at all nor concerning whether either is content with what he has.

This defect in egalitarianism appears plainly in Thomas Nagel's development of the doctrine. According to Nagel: "The essential feature of an egalitarian priority system is that it counts improvements to the welfare of the worse off as more urgent than improvements to the welfare of the better off. . . . What makes a system egalitarian is the priority it gives to the claims of those . . . at the bottom. . . . Each individual with a more urgent claim has priority . . . over each individual with a less urgent claim."[17] And in discussing Rawls's Difference Principle, which he endorses, Nagel says: the Difference Principle "establishes an order of priority among needs and gives preference to the most urgent."[18] But the preference actually assigned by the Difference Principle is not in favor of those whose needs are most urgent; it is in favor of those who are identified as worst off. It is a mere assumption, which Nagel makes without providing any grounds for it whatever, that the worst off individuals have urgent needs. In most societies the people who are economically at the bottom are indeed extremely poor, and they do, as a matter of fact, have urgent needs. But this relationship between low economic status and urgent need is wholly contingent. It can be established only on the basis of empirical data. There is no necessary conceptual connection between a person's relative economic position and whether he has needs of any degree of urgency.[19]

It is possible for those who are worse off not to have more urgent needs or claims than those who are better off because it is possible for them to have no urgent needs or claims at all. The notion of "urgency" has to do with what is *important.* Trivial needs or interests, which have no significant bearing upon the quality of a person's life or upon his readiness to be content with it, cannot properly be construed as being urgent to any degree whatever or as supporting the sort of morally demanding claims to which genuine urgency gives rise. From the fact that a person is at the bottom of some economic order, moreover, it

17. Nagel, p. 118.

18. Ibid., p. 117.

19. What I oppose is the claim that when it comes to justifying attempts to improve the circumstances of those who are economically worst off, a good reason for making the attempt is that it is morally important for people to be as equal as possible with respect to money. The only morally compelling reason for trying to make the worse off better off is, in my judgment, that their lives are in some degree bad lives. The fact that some people have more than enough money suggests a way in which it might be arranged for those who have less than enough to get more, but it is not in itself a good reason for redistribution.

cannot even be inferred that he has *any* unsatisfied needs or claims. After all, it is possible for conditions at the bottom to be quite good; the fact that they are the worst does not in itself entail that they are bad or that they are in any way incompatible with richly fulfilling and enjoyable lives.

Nagel maintains that what underlies the appeal of equality is an "ideal of acceptability to each individual."[20] On his account, this ideal entails that a reasonable person should consider deviations from equality to be acceptable only if they are in his interest in the sense that he would be worse off without them. But a reasonable person might well regard an unequal distribution as entirely acceptable even though he did not presume that any other distribution would benefit him less. For he might believe that the unequal distribution provided him with quite enough, and he might reasonably be unequivocally content with that, with no concern for the possibility that some other arrangement would provide him with more. It is gratuitous to assume that every reasonable person must be seeking to maximize the benefits he can obtain, in a sense requiring that he be endlessly interested in or open to improving his life. A certain deviation from equality might not be *in* someone's interest because it might be that he would in fact be better off without it. But as long as it does not *conflict* with his interest, by obstructing his opportunity to lead the sort of life that it is important for him to lead, the deviation from equality may be quite acceptable. To be wholly satisfied with a certain state of affairs, a reasonable person need not suppose that there is no other available state of affairs in which he would be better off.[21]

Nagel illustrates his thesis concerning the moral appeal of equality by considering a family with two children, one of whom is "normal and quite happy" while the other "suffers from a painful handicap."[22] If this family were to move to the city the handicapped child would benefit from medical and educational opportunities that are unavailable in the suburbs, but the healthy child would have less fun. If the family were to move to the suburbs, on the other hand, the handicapped child would be deprived but the healthy child would enjoy himself more. Nagel stipulates that the gain to the healthy child in moving to the suburbs would be greater than the gain to the handicapped child in moving to the city: in the city the healthy child would find life positively disagreeable, while the handicapped child would not become happy "but only less miserable."

Given these conditions, the egalitarian decision is to move to the city; for "it is more urgent to benefit the [handicapped] child even though the benefit we can give him is less than the benefit we can give the [healthy] child." Nagel explains that this judgment concerning the greater urgency of benefiting the handicapped child "depends on the worse off position of the [handicapped] child. An improvement in his situation is

20. Nagel, p. 123.
21. For further discussion, see Sec. VII below.
22. Quotations from his discussion of this illustration are from Nagel, pp. 123–24.

more important than an equal or somewhat greater improvement in the situation of the [normal] child." But it seems to me that Nagel's analysis of this matter is flawed by an error similar to the one that I attributed above to Dworkin. The fact that it is preferable to help the handicapped child is not due, as Nagel asserts, to the fact that this child is worse off than the other. It is due to the fact that this child, and not the other, suffers from a painful handicap. The handicapped child's claim is important because his condition is *bad*—significantly undesirable—and not merely because he is *less well off* than his sibling.

This does not imply, of course, that Nagel's evaluation of what the family should do is wrong. Rejecting egalitarianism certainly does not mean maintaining that it is always mandatory simply to maximize benefits and that therefore the family should move to the suburbs because the normal child would gain more from that than the handicapped child would gain from a move to the city. However, the most cogent basis for Nagel's judgment in favor of the handicapped child has nothing to do with the alleged urgency of providing people with as much as others. It pertains rather to the urgency of the needs of people who do not have enough.[23]

VII

What does it mean, in the present context, for a person to have enough? One thing it might mean is that any more would be too much: a larger amount would make the person's life unpleasant, or it would be harmful or in some other way unwelcome. This is often what people have in mind when they say such things as "I've had enough!" or "Enough of that!" The idea conveyed by statements like these is that *a limit has been reached,* beyond which it is not desirable to proceed. On the other hand, the assertion that a person has enough may entail only that *a certain requirement or standard has been met,* with no implication that a larger quantity would be bad. This is often what a person intends when he says something like "That should be enough." Statements such as this one characterize the indicated amount as sufficient while leaving open the possibility that a larger amount might also be acceptable.

In the doctrine of sufficiency the use of the notion of "enough" pertains to *meeting a standard* rather than to *reaching a limit.* To say that a person has enough money means that he is content, or that it is reasonable for him to be content, with having no more money than he has. And to say this is, in turn, to say something like the following: the person does not (or cannot reasonably) regard whatever (if anything) is unsatisfying or distressing about his life as due to his having too little money. In other words, if a person is (or ought reasonably to be) content with the amount of money he has, then insofar as he is or has reason to be unhappy

23. The issue of equality or sufficiency that Nagel's illustration raises does not, of course, concern the distribution of *money*.

with the way his life is going, he does not (or cannot reasonably) suppose that money would—either as a sufficient or as a necessary condition—enable him to become (or to have reason to be) significantly less unhappy with it.[24]

It is essential to understand that having enough money differs from merely having enough to get along or enough to make life marginally tolerable. People are not generally content with living on the brink. The point of the doctrine of sufficiency is not that the only morally important distributional consideration with respect to money is whether people have enough to avoid economic misery. A person who might naturally and appropriately be said to have just barely enough does not, by the standard invoked in the doctrine of sufficiency, have enough at all.

There are two distinct kinds of circumstances in which the amount of money a person has is enough—that is, in which more money will not enable him to become significantly less unhappy. On the one hand, it may be that the person is suffering no substantial distress or dissatisfaction with his life. On the other hand, it may be that although the person is unhappy about how his life is going, the difficulties that account for his unhappiness would not be alleviated by more money. Circumstances of this second kind obtain when what is wrong with the person's life has to do with noneconomic goods such as love, a sense that life is meaningful, satisfaction with one's own character, and so on. These are goods that money cannot buy; moreover, they are goods for which none of the things money can buy are even approximately adequate substitutes. Sometimes, to be sure, noneconomic goods are obtainable or enjoyable only (or more easily) by someone who has a certain amount of money. But the person who is distressed with his life while content with his economic situation may already have that much money.

It is possible that someone who is content with the amount of money he has might also be content with an even larger amount of money. Since having enough money does not mean being at a limit beyond which more money would necessarily be undesirable, it would be a mistake to assume that for a person who already has enough the marginal utility of money must be either negative or zero. Although this person is by hypothesis not distressed about his life in virtue of any lack of things which more money would enable him to obtain, nonetheless it remains possible that he would enjoy having some of those things. They would not make him less unhappy, nor would they in any way alter his attitude toward his life or the degree of his contentment with it, but they might bring him pleasure. If that is so, then his life would in this respect be better with more money than without it. The marginal utility for him of money would accordingly remain positive.

24. Within the limits of my discussion it makes no difference which view is taken concerning the very important question of whether what counts is *the attitude a person actually has* or *the attitude it would be reasonable for him to have*. For the sake of brevity, I shall henceforth omit referring to the latter alternative.

To say that a person is content with the amount of money he has does not entail, then, that there would be no point whatever in his having more. Thus someone with enough money might be quite *willing* to accept incremental economic benefits. He might in fact be *pleased* to receive them. Indeed, from the supposition that a person is content with the amount of money he has it cannot even be inferred that he would not *prefer* to have more. And it is even possible that he would actually be prepared to *sacrifice* certain things that he values (e.g., a certain amount of leisure) for the sake of more money.

But how can all this be compatible with saying that the person is content with what he has? What *does* contentment with a given amount of money preclude, if it does not preclude being willing or being pleased or preferring to have more money or even being ready to make sacrifices for more? It precludes his having an *active interest* in getting more. A contented person regards having more money as *inessential* to his being satisfied with his life. The fact that he is content is quite consistent with his recognizing that his economic circumstances could be improved and that his life might as a consequence become better than it is. But this possibility is not important to him. He is simply not much interested in being better off, so far as money goes, than he is. His attention and interest are not vividly engaged by the benefits which would be available to him if he had more money. He is just not very responsive to their appeal. They do not arouse in him any particularly eager or restless concern, although he acknowledges that he would enjoy additional benefits if they were provided to him.

In any event, let us suppose that the level of satisfaction that his present economic circumstances enable him to attain is high enough to meet his expectations of life. This is not fundamentally a matter of how much utility or satisfaction his various activities and experiences provide. Rather, it is most decisively a matter of his attitude toward being provided with that much. The satisfying experiences a person has are one thing. Whether he is satisfied that his life includes just those satisfactions is another. Although it is possible that other feasible circumstances would provide him with greater amounts of satisfaction, it may be that he is wholly satisfied with the amounts of satisfaction that he now enjoys. Even if he knows that he could obtain a greater quantity of satisfaction overall, he does not experience the uneasiness or the ambition that would incline him to seek it. Some people feel that their lives are good enough, and it is not important to them whether their lives are as good as possible.

The fact that a person lacks an active interest in getting something does not mean, of course, that he prefers not to have it. This is why the contented person may without any incoherence accept or welcome improvements in his situation and why he may even be prepared to incur minor costs in order to improve it. The fact that he is contented means only that the possibility of improving his situation is not *important* to him. It only implies, in other words, that he does not resent his circumstances, that he is not anxious or determined to improve them, and that he does not go out of his way or take any significant initiatives to make them better.

It may seem that there can be no reasonable basis for accepting less satisfaction when one could have more, that therefore rationality itself entails maximizing, and, hence, that a person who refuses to maximize the quantity of satisfaction in his life is not being rational. Such a person cannot, of course, offer it as his reason for declining to pursue greater satisfaction that the costs of this pursuit are too high; for if that were his reason then, clearly, he would be attempting to maximize satisfaction after all. But what other good reason could he possibly have for passing up an opportunity for more satisfaction? In fact, he may have a very good reason for this: namely, *that he is satisfied with the amount of satisfaction he already has.* Being satisfied with the way things are is unmistakably an excellent reason for having no great interest in changing them. A person who is indeed satisfied with his life as it is can hardly be criticized, accordingly, on the grounds that he has no good reason for declining to make it better.

He might still be open to criticism on the grounds that he *should not* be satisfied—that it is somehow unreasonable, or unseemly, or in some other mode wrong for him to be satisfied with less satisfaction than he could have. On what basis, however, could *this* criticism be justified? Is there some decisive reason for insisting that a person ought to be so hard to satisfy? Suppose that a man deeply and happily loves a woman who is altogether worthy. We do not ordinarily criticize the man in such a case just because we think he might have done even better. Moreover, our sense that it would be inappropriate to criticize him for that reason need not be due simply to a belief that holding out for a more desirable or worthier woman might end up costing him more than it would be worth. Rather, it may reflect our recognition that the desire to be happy or content or satisfied with life is a desire for a satisfactory amount of satisfaction and is not inherently tantamount to a desire that the quantity of satisfaction be maximized.

Being satisfied with a certain state of affairs is not equivalent to preferring it to all others. If a person is faced with a choice between less and more of something desirable, then no doubt it would be irrational for him to prefer less to more. But a person may be satisfied without having made any such comparisons at all. Nor is it necessarily irrational or unreasonable for a person to omit or to decline to make comparisons between his own state of affairs and possible alternatives. This is not only because making comparisons may be too costly. It is also because if someone is satisfied with the way things are, he may have no motive to consider how else they might be.[25]

25. Compare the sensible adage: "If it's not broken, don't fix it."

Contentment may be a function of excessive dullness or diffidence. The fact that a person is free both of resentment and of ambition may be due to his having a slavish character or to his vitality being muffled by a kind of negligent lassitude. It is possible for someone to be content merely, as it were, by default. But a person who is content with resources providing less utility than he could have may not be irresponsible or indolent or deficient in imagination. On the contrary, his decision to be content with those resources—in other words, to adopt an attitude of willing acceptance toward the fact that he has just that much—may be based upon a conscientiously intelligent and penetrating evaluation of the circumstances of his life.

It is not essential for such an evaluation to include an *extrinsic* comparison of the person's circumstances with alternatives to which he might plausibly aspire, as it would have to do if contentment were reasonable only when based upon a judgment that the enjoyment of possible benefits has been maximized. If someone is less interested in whether his circumstances enable him to live as well as possible than in whether they enable him to live satisfyingly, he may appropriately devote his evaluation entirely to an *intrinsic* appraisal of his life. Then he may recognize that his circumstances do not lead him to be resentful or regretful or drawn to change and that, on the basis of his understanding of himself and of what is important to him, he accedes approvingly to his actual readiness to be content with the way things are. The situation in that case is not so much that he rejects the possibility of improving his circumstances because he thinks there is nothing genuinely to be gained by attempting to improve them. It is rather that this possibility, however feasible it may be, fails as a matter of fact to excite his active attention or to command from him any lively interest.[26]

APPENDIX

Economic egalitarianism is a drily formalistic doctrine. The amounts of money its adherents want for themselves and for others are calculated without regard to anyone's personal characteristics or circumstances. In this formality, egalitarians resemble people who desire to be as rich as possible but who have no idea what they would do with their riches. In neither case are the individual's ambitions, so far as money is concerned, limited or measured according to an understanding of the goals that he intends his money to serve or of the importance of these goals to him.

26. People often adjust their desires to their circumstances. There is a danger that sheer discouragement, or an interest in avoiding frustration and conflict, may lead them to settle for too little. It surely cannot be presumed that someone's life is genuinely fulfilling, or that it is reasonable for the person to be satisfied with it, simply because he does not complain. On the other hand, it also cannot be presumed that when a person has accommodated his desires to his circumstances, this is itself evidence that something has gone wrong.

The desire for unlimited wealth is fetishistic, insofar as it reflects with respect to a *means* an attitude—namely, desiring something for its own sake—that is appropriate only with respect to an *end*. It seems to me that the attitude taken by John Rawls toward what he refers to as "primary goods" ("rights and liberties, opportunities and powers, income and wealth")[27] tends toward fetishism in this sense. The primary goods are "all purpose means," Rawls explains, which people need no matter what other things they want: "Plans differ, since individual abilities, circumstances, and wants differ . . . ; but whatever one's system of ends, primary goods are a necessary means" (Rawls, p. 93). Despite the fact that he identifies the primary goods not as ends but as means, Rawls considers it rational for a person to want as much of them as possible. Thus, he says: "Regardless of what an individual's rational plans are in detail, it is assumed that there are various things which he would prefer more of rather than less. While the persons in the original position do not know their conception of the good, they do know, I assume, that they prefer more rather than less primary goods" (Rawls, pp. 92–93). The assumption that it must always be better to have more of the primary goods rather than less implies that the marginal utility of an additional quantity of a primary good is invariably greater than its cost. It implies, in other words, that the incremental advantage to an individual of possessing a larger quantity of primary goods is never outweighed by corresponding incremental liabilities, incapacities, or burdens.

But this seems quite implausible. Apart from any other consideration, possessing more of a primary good may well require of a responsible individual that he spend more time and effort in managing it and in making decisions concerning its use. These activities are for many people intrinsically unappealing; and they also characteristically involve both a certain amount of anxiety and a degree of distraction from other pursuits. Surely it must not be taken simply for granted that incremental costs of these kinds can never be greater than whatever increased benefits a corresponding additional amount of some primary good would provide.

Individuals in the original position are behind a veil of ignorance. They do not know their own conceptions of the good or their own life plans. Thus it may seem rational for them to choose to possess primary goods in unlimited quantities: since they do not know what to prepare for, perhaps it would be best for them to be prepared for anything. Even in the original position, however, it is possible for people to appreciate that at some point the cost of additional primary goods might exceed the benefits those goods provide. It is true that an individual behind the veil of ignorance cannot know at just what point he would find that an addition to his supply of primary goods costs more than it is worth. But his ignorance of the exact location of that point hardly warrants his acting as though no such point exists at all. Yet that is precisely how he does act if he chooses that the quantity of primary goods he possesses be unlimited.

Rawls acknowledges that additional quantities of primary goods may be, for some individuals, more expensive than they are worth. In his view, however, this does not invalidate the supposition that it is rational for everyone in the original position to want as much of these goods as they can get. Here is how he explains the matter:

27. John Rawls, *A Theory of Justice* (Cambridge, Mass.: Harvard University Press, 1971), p. 92. Additional references to this book appear in parentheses in the text.

> I postulate that they [i.e., the persons in the original position] assume that they would prefer more primary social goods rather than less. Of course, it may turn out, once the veil of ignorance is removed, that some of them for religious or other reasons may not, in fact, want more of these goods. But from the standpoint of the original position, it is rational for the parties to suppose that they do want a larger share, since in any case they are not compelled to accept more if they do not wish to, nor does a person suffer from a greater liberty. [Rawls, pp. 142–43]

I do not find this argument convincing. It neglects the fact that dispensing with or refusing to accept primary goods that have been made available is itself an action that may entail significant costs. Burdensome calculations and deliberations may be required in order for a person to determine whether an increment of some primary good is worth having, and making decisions of this sort may involve responsibilities and risks in virtue of which the person experiences considerable anxiety. What is the basis, moreover, for the claim that no one suffers from a greater liberty? Under a variety of circumstances, it would seem, people may reasonably prefer to have fewer alternatives from which to choose rather than more. Surely liberty, like all other things, has its costs. It is an error to suppose that a person's life is invariably improved, or that it cannot be made worse, when his options are increased.[28]

28. For pertinent discussion of this issue, see Gerald Dworkin, "Is More Choice Better than Less?" in *Midwest Studies in Philosophy*, ed. P. French, T. Uehling, and H. Wettstein (Minneapolis: University of Minnesota Press, 1982), vol. 7, pp. 47–61.

[16]

Madison Powers

Forget About Equality

ABSTRACT. Justice is widely thought to consist in equality. For many theorists, the central question has been: Equality of what? The author argues that the ideal of equality distorts practical reasoning and has deeply counterintuitive implications. Moreover, an alternative view of distributive justice can give a better account of what egalitarians should care about than can any of the competing ideals of equality.

FOR THE LAST 25 YEARS most moral and political philosophers have assumed that justice is a matter of equality. Citing Aristotle for support, the central question for theories of justice has been: Equality of what? (Nagel 1979; Dworkin 1981; Sen 1992; Cohen 1989). Among the answers frequently proposed are equality of welfare, resources, income, opportunity, or some hybrid thereof (Cohen 1989; Dworkin 1981; Fried 1978; Green 1989; Arneson 1989). Other theorists defend the notion of equality of access to specific goods and services, such as education and health care (Daniels 1985; Green 1976; Veatch 1986). Almost all of the major theoretical contenders occupy what Ronald Dworkin has called the egalitarian plateau (Dworkin 1987). The assumption is that we merely disagree over the best interpretation(s) of equality and about the importance of equality in relation to other values or principles. For many, the aim is to identify the core, foundational conception of equality that provides the key to the resolution of all of the practical questions of distributive justice. Some other theorists agree that justice is a matter of equàlity of some sort, but they doubt the existence of any deeper unifying structure. They argue that a plurality of ideals of equality may pull us in different directions and that all are of central importance in moral and political thought, even though they do not yield ready answers to questions of distributive justice (Williams 1962).

I want to challenge the assumption that the central concern of a theory of justice should be equality, at least when discussing the distribution of

KENNEDY INSTITUTE OF ETHICS JOURNAL • JUNE 1996

scarce resources. I offer six arguments for the claim that it is not equality itself, but other moral concerns, that egalitarians ought to care about in the allocation of scarce resources.

THE DISTORTION OF PRACTICAL REASONING

The first argument is that a commitment to equality suffers from a defect analogous to that which many believe plagues utilitarianism, namely, the need to maximize. David Wiggins (1980, p. 232) argues that utilitarians' initial assumption of some prima facie duty to maximize anything distorts one's appreciation of the moral landscape. Taking the duty to maximize as a benchmark, from which all departures from maximization stand in need of justification, distorts the way one should think about the variety of goods to be realized in human action. We should focus first and foremost on the diversity of human goods at stake in each particular context of practical judgment, not on maximizing. The general presumption in favor of maximizing turns practical reasoning upside down and diverts attention away from concrete ends. In short, the overarching aim of maximizing is not a reliable guide to thinking about what we ought to do, even as a generalization.

We can say the same to the egalitarian. There is no prima facie duty to equalize anything. Like Wiggins, I take my cue from Aristotle. Contrary to a standard reading of Aristotle, I emphasize that when he says that justice must be some form of equality, he is reporting what he takes to be the "general opinion." However, I think that Aristotle's *Politics* is best read as a refutation of general opinion. Because Aristotle's arguments are complex, a full defense of any reading of the *Politics* would require a much more detailed historical treatment than I can supply here. Nonetheless, I shall argue that Aristotle offers three arguments that, taken together, suggest an approach to equality that is much different from the currently dominant view.

First, Aristotle argues that often the relevant form of equality one should seek is not strict equality in distribution, but a rather different notion, which he calls proportional equality (*Politics* 1282b–1283a). Proportional equality directs us to distribute goods according to the specific aims of an activity and in light of the ways that individuals conceive of the purposes of association that underlie our various institutions and social practices (Walzer 1983; Taylor 1985). Such aims and purposes, of course, are matters of interpretation and potential disagreement, but Aristotle's central point is clear: the relevant principle of distribution will vary according to the type of institution or activity. In most contexts, for example, we do not seek strict equality of distribution among potential recipients of goods. Instead, we distribute in proportion to the contribution each recipient makes to the activity under consideration. We award the prize for the race to the swift. We distribute flutes to those with musical ability, not because they are well-born. So a strict adherence to equal distribution is but one of many principles of distributive justice that may be relevant to any given activity, association, or institution.

Second, although Aristotle maintains that strict equality may be a good at which one sometimes should aim, often it is not, primarily because its moral importance depends on the other goods at stake in a particular context. The circumstances in which one has reason to favor equality in the possession of a good are ones in which the reduction of inequality is instrumental in the achievement of some other good. For example, Aristotle considers the value of equality of property in securing social cohesion and maintaining the stability of political institutions (*Politics* 1266b–1267b). He admits that there is some merit in the argument that inequality contributes to civil discord, but he says that a lack of necessities is not the sole cause of crime and social disintegration. He famously notes that "Men do not become tyrants in order to avoid exposure to the cold" (*Politics* 1267a). Moreover, strict equality itself can be a source of civil discord. For strict equality ignores the claims those who contribute more to the community, and so dissatisfaction arises among those who feel that "they deserved something more than mere equality" (*Politics* 1267a). If equality of property contributes to both the lessening and the increase of social tensions that influence political cohesion and communal solidarity, then the pursuit of strict equality appears to be an inherently unstable social solution. We therefore need to look elsewhere for a better account of justice.

Aristotle's third argument against the notion of justice as equality proceeds from a consideration of the question he addresses in the *Ethics*: What is the best way of life for the majority of individuals and states? For Aristotle, a good life is a life free of impediments. It is important for a good life that one have a sufficient amount of material resources. However, individuals who enjoy either too few or too many advantages are not prepared to listen to reason. Those with too little are far too mean and poor-spirited, while those nurtured in luxury never acquire the habit of discipline. With great extremes in wealth, the state will be characterized by "envy on the one side and on the other contempt . . . . Men will not even share the same path" *(Ethics* 1295b). The claim is not that inequality itself is bad, but that a life lived at either extreme is not a good life. Extreme inequality, therefore, ought to be reduced, not because precise equality is an ideal to be sought, but because neither too much nor too little permits human flourishing.

In effect, Aristotle offers three closely related arguments. First, justice is too complex to be seen as an all-purpose principle specifying the kinds of inequalities that are in need of justification. More often than not the distributive principles appropriate to a specific context have nothing to do with equalizing anything. Second, the fact that strict equality can easily promote bad consequences as well as good shows that it should not be taken as prima facie ideal. Its goodness or badness depends on its instrumental role in achieving good lives for individuals and states. Third, the claim that either having too much or too little is detrimental to human flourishing does not provide a rationale for the pursuit of equality. That both extremes are inconsistent with human well-being does, however, provide an indirect argument for the compression of the range of inequality within society in order that all citizens can flourish.

COUNTERINTUITIVE IMPLICATIONS

Despite the preceding arguments, the prospect that equality loses its luster in particular contexts will not provide sufficient reason for some egalitarians to abandon their commitment to it as an ideal, only reason to limit its scope. However, my second argument is that a commitment to the intrinsic value of equality, even in prima facie form, has counterintuitive implications. If equality is intrinsically valuable, or a moral ideal of independent significance, then its proponents must admit the existence of at least *some* moral reason to favor a state of affairs in which everyone is more nearly equal, even if no one is made better off and some are made much worse off. But we may ask, why should anyone think that equality itself has value when there is no one for whom life is made better? Arguments of this sort against equality have been around for some time, although sometimes only as implicit assumptions behind other arguments (Parfit 1989; Tempkin 1993, pp. 248–64).

Other anti-egalitarian arguments have been more direct. J. R. Lucas (1965, p. 303) claims that the "acid test" of a true egalitarian is to find it morally objectionable that some may have more than others and to be prepared on principle to reduce the level of well-being of one individual, even if doing so would provide negligible or no benefit to others. Joseph

Raz (1986, p. 235) similarly notes that egalitarian principles are "indifferent between taking away from those who have and giving to those who have not." Such indifference seems to involve an irrational attitude toward waste. As Robert Goodin (1995, p. 248) observes, there is something "undeniably paradoxical about wantonly destroying needed resources" for the sake of obtaining equality. Equality can be achieved by leveling down as well as by leveling up, and the commitment to equality itself does not discriminate between the two (Goodin 1995, p. 248).

Some egalitarians accept that a commitment to the intrinsic importance of equality may have somewhat counterintuitive implications. However, they argue that only the belief that inequality itself is bad can account for the guilt or shame that egalitarians feel in its presence. Their claim is that moral concerns about equality are essentially comparative judgments about how one fares relative to others, not concerns merely about the absolute low level of welfare of the worse-off (Tempkin 1993, p. 246). Even if a prima facie commitment to equality commits one to leveling down on some occasions, some egalitarians argue that equality has an independent moral significance that cannot be explained away (Nagel 1991, p. 107). Moreover, defenders of equality claim that the leveling-down objection can be answered in the same way that other prima facie ideals are defended. Arguments based on the independent moral significance of equality may be outweighed by other ideals, which would, "if exclusively pursued, have terrible implications" (Tempkin 1993, p. 282). The egalitarians therefore claim that the counterintuitiveness of leveling down in the name of equality is no more an objection to equality than it is an objection to the ideal of liberty that in some instances its pursuit may undermine other values, including equality.

Nevertheless, I think that the charge of counterintuitiveness can be made to stick, especially if an alternative account can be given of the motivation for the concern many individuals experience in the face of inequality. The development of this alternative is the thrust of the third and fourth arguments.

TOWARD AN EGALITARIAN ALTERNATIVE

A third argument against the primacy of equality may make abandonment of the ideal more palatable to egalitarian sensibilities. Suppose that we reject any reason to equalize, except when the move to equality benefits the worse-off. Derek Parfit (1989) calls this the priority view. In its most general form, the priority view simply holds that one should give some priority to the worse-off. Some treat the priority view as a weak form of egalitarianism. However, the rationale underlying this view is benefit to the worse-off, rather than the assumption of something special about equality itself. Thus, while the priority view's focus on the worse-off makes it sensitive to moral concerns associated with inequality, it is not committed to the importance of equality per se that is the hallmark of the pure egalitarian position.

Priority views can come in varying strengths. The version defended by Rawls, for example, is especially stringent. His distributive principle gives strict priority to improving the condition of the worst-off. No improvement in the condition of the better-off is permitted unless it benefits, or at least does not worsen, the condition of the worst-off (Rawls 1971, p. 151). Moreover, the priority view only justifies benefits to the better-off when they make the condition of the worst-off as good as possible under any alternative, morally acceptable social arrangement. It would not allow, say, a benefit to the better-off, even if the change provided a small benefit to the worst-off, if some other alternative would provide still greater benefits to the worst-off without violating other moral commitments, such as equal liberties and equal opportunity.

At first glance, the priority view appears to embody an implicit commitment to equality (Nagel 1979, p. 107). Such a view narrows the conditions under which inequalities may be morally justified and, under certain conditions, can compress the range of inequality within a society. But two arguments show that it is incorrect to treat the priority view and a commitment to the independent moral importance of equality as equivalent. First, the priority view does not permit taking goods away from the better-off when doing so would produce no benefit for the worse-off. Consequently, the priority view is interested solely in benefiting the worse-off, not in equalizing anything. Second, although the priority view tends toward equality, it does so only contingently. The priority view permits the better-off to benefit greatly while the worse-off benefit only slightly, as long as the worse-off benefit at least as much as they would under any other available, morally acceptable social option. Thus, the priority view, even in its most stringent form, clearly does not care about equality for its own sake, for it is in principle compatible with extreme inequalities (Nagel 1991, p. 107; Parfit 1989).

What then is the rationale for giving priority to the worse-off? The weak egalitarians who adopt some sort of priority view identify the unfortunate condition of the worse-off as the morally salient concern.

Joseph Raz (1986, p. 240), for example, claims that "their hunger is greater, their need more pressing, their suffering more hurtful, and therefore our concern for the hungry, the needy, the suffering, not our concern for equality, makes us give them priority." This account of the rationale for the heightened moral concern triggered by the experience of inequality avoids the counterintuitive implications of the pure egalitarian position. In this case, the only prima facie reason to prefer a movement toward equality is the urgency of unmet needs.

Although the shift to a priority view avoids some of the difficulties associated with the treatment of equality as intrinsically valuable, it does not respond to all of the egalitarian's concerns. The reason that equality still retains its appeal for many egalitarians is their response to more than the low *absolute* level of well-being of the worse-off. They also attach moral significance to the *comparative* or *relative* differences in levels of well-being. This may seem to provide the pure egalitarian with renewed hope: In some cases, equality seems to be what people care about after all. If some version of the priority view is to be shown preferable to a pure egalitarian view, we need a fourth argument that can take account of these residual concerns without encountering new objections.

WHICH INEQUALITIES MATTER?

Consider first the implications of the strict priority view in a variation of an example offered by Harry Frankfurt. Suppose that there is enough medicine at the appropriate dosage to enable some but not all members of a population to survive. According to the priority view, no prima facie reason exists to prefer an equal distribution of the medicine, which would guarantee the death of all. However, suppose that the drug is only fully effective in eradicating the illness in persons who are otherwise healthy; for those who are less healthy, the drug will only sustain their lives for a few months. Suppose also that the less healthy half of the population have lived what everyone would agree are worse lives overall, say, lives that they themselves would subjectively judge to be barely worth living. Such individuals are clearly worse-off in anyone's view. They are worse-off at the present moment, they have been among the worse-off over a lifetime, and they are worse-off according to their own estimates. The strict priority view seems to recommend giving the drug to those worse-off individuals, even though doing so would mean that the otherwise healthy half of the population will die immediately and the worse-off half will die after a few more months of life barely worth living.

In such cases, a strict priority principle will appear irrational to many, and, indeed, the defender of some version of the priority claim may be moved to adopt a more flexible priority principle, for example, one that gives *some*, but not absolute, priority to the worse-off, just because they are worse-off (Nagel 1991, p. 73). A view that gives flexible priority to the worse-off remains sensitive to moral differences that are associated with relative inequalities. However, the justification for sometimes benefiting the better-off is no longer conditional upon its benefiting, or at least not harming, the worse-off. Accordingly, a flexible priority principle can avoid the charge of irrationality in cases in which giving priority to the worse-off has dire consequences for some and little or no clear benefit for others.

If we retreat from a strict priority view to some weaker principle that accords the needs of the worse-off some indeterminate greater weight in distribution than the needs of the better-off, we naturally may ask whether we can further delineate which needs most merit the assignment of greater weight. Frankfurt's alternative to pure egalitarian principles provides one answer. In his view, it is morally important that everyone has enough, or a sufficient amount, of what contributes to a good life, not that everyone has an equal share. The true concern, as Frankfurt has argued, is with a principle of sufficiency. For example, he claims that it is not that "some of the individuals . . . have *less* money than others but the fact that those with less have *too little*" (Frankfurt 1987, p. 32).

Frankfurt's sufficiency principle differs from the strict priority principle as much as it departs from pure egalitarianism. Seeing to it that all (or as many as possible) have enough is not equivalent to giving strict priority to the worse-off. On Frankfurt's view, the needs of the worse-off are not more urgent simply because they are the needs of the comparatively worse-off. He claims that, "[i]f everyone had enough, it would be of no moral consequence whether some had more than others" (Frankfurt 1987, p. 21). Hence, the only needs of the worse-off that are morally more urgent are those that cause persons to fall below the level of sufficiency. Above that level, the fact that some are worse-off does not matter morally. Whatever priority we accord to the worse-off in the sufficiency view, we do so only to the extent that their condition falls below some minimum absolute level of well-being.

It is important to see clearly how the strict priority view and the sufficiency view differ. On the strict priority view, concern with the condition of the worse-off never ends. As long as there is some group that is at least somewhat worse-off, the priority view struggles to close the gap. So even though the strict priority view does not take equality as important in itself, its moral concern extends until all inequalities are eliminated. By contrast, the moral concern of the sufficiency view stops at the point at which everyone has enough. Consequently, the sufficiency view appears to be less able than the strict priority view to accommodate the egalitarian's concerns about relative inequality.

If, as I have argued, the pure egalitarian view is counterintuitive, the strict priority view is irrational in some cases, and the sufficiency view fails to address the egalitarian's concerns, where does that leave us? I shall argue that the best option is to defend a revised version of the sufficiency view. However, the defense of the revised sufficiency principle must explain the felt moral significance of relative inequality without capitulating to the egalitarian's claim that equality itself is what matters. In order to steer between a position that exhibits no concern about relative inequality and one that takes all inequalities as bad in themselves, we must defend a view that responds only to cases in which inequalities are unjustified. I argue that such cases are those in which "everyone's having enough" is incompatible with some having too much.

The requirements for a good life include the satisfaction of needs that Robert Goodin calls relative needs. Relative needs are ones for which the benchmark for satisfaction is tied to how well everyone else in society fares. Developing insights originated by Rousseau, Goodin points to cases of what he calls competitive utilization. He argues that the measure of how useful goods such as money are "to you in pursuing your ends depends on how much others have and use of it in bidding against you for scarce resources that you both desire but cannot both simultaneously enjoy" (Goodin 1995, p. 250). Thus, one cannot say when someone has sufficient money to secure enough of what contributes to a good life in the absence of an account of how much others have.

Money is not the only good for which sufficiency can only be judged in light of the amount others have. That some individuals can outbid others who have less money in competitive struggles for scarce goods is not the only way that someone's life may be made worse-off as a by-product of inequality. There are many goods from which the possessors gain advantage only when their possession works to the detriment of others (Rousseau [1755] 1988, pp. 9, 32). Honor and distinction, and power and influence are examples. The more others have of them, the less good mine will do me (Goodin 1995, p. 251). So for some goods, the claim that some individuals have *too little* must mean that they have an insufficient amount necessary to pursue the kinds of lives generally available to others in their community (Lucas 1965, p. 302). Those who command a substantially greater share of the community's goods are better-off, not simply by virtue of having a higher standard of living, but also by having disproportionate influence in public affairs and augmented bargaining power in private transactions. In addition to lacking the resources needed to outbid the better-off for scarce goods, the worse-off may have little option but to do the hardest work, with the greatest threat to health and safety, at the least convenient times, with the greatest risk of ruin by a turn of bad fortune, and for wages that can never raise them above their current conditions. What the better-off have that the worse-off lack under conditions of extreme inequality is the option of saying no, or at least the option of holding out for better terms in both private and public spheres.

Contrary to the pure egalitarian's claim that it is inequality itself that accounts for the shame and guilt that many experience in the face of it, I claim that some inequalities are morally troubling precisely because they indicate that some enjoy benefits that may only be obtained by acting to the detriment of others. The extreme inequalities that give us the most cause for moral concern, and thus a justified feeling of shame and guilt, are the ones that subordinate others. Such inequalities virtually guarantee that others will have diminished ability to exercise control over their own individual lives, that the terms of social cooperation will be established largely by those who have disproportionate influence in the public sphere, and that the prospects for the worse-off to improve their own lot through their own efforts are stymied by a lack of bargaining power.

The true object of egalitarian moral concern, therefore, must be those inequalities that permit some to benefit greatly only by making others substantially worse-off, leaving them with little possibility of improving their fortune. Inequalities that result in the subordination of one group to the interests of another undermine social solidarity, political sustainability, and the conditions necessary for individuals to lead autonomous lives.

We can now summarize how the revised sufficiency principle differs from its two main competitors. Pure egalitarians are too indiscriminate in their approach to inequality. They care equally about all inequalities. The strict priority view maintains permanent vigilance against inequality, but it permits inequalities that can result in the subordination of the worse-off. Both the pure egalitarian and the priority theorist falsely assume that a closer approximation to equality necessarily represents a moral improvement. A false precision in pursuit of some metric of equality can lead us to ignore the deeper sources of our moral objections to inequality. Once we see that having enough goods to lead a good life sometimes requires that others not have too much, the principle of sufficiency can offer a more plausible account of which inequalities to notice than can either the pure egalitarian or the strict priority theorist.

THE RECONSTRUCTION OF COMMONSENSE MORALITY

My fifth argument is that neither strict equality nor Rawls's strict priority view provides a plausible reconstruction of the considered judgments of commonsense morality. For some, this objection will not much matter, but for those who accept something like Rawls's account of reflective equilibrium, this will be a serious criticism. The plausibility of a theory itself depends on some explanatory connection to our most carefully considered moral judgments. Of course, what we take as our considered judgments is highly contestable. Nonetheless, I offer two examples of views that I think defenders of various principles of distributive justice ought to take seriously in any process of reaching reflective equilibrium.

First, a great deal of empirical work on attitudes toward justice and distribution reveals that the distributive principle chosen in specific contexts typically depends on the goals of the institution and on the nature of the relationships among the participants. In a survey of such empirical evidence, David Miller (1992) offers a number of examples that support both the account of sufficiency as a societal ideal and the claim that a plurality of context-sensitive distributive principles will be needed to address specific distributional questions. In cases where the main goal is efficiency and the relations among persons are more distant, the most widely supported principle of distribution is reward for desert or merit. But where cooperation is prized and the preservation of social solidarity and trust is a central aim among intimates or like-minded persons, a principle of roughly equal division is more likely to be favored. In other words, people respond exactly in the way that Aristotle predicted. The respondents favored the distribution of goods in accordance with how they understood the specific aims of an activity and in light of their views of the purposes of the association.

Experimental evidence about attitudes toward societal distributive principles also shows virtually no support for either strict egalitarian distributions or for Rawls's priority view. The most frequently endorsed view is a principle of maximization, subject to a welfare floor—i.e., a distributional constraint that guarantees a basic level of well-being to all persons. Most individuals also endorse the compression of the range of inequality, say, when great inequalities threaten social solidarity or political sustainability.

Second, Jon Elster (1992) summarizes conclusions about "the commonsense theory of justice" that he has identified as pervasive among institutional problem solvers. Elster's account reinforces a number of the points developed in Miller's survey of attitudes toward distributive principles. The central claim is that one should maximize aggregate welfare subject to a floor constraint on individual welfare (Elster 1992, p. 238), and other principles come into play in a rough hierarchy, depending on the context and the institutional goals. Elster's account of justice is:

> 1. Maximize total welfare. 2. Deviate from that goal if necessary to ensure that all achieve a decent minimum level of welfare. 3. Deviate from the requirement of a decent minimum level of welfare in the case of persons who fall below it because of their own choices. 4. Deviate from the principle of not supporting the persons identified in 3 if their failure to plan ahead and react to incentives is due to severe poverty and deprivation. (Elster 1992, p. 240)

Although Elster's account is somewhat more systematic than either the one I have defended or the picture that emerges from Miller's survey, it reinforces the claim that the choice of distributive principles depends on the relative importance of other contextual goals. These goals might include efficiency, incentives and rewards for contribution, the need for cooperation, and the drive to preserve the solidarity of a social group. Elster's account also reinforces my claim that a principle of sufficiency explains more of our considered judgments about societal justice than can either pure egalitarianism or strict priority views.

THE APPEAL TO ABSTRACT EQUALITY

The sixth and final argument against the ideal of equality is that whatever merit there may be in the claim that all moral and political theories ultimately rest on some abstract conception of equality, it has few implications for how one determines the best-justified principles of distributive justice. Even if a foundational concept of equality is the warp out of which principles of distributive justice are woven, those principles need not produce the strong egalitarian pattern that involves equalizing anything.

One problem is that the abstract account of equality that undergirds all moral and political theories arguably is not a theory of equality in the same sense that egalitarians claim equality for their distributive principles. Perhaps all plausible theories aspire to articulate principles that apply to everyone *equally*, but equal application does not entail a substantive requirement to treat everyone equally in any particular respect. Any commitment to an equality that means more than the *universality* of application remains in need of further argument (Lucas 1965, p. 298; Raz 1986, p. 228).

One such argument is that the very idea of impartial morality implies a commitment to equality (Nagel 1991, p. 65). However, impartiality is no less controversial a concept than equality. For some, the ideal of impartiality is the view that all persons are to be treated without distinction, say, without regard to whether they are friends, clients, or strangers. But for others, the core conception of impartiality is more like the Humean ideal in which each is exhorted "to depart from his private and particular situation, . . . and chuse a view, common to him with others" (Hume [1777] 1983, p. 75). In this second account, impartiality functions more like a comprehensive, deliberative framework in which the interests of all are taken into account and all interests are given adequate weight (Powers 1993). Thomas Scanlon, for example, argues that, "to judge impartially that a principle is acceptable is . . . to judge that it is one which you would accept no matter who you are" (Scanlon 1982, p. 120). It therefore remains an open question whether a comprehensive, impartial framework demands equality of treatment. It is possible that some other principle might be made acceptable to all.

Not surprisingly, some argue that the very idea of the requirement of unanimous agreement embodied in contractualist moral theories leads one to accept egalitarian distributive principles (Nagel 1979, p. 123; 1991, pp. 63–70). But as Frankfurt (1987, p. 36) rightly observes, it is not a necessary feature of reasonable agreement that individuals would favor those social arrangements that deviate from strict equality only when they would be made better off thereby. Some proponents of the contractualist model argue that no unique set of principles will be justified by unanimous agreement, and the explanation for the possible divergence in what reasonable people may agree upon is that reasonableness itself is indeterminate (Scanlon 1982, 112). While the bounds of reasonableness may be adequate to exclude many options, they may not be precise enough to identify the best among the remaining alternatives. Indeed, virtually every conceivable distributive principle has been defended in terms of what reasonable agreement requires, but the problem with this defense is that it confuses principles to which reasonable people may agree with principles that are uniquely justified (Powers 1994).

Moreover, one can imagine a wide range of competing social aims that reasonable people would rank as more important than the preservation of strict equality. Conceiving the political task of acting justly to be an equalizing of the condition of all involves a number of distortions in our understanding of ourselves as moral agents. The preoccupation with the pursuit of equality for its own sake directs our attention away from a proper consideration of the kinds of lives we want to live and of what each of us needs to flourish in light of what we most care about (Frankfurt 1987, pp. 22–23). Our aim should be to enable people to form and revise their own conceptions of the good, not for government to reshape them to get everyone to some targeted—i.e., equal—level of welfare (Daniels 1990).

In sum, I have argued: (1) that the ideal of equality is a poor guide to what our moral duties are in general; (2) that equality has counterintuitive implications when taken to be intrinsically valuable; (3) that there is a better alternative account of the real concerns of the egalitarian; (4) that this alternative to egalitarianism also offers a better explanation of when some inequalities have derivative or instrumental moral significance; (5) that the alternative offers a more plausible reconstruction of the considered judgments of commonsense morality; and (6) that even if equality provides a part of the foundation for any plausible moral or political theory, there are strong reasons to think that it need not, and often should not, lead to the acceptance of an egalitarian principle of distributive justice. In other words, as we move forward into the next century, we should forget about equality and the kinds of questions that have been at the center of theories of distributive justice for the last generation. We have to shift the terms of the discussion to a debate about the purposes of association and the ends of specific activities, such as education and health care delivery, to get a fix on the best-justified distributive principles for those arenas. However, the principles appropriate to each local distributive context should be evaluated with an eye to their cumulative effect. Distributive principles that seem fair when viewed in isolation may be deemed unjust on a global scale when our institutions of allocation are evaluated in their totality (Elster 1992). I have argued that the most plausible global principle of distributive justice is a version of the sufficiency principle. It understands "each having enough" in terms of what is necessary for each to pursue the kinds of lives generally available to others in their community. Sufficiency, though difficult to formulate precisely, must mean enough to prevent the subordination of some groups to the will of others.

Support for this research has been provided by the Robert Wood Johnson Foundation, Investigator Awards in Health Policy Research, Grant # 26418.

REFERENCES

Arneson, Richard. 1989. Equality and Equal Opportunity for Welfare. *Philosophical Studies* 56: 73–93.

Cohen, G. A. 1989. On the Currency of Egalitarian Justice. *Ethics* 99: 906–44.

Daniels, Norman. 1985. *Just Health Care.* Cambridge: Cambridge University Press.

———. 1990. Equality of What: Welfare, Resources, or Capabilities? *Philosophy and Phenomenological Research* 50: 273–96.

Dworkin, Ronald. 1981. What Is Equality? Part 2: Equality of Resources. *Philosophy and Public Affairs* 10: 283–345.

———. 1987. What Is Equality? Part 3: The Place of Liberty. *Iowa Law Review* 73: 1–54.

Elster, Jon. 1992. *Local Justice.* NY: Russell Sage Foundation.

Frankfurt, Harry. 1987. Equality as a Moral Ideal. *Ethics* 98: 21–43.

Fried, Charles. 1978. *Right and Wrong.* Cambridge, MA: Harvard University Press.

Goodin, Robert. 1995. *Utilitarianism as a Public Philosophy.* Cambridge: Cambridge University Press.

Green, Ronald. 1976. Health Care and Justice in Contract Theory Perspective. In *Ethics and Health Policy,* ed. Robert Veatch and Roy Branson, pp. 111–26. Cambridge, MA: Ballinger Publishing Co.

Green, S. J. D. 1989. Competitive Equality of Opportunity: A Defense. *Ethics* 100: 5–32.

Hume, David. [1777] 1983. *An Enquiry Concerning the Principles of Morals,* ed. J. Schneewind. Indianapolis: Hackett Publishing Company.

Lucas, J. R. 1965. Against Equality. *Philosophy* 40: 296–307.

Miller, David. 1992. Distributive Justice: What People Think. *Ethics* 102: 555–93.

Nagel, Thomas. 1979. Equality. Reprinted in his *Mortal Questions*, pp. 106–27. Cambridge: Cambridge University Press.

———. 1991. *Equality and Partiality*. NY: Oxford University Press.

Parfit, Derek. 1989. On Giving Priority to the Worse-Off, unpublished manuscript.

Powers, Madison. 1993. Contractualist Impartiality and Personal Commitments. *American Philosophical Quarterly* 30: 63–71.

———. 1994. Hypothetical Choice Approaches to Health Care Allocation. In *Biomedical Ethics Reviews, 1994: Allocating Health Resources*, ed. James M. Humber and Robert F. Almeder, pp. 147–76. Clifton, NJ: Humana Press.

Rawls, John. 1971. *A Theory of Justice*. Cambridge, MA: Harvard University Press.

Raz, Joseph. 1986. *The Morality of Freedom*. Oxford: Clarendon Press.

Rousseau, Jean-Jacques. [1777] 1988. *Discourse on the Origin of Inequality*. Reprinted in *Rousseau's Political Writings*, ed. Alan Ritter, trans. Julia Conway. New York: W. W. Norton and Company.

Scanlon, T. M. 1982. Contractualism and Utilitarianism. In *Utilitarianism and Beyond*, ed. A. Sen and B. Williams, pp. 103–28. Cambridge: Cambridge University Press.

Sen, Amartya. 1992. *Inequality Reexamined*. NY: Russell Sage Foundation.

Taylor, Charles. 1985. The Nature and Scope of Distributive Justice. Reprinted in his *Philosophy and the Human Sciences, Philosophical Papers*, 2, pp. 289–317. Cambridge: Cambridge University Press.

Tempkin, Larry. 1993. *Inequality*. Oxford: Oxford University Press.

Veatch, Robert. 1986. *The Foundations of Justice: Why the Retarded and the Rest of Us have Claims to Equality*. NY: Oxford University Press.

Walzer, Michael. 1983. *Spheres of Justice*. NY: Basic Books.

Wiggins, David. 1980. Deliberation and Practical Reason. Reprinted in *Essays on Aristotle's Ethics*, ed. Amelie Rorty, pp. 221–40. Berkeley: University of California Press.

Williams, Bernard. 1962. The Idea of Equality. In *Philosophy, Politics and Society* (Second Series), ed. Peter Laslett and W. G. Runciman, pp. 110–31. Oxford: Basil Blackwell.

Part V
New Distinctions within Egalitarianism

[17]

Equality and Time*

Dennis McKerlie

An egalitarian moral view must make certain choices in formulating principles to explain our particular egalitarian judgments. The first choice concerns the form of the egalitarian principle, the particular way in which it differs from a principle simply telling us to maximize what is good across all lives. One kind of egalitarianism aims at equality between different lives. It uses a measure of inequality and requires us to minimize the inequality recorded by the measure. A second kind of egalitarianism gives priority to helping those who are badly off. This view has a tendency toward equality, but not because it believes that equality itself has value. It furthers equality because that will be the result of helping the badly off, and it believes that improving a bad life takes priority over improving better lives. I will use "maximin egalitarianism" as a general name for this kind of view, even though some of the principles it covers are not exclusively concerned with the interests of the very worst off. A third version of egalitarianism believes that everyone should receive at least a specified minimum share of advantages or benefits. It could be called "minimum entitlement egalitarianism."

Each view puts constraints on the distribution of good and bad things across different lives. An egalitarian theory must also specify the things to which the constraints apply. To take the first kind of egalitarianism as an example, which things should be distributed equally across different lives? Some views apply the requirement of equality to possessions and services like wealth, property, and medical care. In Ronald Dworkin's terms these views aim at equality in resources.[1] Other views say that lives should be equal in terms of happiness or the extent to which people's desires are satisfied—equality of welfare in Dworkin's terms. Yet other views start with a list of objective features that give lives value—knowledge, achievements, relationships with others, and so on—and say that we should try to make lives equal in these value-conferring respects.

These two choices have been extensively discussed. A third question is less familiar. As well as specifying the things to be distributed, an

* This paper has benefited greatly from the suggestions and criticisms of the reviewers and editors of *Ethics.* And I owe a special debt to Thomas Hurka.

1. Ronald Dworkin, "What Is Equality? Part 1: Equality of Welfare," *Philosophy and Public Affairs* 10 (1981): 185–246, and "What Is Equality? Part 2: Equality of Resources," *Philosophy and Public Affairs* 10 (1981): 283–345.

Ethics 99 (April 1989): 475–491

egalitarian theory must also specify the units across which the distribution is to take place, or the items with a claim to an equal share of what is distributed. For almost all egalitarians people are the units of distribution. The things to be distributed, whatever they are, should be divided equally among the lives of different people. But our lives are lived through time. Because of this fact there are different ways in which the egalitarian requirement could be applied to us.

These different possibilities have been neglected because egalitarians seem to agree on a particular view about how the egalitarian constraint should be applied. I will call it the "complete lives view."[2] It says that different people's shares of resources, or welfare, should be equal when we consider the total amounts of those things that they receive over the complete course of their lives. To apply this view we would begin by estimating the size of a person's share of the relevant good things at each temporal stage of that person's life. Then we would add these figures together to determine that person's share in terms of a complete life. Finally this share would be compared with the shares of other people over their complete lives in order to test for equality.

The complete lives view has two features. It applies distributional constraints only to shares over complete lives, and it considers the total amounts of the relevant good and bad things that different lives contain. I think that both features can be questioned. Most of this paper will be concerned with the first feature, but I will begin with a brief discussion of the second feature.

It is well known that utilitarians must choose between maximizing the total sum of happiness across all lives or maximizing the average happiness of lives. These two goals can differ if the population size is not constant. Egalitarians face a similar choice. Should we equalize the average happiness of different lives, where the average happiness of a life is the total amount of happiness that it contains divided by its length, or should we equalize the total amounts of happiness that lives contain (assuming, for the sake of simplicity, that happiness is the good which should be distributed equally)? These two goals can differ when lives have different lengths. If one life will be longer than another, is it fair if they have the same average happiness per year, or would it be fairer if the shorter life had a higher average so that the total amounts of happiness in the two lives would be the same? If someone must die in ten years of an incurable disease should egalitarianism say that this person is entitled to a higher average happiness during the ten years?

2. For example, Rawls describes the hypothetical contractors in the original position as considering the life prospects and long-term expectations of representative people from different social classes, so it seems that his difference principle would deal with shares of primary goods over complete lives (John Rawls, *A Theory of Justice* [Cambridge, Mass.: Harvard University Press, 1971], pp. 78, 178). Dworkin explicitly says that his egalitarian principle requires equal shares of resources over complete lives ("What Is Equality? Part 2: Equality of Resources," pp. 304–5).

I think that most people would find a strict version of the average view implausible. If one person has a happy life of an ordinary length while someone else dies at thirty, this does seem a natural unfairness that egalitarianism should at least regret, even though the average happiness of the two lives might be the same. But it might also be thought unfair if one person lives a very long life with a miserably low average happiness—for example, an octogenarian peasant in a Third World country—even though the total sum of happiness in that life equals that of a middle-class North American who dies at fifty-five. So it may seem that the best view will be a compromise between the total view and the average view.

Perhaps we can at least conclude that this is not a question for egalitarianism as such to answer. We must decide what we value in lives, choosing between such things as a high average happiness and a large total amount of happiness. This question can be answered by one person thinking about his or her own life. Once an answer is obtained, egalitarians will use it in ensuring that different lives go equally well. But the choice between the average view and the total view will not be made for distinctively egalitarian reasons, reasons concerned with the relationship that should hold between different lives.

The average view and the total view agree in only considering shares of resources or happiness over complete lives, and this is the feature of complete lives egalitarianism that I will criticize. I will begin with examples where the complete lives view draws conclusions that I find implausible. I will then describe some alternative views that would assess the examples differently. Finally, I will briefly discuss the philosophical issues raised by what seems to me the most attractive alternative view.[3]

If egalitarianism takes complete lives as the units across which we should distribute equally, the past will be important for egalitarianism in a way that it is not important for moral views like utilitarianism. The past does not matter for those views because it cannot be changed. We need only consider the consequences of our actions in the present and the future.[4] But if we aim at equality between complete lives the past history of those lives will partly determine which present action would do best at achieving equality.

Derek Parfit has discussed an example that illustrates this feature of complete lives egalitarianism.[5] A doctor has two patients feeling pain.

3. This paper mainly discusses the kind of egalitarianism that aims at equality and how it might give weight to temporal considerations. But I will also briefly discuss how the other kinds of egalitarianism might handle the issue. My discussion is intended to be neutral between different views about which goods should be equally distributed.

4. Utilitarians could give importance to satisfying past desires directed at the future (past desires are discussed by Derek Parfit in *Reasons and Persons* [Oxford: Clarendon Press, 1984], pp. 149–58). But the fact that such a desire is satisfied is not simply a fact about the past.

5. Derek Parfit, "Comments," *Ethics* 96 (1986): 832–72, esp. pp. 869–70.

Patient A's suffering is not as severe as the suffering of B, but in the past A has suffered much more than B. The doctor can only help one patient, and the treatment would relieve more suffering if it were given to B.

Complete lives egalitarianism would tell the doctor to help A. It makes this choice because of the past even though there would be less suffering if B were helped instead. Many people would disagree with this conclusion. But the example is not a decisive objection against the complete lives view. If we combine complete lives egalitarianism with a principle of minimizing suffering we are not forced to agree that, all things considered, helping A would be best. We might decide that the good done by relieving more suffering outweighs the loss in terms of equality.[6]

There are simpler examples in which the complete lives view makes the past matter, and I think that many egalitarians would doubt the conclusions it draws. If we judge equality in the complete lives way we will think that past inequality calls for a certain response. If in the past A's life has been worse than B's life, complete lives egalitarianism will tell us not merely to end the inequality but to reverse it—to create a new inequality in A's favor so that their complete lives will be equal. Suppose that one person's childhood was very unhappy, involving the loss of parents, poverty, and suffering. But these tragedies left no scars, and today that person is happy with a bright future. A second person has equally good prospects after a happy childhood. Complete lives egalitarianism will say that it would be a good thing if the first person's future went better than that of the second person, in order to compensate for the difference in their childhoods. I think that many egalitarians would resist that conclusion.

The question raised by the example is whether we should always compensate for past inequality. This policy seems especially inappropriate in the example of the two childhoods. It might be suggested that the example can be handled by slightly modifying the complete lives view. We should aim at equality between complete lives, but we may ignore inequalities in the distant past. But the restriction seems arbitrary. If we only care about inequality between complete lives, why should the fact that the inequality was in the distant past matter? If we think that compensation is not appropriate we should look for a different kind of egalitarian view to explain our judgment.[7]

6. Parfit also discusses a version of the example in which the medicine would relieve more suffering in the long run if it were given to A, although B's current suffering is more intense than A's suffering will ever be ("Comments," p. 870). Parfit thinks that it would not be absurd to decide to help B. This conclusion could not be explained by a combination of complete lives egalitarianism and the principle of minimizing suffering. If the conclusion is an egalitarian judgment it must be based on what Parfit calls a "time-relative" principle of equality.

7. The question of the importance of the past also arises for maximin egalitarianism. Suppose that one group of people began with lives of extreme poverty (perhaps they were originally penniless refugees), but they have now reached a reasonable level of happiness.

The complete lives view has another consequence. It does not see any disvalue in inequality between parts of lives as long as the inequality is compensated for at earlier or later times so that there is no inequality between complete lives. This enables us to imagine a new kind of egalitarian society. It contains great inequality, with happier lives attached to certain social positions. But at a fixed time people change places and switch from a superior position to an inferior one or vice versa. One example would be a feudal society in which peasants and nobles exchange roles every ten years. The result is that people's lives as wholes are equally happy. Nevertheless during a given time period the society contains great inequality, and in one sense this always remains true. I will call this system "changing places egalitarianism." If equality between complete lives were all that mattered, an egalitarian could not object to it. But I think that many egalitarians would find it objectionable.[8]

A realistic example is in some respects analogous to changing places egalitarianism. In our society the elderly generally have worse lives than those who are younger.[9] Perhaps that has always been so, but today it is a more serious problem. Retirement policies and medical advances mean that for most people old age begins sooner and lasts longer than it did in the past. Consider the different futures of a thirty-five-year-old and someone who has just retired at the age of sixty-five. The first life might be happy and affluent, the second life might be equivalent to that of a welfare recipient. This inequality is very great, and it might last for as long as twenty-five years.

Some people think the inequality is an injustice and should be mitigated. But according to the complete lives view it is not necessarily unjust. The inequality is not objectionable unless the two lives as wholes differ in quality. If the earlier years of the person who is now old were just as happy as the current life of the middle-aged person, their complete lives might not differ in quality. If the appropriate things happen before it and afterward, the twenty-five-year inequality has no moral importance.[10]

A second group have lives that are better than the lives of the first group used to be, but not as happy as their lives currently are. The first group have the better present and future, but it may be true that because of the past their lives viewed as wholes are less attractive than the lives of the second group. Which group should a maximin egalitarian give first priority to helping?

8. Some would only object to the changing places system if the inequality were greater than some specified degree or if the quality of life of the peasants were below some specified minimum level. But they are still applying egalitarian principles, although different egalitarian principles, to temporal parts of lives. However, if we only object to the example because we suppose that the peasants' rights are violated, then the example will not serve my purposes. A rights view will rule out any violation at any time of the rights it recognizes, so it will not take temporal considerations into account in any important way.

9. This claim is disputable if it is applied to all those over sixty-five. But if it is restricted to the 'old old'—those seventy-five or older—it has more force.

10. The question whether to give importance to temporal segments of lives also arises for maximin egalitarianism. Ordinarily maximin views tell us to make the worst complete

I would not claim that these examples should destroy our confidence in the complete lives view. But I think they are strong enough to motivate an examination of alternative versions of egalitarianism that would reach different conclusions about the examples.

In the example of the two childhoods the complete lives view makes the past important. An obvious alternative is what might be called "future-directed egalitarianism." It ignores the past and tells us to aim at equality in the present and future.

This view makes the present a dividing line for moral significance. But the present is a boundary that moves. Consequently, the view faces a formal problem. It gives us the goal "equality from now on." But this is not a single goal. It changes as the reference of "now" changes. And if we pursue different goals at different times our effort to achieve our goal at one time may interfere with our effort to achieve another goal at another time.

For example, at T1 we have the goal of equality from T1 on. We may decide that the best way to achieve this goal is to make one person, A, bear a burden until a later time T2 while another person, B, enjoys a benefit until T2. After T2 A will receive the benefit and B will bear the burden. But when it is T2 we have a new goal, equality from T2 on. A's burden and B's benefit are now in the past and do not count. So at T2 we might abandon the policy of switching, with the result that A will go on bearing the burden and B will continue to enjoy the benefit. I do not think an egalitarian would accept these consequences of future-directed egalitarianism.

It is true that if at T1 we anticipate our later decision at T2 future-directed egalitarianism would not tell us to begin the policy at T1. The example is not one in which the view, applied at different times with full knowledge of the facts (including facts about the future), has clearly unacceptable consequences. But future-directed egalitarianism might tell us to begin the policy at T1 if we can also take measures that will make it very difficult to reverse the policy at T2. And then at T2 it will tell us to do all we can to defeat those measures, with the aim of allowing the inequality between A and B to continue. Is it reasonable to adopt a view which gives us this advice about how to achieve equality in the lives of A and B?

Future-directed egalitarianism takes time into account by supposing that there is a fundamental difference in moral importance between the

---

life as happy as possible. Suppose that a life which is not the worst complete life contains a significant segment—perhaps, a twenty-five-year stretch—whose quality is far worse than the quality of the worst segment of the worst complete life. Some egalitarians might think that improving this segment has a priority that competes with the priority of improving the worst complete life. Applying maximin principles to parts of lives would have implications for the treatment of the elderly. When economic disadvantages are combined with the other undesirable effects of aging, their current lives would be among the worst segments of lives. So this kind of egalitarianism would give priority to helping them.

past on the one hand and the present and future on the other. The second way for egalitarianism to give weight to facts about time is to apply the requirement of equality to temporal parts of lives, not just to complete lives.

Consider the lives of A and B during the time periods T1, T2, and T3.

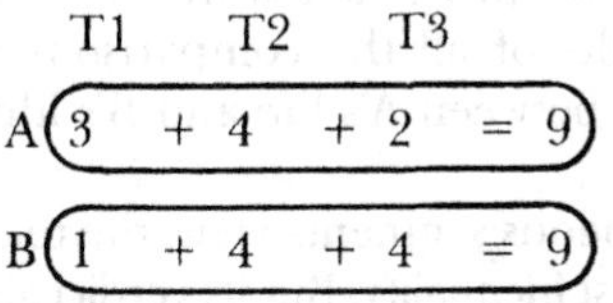

The numbers measure the happiness of A and B during those periods. The complete lives view measures inequality in the following way: it first adds the numbers to measure the happiness of the complete lives of A and B and then compares them to determine the inequality. In this example the inequality is zero.

But there is another way of measuring the inequality. It begins by measuring the inequality across lives within each time period and then adds these figures to measure the total inequality through time.

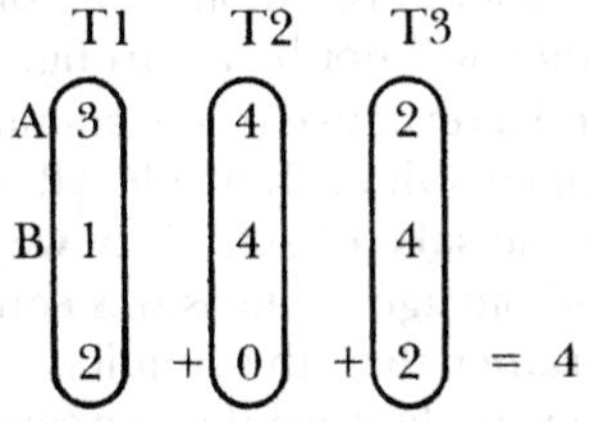

The two ways of measuring inequality can lead to different conclusions. For example the first method would prefer A = 2 and B = 4 during T3 to A = 3 and B = 3, but the second method would reach the opposite conclusion.

The second method tells us to minimize the sum total of inequality-at-a-time. I will call it the "simultaneous segments view." It would not compensate for past inequality. It does not see A = 2 and B = 4 during T3 as compensating for the inequality during T1 but rather as adding to the inequality between the lives of A and B. What we do now cannot change the inequality at past times, so this view would agree with future-directed egalitarianism about the example of the two childhoods. It rejects changing places egalitarianism because that system would involve a very high sum total of inequality-at-a-time. The view would recognize a claim to equality between the simultaneously lived final twenty-five years of one person's life and the middle twenty-five years of another person's life. It would see this inequality as something regrettable in itself, and as something whose importance could not be entirely erased by what happens before or after the twenty-five years. The view escapes the formal problem because it gives us the same goal—minimizing the sum total of inequality-at-a-time—at every time.[11]

There are two more egalitarian views that use parts of lives in judging equality. The first begins with a segment of A's life and compares it, not just with the simultaneous segment of B's life, but with every segment of B's life. It repeats this procedure with every other segment of A's life and it adds the results of all the comparisons to determine the total amount of inequality between A's life and B's life. I will call it the "total segments view."

Like the simultaneous segments view, the total segments view would not compensate for past inequality. But its verdict about a present inequality favoring B depends on comparing this segment of B's life with the past segments of A's life, not just with the simultaneous segment of A's life. If A has been better off than B in the past, a present inequality favoring B might do best at minimizing the inequality among all the parts of their lives. Because it values equality in all segments of all lives, the total segments view does recognize a claim to equality between the simultaneously lived final twenty-five years of one life and the middle twenty-five years of another life. However, it does not see any importance in the fact that the unequal parts are simultaneous. So the view's consequences for the treatment of the elderly should resemble those of the simultaneous segments view, although they will not be as strong.

I will call the second alternative to the simultaneous segments view the "corresponding segments view." It divides all lives into the same series of temporal parts. For the sake of simplicity we can suppose that they are youth, maturity, and old age. It measures equality by comparing the corresponding stages, rather than the simultaneous stages, of different lives. In the following example it would compare A's middle years with B's middle years, while the simultaneous segments view compares A's middle years with B's youth.

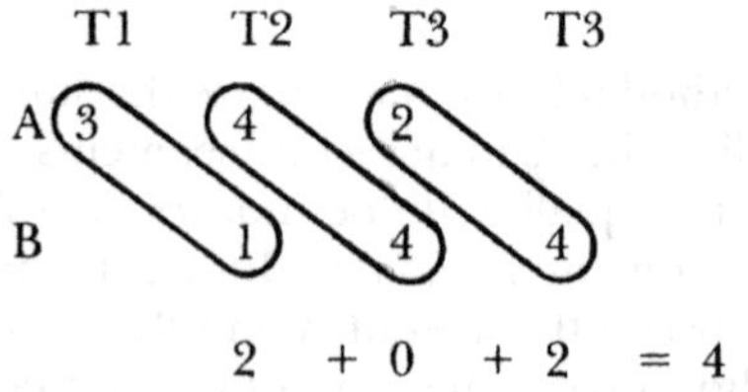

11. According to Amartya Sen ("Utilitarianism and Welfarism," *Journal of Philosophy* 76 [1979]: 463–89, esp. p. 471), when economists measure inequality they use a time-slice cut across people's lives instead of considering shares of resources or wealth over complete lives. This method seems very close to the simultaneous segments view. But it is not clear from what Sen says whether economists adopt the method because it is convenient, or whether they believe that adding time-slice inequalities is a better measure of inequality over time.

Concerning the elderly, the corresponding segments view requires that their lives should be as happy as ours will be when we reach their age, but it does not object to inequalities between their final years and the middle-age of others.

In the examples in which we seem to care about inequality between parts of lives I think that we care specifically about inequality in the simultaneous parts of different lives. We are disturbed because the elderly are poor and unhappy while others prosper. So I will concentrate on the simultaneous segments view in the next part of the paper, although similar points could be made about the other segments views.

How should the simultaneous segments view divide lives into parts to measure inequality? The view will seem implausible if the time periods within which the inequality is measured are too short. If two people will see a dentist tomorrow, it would tell them to schedule simultaneous appointments so that there will be equality in suffering at that time. Are there serious egalitarian reasons for preferring two 10:30 appointments to an appointment at 10 and an appointment at 11?

We might claim that the inequality in the dental appointments example does have some small importance. This claim would not have implausible consequences since the inequality would not count for much in the context of all the other simultaneous parts of the lives of the people concerned. But the view would still face a difficult example. Suppose that the lives of A and B consist of very short periods of enjoyment alternating with very short periods of intense pain. Perhaps they are the prisoners of a sadistic regime and are alternately tortured and pampered. If their periods of suffering are not synchronized, their lives will involve a very high total sum of inequality-at-a-time. Is it really important that their periods of suffering should be synchronized?

To defend the simultaneous segments view we might claim that, given the nature of the relevant goods and evils, lives cannot change radically within short periods of time in the respects that matter for egalitarianism. For example, if the egalitarian principle requires equality in happiness, perhaps it is not possible for a life to change from happy to unhappy in an afternoon. However, some people may not be convinced by these claims about happiness or welfare, and it would be desirable if the simultaneous segments view were independent of particular views about which good things should be equally distributed.

A different reply claims that an inequality must include a significant portion of the lives in question to be morally important. This reply also faces difficulties. It is difficult to suggest a plausible threshold to pick out the inequalities which last long enough to matter. If the time periods used to measure inequality are too long, there could be intuitively objectionable changing places arrangements inside them. The threshold must be low enough to avoid this problem but high enough to avoid the dental appointments example. And if we are sympathetic to the simultaneous segments view it is difficult to understand why short-term inequalities should not count at all, as opposed to having some slight disvalue. These difficulties are serious, but I do not think the problem of choosing time periods is a decisive objection against the simultaneous segments view.

How persuasive is the simultaneous segments view as an alternative to complete lives egalitarianism? Many people will reject it because it does not give any weight to equality between complete lives. It will not compensate for past inequality, no matter how long the inequality has lasted and no matter how extreme it has been. If we believe that sometimes we owe those who have been badly off in the past more than just equality with others in the future, we will see this as an objection to the view. And if for some reason there must be inequality in the future, the simultaneous segments view does not distinguish between an inequality favoring those who previously were badly off and an inequality favoring the people who had better lives in the past.

On the other hand, we may be persuaded by the examples which seem to show us caring about inequality between parts of lives. A society in which peasants and nobles change places after twenty-five years does seem worse than a society without gross inequality between large parts of people's lives. Even if the second society contained slightly more inequality between complete lives than changing places egalitarianism, I think that egalitarians should prefer it.

In the case of the elderly, the complete lives view could in principle justify very great inequality between the simultaneous parts of different lives. I would support reducing the current inequality even if this means that those who are now old will have better complete lives than those who are now younger. If it is plausible to think that resources would do more good in utilitarian terms if they were given to younger people (because of the intractable nature of their problems the elderly, like the handicapped, are difficult to help), this conclusion could not be explained by a mixture of complete lives egalitarianism and utilitarianism.

One possibility is that we care about equality between complete lives, but we also accept some principle that gives weight to inequalities between parts of lives. Faced with a choice about whether to compensate for past inequality, we would have to balance the gain in equality between complete lives against the loss in terms of equality between parts of lives. In some examples we might think that one consideration was stronger, in different examples the other consideration.[12]

I have considered some alternatives to complete lives egalitarianism, and I have described their implications in a series of examples. I will now discuss some questions about the ways of thinking about our lives and time that might lie behind the alternative views.

12. If we also value equality between complete lives we can recognize a claim to equality between members of different generations whose lives have no simultaneous parts. In this case there would be no egalitarian constraints applying to the parts of their lives.

The simultaneous segments view values equality in temporal segments of lives. It might be thought that any such view must depend on revising the ordinary view of personal identity. If we believe that inequality between the later years of A and B does not morally cancel an inequality in their youth, it must be because we believe that in some sense A and B are not the same people in age that they were in youth. Some theories of personal identity divide the life of a person into the lives of a series of different but related 'selves.' The suggestion I will consider is that the plausibility of this kind of egalitarianism depends on the plausibility of some such way of thinking of personal identity.

If the egalitarian views did have this basis certain consequences would follow. First, the views should require equality between the different parts of a single person's life. If an inequality disadvantaging A as a youth cannot be compensated for by an inequality benefiting the elderly A, because the young A and the elderly A are like different persons, then there should also be a claim to equality between the young A and the elderly A.

Second, the theory of personal identity would change our prudential judgments as well as our moral judgments. If the young A and the elderly A are like different people, then A as a youth should not believe that happiness in the distant future would reward him for sacrifices made now.

In fact this view of personal identity would not support the simultaneous segments version of egalitarianism. If different selves should be treated as we now treat different persons, it is natural to suppose that we should aim at minimizing the inequality among all of those distinct selves. The total segments view would give us this goal, once it had been extended to apply to the parts of a single life. But the simultaneous segments view has a different goal. To use the language of selves, it tells us to measure the inequality among each simultaneously existing group of selves and then requires us to minimize the sum total of these inequalities through time.

The difference between the two views is shown by the example of gradual progress. A and B share lives of poverty and sacrifice during T1, modest prosperity during T2, and considerable happiness during T3.

	T1	T2	T3
A	1	3	6
B	1	3	6

The simultaneous segments view (and the complete lives view) would not see any objectionable inequality in this example. But if we aim at minimizing inequality among selves, we will object (for example) to the inequality between A during T1 and B during T2.

I do not think the inequality in the example is objectionable (this is one reason for preferring the simultaneous segments view to the total segments view). And I am not convinced that there is a claim to equality between the different stages of a single life. In the examples where I disagree with complete lives egalitarianism, I do not think that the disagreement results from a special view of personal identity. Objecting to changing places egalitarianism does not require believing that there is some kind of weakening of personal identity in the lives of the people who are in turn peasants and nobles. So I would deny that the plausibility of the simultaneous segments view depends on revising the ordinary view of personal identity.[13]

If it is not supported by a new way of thinking of personal identity, the simultaneous segments view might seem open to serious objections. I will discuss them using the simple example of changing places egalitarianism in which the quality of the lives of A and B is reversed at the midpoint of those lives.

	T1	T2
A	5	2
B	2	5

13. Derek Parfit has discussed versions of egalitarianism that deal with temporal parts of lives and how they might be related to views about personal identity (*Reasons and Persons*, pp. 339–45, and "Comments," pp. 837–43, 871–72). Parfit explains personal identity in terms of psychological connections between experiences rather than relations between selves, and he discusses maximin egalitarian views. He considers what he calls the "extreme view" which makes the units over which the egalitarian principle operates the experiences that people have at particular times. If the extreme view were concerned with minimizing inequality instead of relieving the very worst experiences, it would tell us to minimize inequality among all of the momentary stages of people's lives, and so it would be roughly equivalent to the total segments view. Parfit's suggestion that the extreme view is close to negative utilitarianism (*Reasons and Persons*, p. 345, "Comments," p. 841) is misleading: negative utilitarianism tells us to reduce the total amount of suffering across all lives and times while the extreme view tells us to relieve the worst suffering felt by any person at any time, so the two views would only rarely recommend the same action. In *Reasons and Persons*, Parfit says only that his account of personal identity might be thought to support the extreme view, but in his "Comments," he suggests that egalitarians who accept his account of personal identity should hold the extreme view. Using his example of a double brain transplant, Parfit argues that since we would not compensate the two resulting people for the hardships suffered by the original person we must believe that the further fact of identity is essential for there to be compensation for burdens within a single life. If we agree with Parfit's reductionist view of personal identity we will think that the further fact is never present, and so we will deny that intrapersonal compensation for burdens is possible even within ordinary lives. A difficulty with his argument is that we may judge the splitting examples in the way we do partly because we believe that there is a further fact to identity. If we were fully convinced of the reductionist view we might think that the two resulting people should be compensated for harms experienced by the original person. Parfit does not argue that an egalitarian principle concerned with temporal parts of lives must be based on a special way of thinking of personal identity. But he seems to believe that these principles must deny that benefits at one time can compensate a person for burdens at some other time, and this denial might require support from a revised view of personal identity. I will suggest that we can hold the simultaneous segments view while conceding that there can be intrapersonal compensation for burdens.

If A and B are the same people during T1 and T2, why doesn't the inequality in T2 morally cancel the inequality in T1? The connections between A during T1 and A during T2 are stronger than the connections between A during T1 and B during T1, so that it seems more appropriate to group the segments of lives horizontally rather than vertically in order to measure the inequality.

Moreover, the simultaneous segments view might seem to be ruled out by the judgments we make about single lives. We think that what fundamentally matters for a single life is the overall quality of the complete life. The quality of a temporal part of the life matters only to the extent that it contributes to the quality of the complete life. It would be unreasonable for B to choose a better next twenty years at the cost of a worse complete life. But when it compares lives the simultaneous segments view seems to give independent importance to what happens in temporal parts of lives. It objects to inequalities between parts of lives even if these inequalities are made good over complete lives. We also believe that the timing of benefits and harms does not affect the quality of a life. B should not reject a life because its good features would be concentrated in one temporal part while its bad features would be concentrated in a different temporal part. If the timing of benefits and harms is not important in considering a single life, why should it matter in making egalitarian judgments about different lives? Finally, according to prudence a harm at one temporal stage of a life can be outweighed by benefits at other times. Prudence might judge that B's unhappiness in T1 is outweighed by B's happiness in T2. But if B's unhappiness in T1 is compensated for by later events in B's life, why should we give it any weight when we compare B's life with A's life? It might seem that if we agree with the way in which prudential reasoning assesses single lives we must reject simultaneous segments egalitarianism.

However, I think the simultaneous segments view can be explained in a way that answers these objections. Egalitarianism cares about minimizing the inequality between different lives. Because those lives extend through time there can be more than one view about how to do this. The complete lives view and the simultaneous segments view can both be regarded as built around ways of measuring the inequality between lives. Understood in this way the simultaneous segments view is also concerned with achieving equality between complete lives. But it supposes that the inequality between two lives equals the sum of the inequalities between the simultaneous parts of those lives. Thinking about the example of changing places egalitarianism, it denies that the answer to the question, "How much inequality was there in all between A's life and B's life?" is "Absolutely none." According to it, A and B live all of their lives as unequals. And it claims that the sense in which there was inequality is morally important.

That is why the view does not need to deny the fact of personal identity through time or question the importance of this fact. It does not reach its conclusion by refusing to compare B's life in T1 with B's life in T2.

It applies the requirement of equality to the temporal parts of the lives of A and B. It objects to the inequality in T1, and it objects because of the difference in A's life and B's life during T1. According to it, following T1 by an inequality in T2 favoring B could at most be a second-best alternative to not allowing the inequality in T1 in the first place.

If it is understood in this way the view is not inconsistent with prudence. It does not give independent importance to parts of lives. Like the complete lives view, its fundamental concern is equality between complete lives, but it takes a different view of how inequality in parts of lives contributes to inequality between complete lives. It treats time as important, but it does not claim that the timing of benefits and harms influences the quality of the individual lives which contain them. Time becomes important in making egalitarian judgments about the relationship between different lives, and it is important because of the way in which the simultaneous segments view measures inequality. The view does not depend on thinking that benefits and harms cannot outweigh one another across time. It can agree that B's happiness in T2 might outweigh B's unhappiness in T1 if we are considering B's life. But it would claim that this does not mean that what happens in T1 cannot have any kind of importance. It might matter when we make egalitarian judgments which compare B's life with A's life.

Explained in this way the simultaneous segments view does not clash with prudential judgments about single lives. But the explanation suggests a different objection. The simultaneous segments view seems to undermine the connection between egalitarianism and the quality of individual lives. It objects to inequalities between the temporal parts of the lives of A and B even though the inequalities are compensated for at earlier or later times. But if the inequalities are compensated for at other times, they do not make the lives of A and B worse than they would have been without the inequalities. (We could say that they make parts of the lives of A and B worse, but this reply would suggest that the view did after all care independently about segments of lives.) It might even be true that both A and B have better complete lives because the inequalities exist. Why should we believe that inequalities that do not make people's lives worse are morally objectionable?

However, it is important to remember that I have been discussing a particular kind of egalitarianism, an egalitarian view that values equality as a relationship holding between different lives. It does not object to the poor quality of B's life during T1, it objects to the difference between A's life and B's life. Because its judgments are not based directly on the contents of the lives of A and B, it is easier to understand why there can be differences between judgments about the quality of individual lives and egalitarian judgments. This kind of egalitarianism might object to inequalities even if they did not make people's lives worse.[14]

14. The same issue arises for maximin or minimum entitlement views applied to parts of lives. Prudence tells us to endure ten years of misery if it will be outweighed by enough

If we are forced to treat the simultaneous segments view as a time-relative version of this kind of egalitarianism, some people might take this to be a reason for rejecting the view. But I am not convinced that the view has unacceptable consequences. For example, it is arguable that a health care system and a social security system providing only minimal support for the very old would increase equality over complete lives and use resources in the most efficient way. It is even possible that we would all be better off if we lived our complete lives under these institutions. But this system would create gross inequality between the very old and their younger contemporaries, and I would object to it for that reason despite its advantages.[15]

The simultaneous segments view must also claim that the fact of simultaneity can itself have moral importance.[16] It registers the inequality between A's life during T1 and B's life during T1 as something that should be avoided. But it would not have said the same about an inequality between A's life during T1 and B's life during T2, because those segments of the two lives are not simultaneous. The example of gradual progress, which the view finds unobjectionable, would turn into an objectionable case of changing places egalitarianism if we imagine that B had instead been born twenty-five years before A, so that the happy and unhappy periods of their lives would not be simultaneous.

---

happiness in the further future. Suppose we accept an egalitarian view which gives first priority to improving the worst significant part of a life rather than to improving the worst complete life. Should we revise prudence by claiming that the person whose life it is should also give first priority to improving that segment of the life, or should we say that prudence and maximin egalitarianism can understandably reach different conclusions because of their different standpoints—prudence is concerned with one life while maximin egalitarianism considers the relationship between lives? I have suggested that the kind of egalitarianism which aims at equality should give the second reply. I think that the maximin view should give the first reply. It is motivated by the special importance of meeting certain needs or ending certain kinds of suffering. Since its judgments are based directly on the contents of lives, there is less reason to think that it can reasonably differ from prudence.

15. Comments by Richard Sikora brought home to me that in these circumstances the simultaneous segments view would choose the outcome in which everyone is worse off in terms of complete lives. If everyone would be happier under the alternative system, people might unanimously choose it and stick to their choice even during the periods of their lives when they themselves are disadvantaged by the inequality. What should the simultaneous segments view say about this possibility? It might claim that there is still a moral objection to the system even if it is unanimously chosen. Alternatively it might say that inequalities are not objectionable if they are voluntarily chosen by the people who will be worst off given the inequality. Complete lives egalitarianism could also allow people to choose to have less than an equal share, although this possibility is not usually considered by egalitarian writers. If someone would prefer to have less for the sake of others, should we require her to accept an equal share of resources or happiness?

16. Some versions of complete lives egalitarianism also make simultaneity important. These views hold that the claim to equality between lives that belong to the same generation is stronger than the claim to equality between lives belonging to generations widely separated in time. If social cooperation were a necessary condition for the applicability of egalitarian principles we could perhaps reach the same conclusion without attributing importance to simultaneity. However, I do not think egalitarian principles are subject to this condition of applicability, and I do think that the claim to equality is stronger among contemporaries.

Many people would reject this claim about the importance of simultaneity. Could a mere difference in the timing of B's birth have these moral consequences? They might base their objection on the general claim that facts about time do not have intrinsic moral significance. What matters for morality is what happens, not when it happens. If something good happens it would have had the same value if it had occurred instead at some other time. The view that time is morally irrelevant has apparently been held by many moral philosophers. If simultaneous segments egalitarianism contradicts their view this might seem a sufficient reason for rejecting it.

But it is not clear that those writers really do think that no fact about timing can have any kind of moral importance. They discuss particular ways of making time important. Sidgwick says that the consciousness of one moment is not more important than the consciousness of any other moment.[17] Rawls says that in making decisions about our own lives, or other people's lives, we should not be influenced by a pure time preference which gives more weight to the near future than the distant future.[18]

The simultaneous segments view does not make the mistakes that Sidgwick and Rawls warn us against. It objects to inequality in the simultaneous parts of different lives. But it makes the same judgment about this kind of inequality whenever it occurs, and whether it belongs to the past, present, or future. The view can agree that good and bad things have the same value whenever they occur.

The view gives importance to relative temporal position, not to absolute temporal position. For this reason I think it is appropriate to compare it to the view that punishment should follow and not precede an offense. People might disagree over this claim about punishment, depending on their general views about the justification for punishment. But it would be unreasonable for those who believe it to give it up because they are told that facts about time cannot be morally important.

The situation is analogous with the simultaneous segments view. If we find the view attractive it is because we think that inequality in the simultaneous parts of lives is significant, even if the inequality is compensated for at earlier or later times. This view gives moral importance to the relative timing of benefits and harms in different lives. But if these facts do seem important, we should not be dissuaded by the general claim that time is morally irrelevant.[19]

17. Henry Sidgwick, *The Methods of Ethics* (London: Macmillan, 1907), p. 381.

18. Rawls, pp. 293–94.

19. If we apply maximin egalitarianism or minimum entitlement egalitarianism to parts of lives, simultaneity will not have this importance. The minimum entitlement view would take the form of requiring that every part of every life should receive the minimum entitlement. With the maximin view there are two choices: we could give priority to helping the worst part among all the parts of all lives (this is equivalent to Parfit's extreme view), or we could during each time period give priority to helping the worst part included within that time period. The second alternative is intuitively implausible, and it does not make

I have discussed some examples in which I disagree with the conclusions drawn by the complete lives version of egalitarianism. There are two ways in which the complete lives view could be changed: by denying importance to the past or by applying egalitarian constraints directly to parts of lives. I have argued that the second change is preferable.

I have argued that the most plausible time-relative view about equality will require equality in the simultaneous segments of different lives. Our lives are lived serially through time, and the simultaneous segments view responds to this fact by valuing equality in the simultaneous parts of lives rather than by merely requiring that lives should be equal when viewed timelessly as completed wholes. If we find this view persuasive we will probably care both about equality assessed in the complete lives way and equality in the temporal parts of lives. Someone might care about both and still believe that a gain in equality in terms of complete lives is always more important than a gain in equality between the parts of lives. But if we do care about equality in the simultaneous segments way, I think we should believe that a large gain in equality between parts of lives would outweigh a small gain in equality between complete lives.

I have argued that the plausibility of the simultaneous segments view is independent of theories about personal identity. And I have claimed that the view does not conflict with a requirement of timelessness when we are considering good and bad features within one life, or with the thought that what fundamentally matters for a single life is the overall quality of the complete life. I have applied the view to the practical problem of the treatment of the elderly, where I find its consequences intuitively appealing.

simultaneity important in the way that the simultaneous segments view does. The simultaneous segments view does not say that during a time period we should only be concerned with inequalities within that time period; it says that we should be concerned with minimizing the inequality between lives whenever it might occur.

[18]

Ratio (new series) X 3 December 1997 0034–0006

EQUALITY AND PRIORITY [1]

Derek Parfit

In his article 'Equality', Nagel imagines that he has two children, one healthy and happy, the other suffering from some painful handicap. Nagel's family could either move to a city where the second child could receive special treatment, or move to a suburb where the first child would flourish. Nagel writes:

> This is a difficult choice on any view. To make it a test for the value of equality, I want to suppose that the case has the following feature: the gain to the first child of moving to the suburb is substantially greater than the gain to the second child of moving to the city.

He then comments:

> If one chose to move to the city, it would be an egalitarian decision. It is more urgent to benefit the second child, even though the benefit we can give him is less than the benefit we can give to the first child.[2]

My aim, in this paper, is to discuss this kind of reasoning.

1

Nagel's decision turns on the relative importance of two facts: he could give one child a greater benefit, but the other child is worse off. There are countless cases of this kind. In these cases, when we are choosing between two acts or policies, one relevant fact is how great the resulting benefits would be. For Utilitarians, that is all that matters. On their view, we should always aim for the greatest

[1] This paper is a greatly shortened version of my Lindley Lecture 'Equality or Priority?' (42 pp.), published by the University of Kansas in 1995. That lecture owes much to the ideas of, or comments from, Brian Barry, David Brink, John Broome, Jerry Cohen, Robert Goodin, James Griffin, Shelly Kagan, Dennis McKerlie, David Miller, Thomas Nagel, Robert Nozick, Richard Norman, Ingmar Persson, Janet Radcliffe Richards, Joseph Raz, Thomas Scanlon, and Larry Temkin.

[2] Thomas Nagel, *Mortal Question*, (Cambridge: Cambridge University Press, 1979), pages 123–4. See also Nagel's *Equality and Partiality* (New York: Oxford University Press, 1991).

sum of benefits. But, for egalitarians, it also matters how well off the beneficiaries would be. We should sometimes choose a smaller sum of benefits, for the sake of a better distribution.

Should we aim for a better distribution? If so, when and how? These are difficult questions, but their subject matter is, in a way, simple. It is enough to consider different possible states of affairs, or outcomes, each involving the same set of people. We imagine knowing how well off, in these outcomes, these people would be. We then ask whether either outcome would be better, or would be the outcome that we ought to bring about.

Some writers reject these questions. Nozick objects, for example, that these questions wrongly assume that there is something to be distributed. Most goods, he argues, are not up for distribution, or redistribution.[3] They are goods to which particular people already have entitlements, or special claims. Others make similar claims about desert.

These objections we can set aside. We can assume that, in the cases we are considering, no one deserves to be better off than anyone else; nor does anyone have special claims to whatever we are distributing. Since there are *some* cases of this kind, we have a subject. If we can reach conclusions, we can then consider how widely these apply. Like Rawls and others, I believe that, at the fundamental level, most cases are of this kind.

To ask my questions, we need only two assumptions. First, some people can be worse off than others, in ways that are morally relevant. Second, these differences can be matters of degree. To describe my imagined cases, I shall use figures. Nagel's choice, for example, can be shown as follows:

	The first child	The second child
Move to the city:	20	10
Move to the suburb:	25	9

Such figures misleadingly suggest precision. Even in principle, I believe, there could not be precise differences between how well off different people are. I intend these figures to show only that the choice between these outcomes makes much more difference to Nagel's first child, but that, in both outcomes, the second child would be much worse off.

[3] Robert Nozick, *Anarchy, State, and Utopia* (New York: Basic Books, 1974), pages 149–50.

One point about my figures is important. Each unit is a roughly equal benefit, however well off the person is who receives it. If someone rises from 99 to 100, this person benefits as much as someone else who rises from 9 to 10. Without this assumption we cannot ask some of our questions. Thus we cannot ask whether some benefit would matter more if it came to someone who was worse off.

Since each extra unit is an equal benefit, however well off the recipient is, these units should not be thought of as equal quantities of resources. The same increase in resources usually brings greater benefits to those who are worse off. But these benefits need not be thought of in Utilitarian terms, as involving greater happiness, or desire-fulfilment. They might be improvements in health, or length of life, or education, or range of opportunities, or involve any other goods that we take to be morally important.[4]

2

Most of us believe in some kind of equality. We believe in political equality, or equality before the law, or we believe that everyone has equal rights, or that everyone's interests should be given equal weight. Though these kinds of equality are of great importance, they are not my subject here. I am concerned with people's being *equally well off*. To be egalitarians, in my sense, this is the kind of equality in which we must believe.

Some egalitarians believe that, if people were equally well off, that would be a better state of affairs. If we hold this view, we can be called *Teleological* – or, for short, *Telic* – Egalitarians. We accept

> *The Principle of Equality*: It is in itself bad if some people are worse off than others.[5]

Suppose that the people in some community could all be either equally well off, or equally badly off. The Principle of Equality

[4] For two such broader accounts of well-being, see Amartya Sen, 'Capability and Well-Being', in *The Quality of Life*, edited by Martha Nussbaum and Amartya Sen (Oxford, Oxford University Press, 1993), Amartya Sen, *Inequality Reexamined* (Oxford: Oxford University Press, 1992), Chapter 3; and Thomas Scanlon, 'Value, Desire, and the Quality of Life', in Nussbaum and Sen, *op. cit.*

[5] We might add, 'through no fault or choice of theirs'. In a fuller statement of this principle, we would need to assess the relative badness of different patterns of inequality. But we can here ignore these complications. They are well discussed in Larry Temkin's *Inequality* (New York: Oxford University Press, 1993).

does not tell us that the second would be worse. To explain that obvious truth, we might appeal to

The Principle of Utility: It is in itself better if people are better off.

When people would be on average better off, or would receive a greater sum of benefits, we can say, for brevity, that there would be more *utility*.

If we cared only about equality, we would be *Pure* Egalitarians. If we cared only about utility, we would be Utilitarians. Most of us accept a *pluralist* view: one that appeals to more than one principle or value. According to *Pluralist Egalitarians*, it would be better both if there was more equality, and if there was more utility. In deciding which of two outcomes would be better, we give weight to both these values.

These values may conflict. One of two outcomes may be in one way worse, because there would be more inequality, but in another way better, because there would be more utility. We must then decide which of these two facts would be more important. Consider, for example, the following possibilities:

(1) Everyone at 150
(2) Half at 199 Half at 200
(3) Half at 101 Half at 200

For Pure Egalitarians, (1) is the best outcome, since it contains the least inequality. For Utilitarians, (1) is the worst outcome, since it contains the least utility. For most Pluralist Egalitarians, (1) would be neither the best nor the worst of these outcomes. (1) would be, on balance, worse than (2), since it would be *much* worse in terms of utility, and only *slightly* better in terms of equality. Similarly, (1) would be better than (3), since it would be much better in terms of equality, and only slightly worse in terms of utility.

In many cases the Pluralist View is harder to apply. Compare

(1) Everyone at 150

with

(4) Half at *N* Half at 200.

If we are Pluralist Egalitarians, for which values of *N* would we believe (1) to be worse than (4)? For some range of values – such as 120 to 150 – we may find this question hard to answer. And it may not have an answer. The relative importance of equality and utility may be, even in principle, imprecise.

We should next distinguish two kinds of value. If we claim that equality is good, we may mean only that it has good effects. If people are unequal, for example, that can produce conflict, or damage the self-respect of those who are worst off, or put some people in the power of others. If we care about equality because we are concerned with such effects, we believe that equality has *instrumental* value, or is good as a means. But I am concerned with a different idea. For true Egalitarians, equality has *intrinsic* value, or is in itself good.

This distinction is important. If we believe that, besides having bad effects, inequality is in itself bad, we shall think it to be worse. And we shall think it bad even when it has no bad effects.

To illustrate this second point, consider what I shall call *the Divided World.* The two halves of the world's population are, we can suppose, unaware of each other's existence. Perhaps the Atlantic has not yet been crossed. Consider next two possible states of affairs:

(1) Half at 100 Half at 200
(2) Everyone at 145

Of these two states, (1) is in one way better than (2), since people are on average better off. But we may believe that, all things considered, (1) is worse than (2). How could we explain this view?

If we are Telic Egalitarians, our explanation would be this. While it is good that, in (1), people are on average better off, it is bad that some people are worse off than others. The badness of this inequality morally outweighs the extra benefits.

In making such a claim, we could not appeal to inequality's bad effects. Since the two halves of the world's population are quite unconnected, this inequality has no effects. If we are to claim that (1) is worse because of its inequality, we must claim that this inequality is in itself bad.[6]

[6] In his paper in this volume, which I cannot properly discuss here, Richard Norman writes: '[Parfit] asks us whether (1) is worse that (2). I have to confess that I do not know how to answer that question, and I do not think that this is simply a personal confession on my part. . . . I want to say of Parfit's Divided world example that when you abstract the question from the social context in which we make judgements about equality and inequality, it is no longer clear how to answer it' (pp. 240–1 below). It is, I agree, not obvious whether the inequality in (1) is bad. But that it is not because we cannot make value judgments about such examples. It is clear that (1) would be better than

(3) Half at 100, Half at 50,

but worse than

(4) Everyone at 200.

3

We can now turn to a different kind of egalitarian view. According to *Deontic Egalitarians*, though we should sometimes aim for equality, that is *not* because we would thereby make the outcome better. On this view, it is not in itself bad if some people are worse off than others. When we ought to aim for equality, that is always for some other moral reason.

Such a view typically appeals to claims about *comparative* justice. Whether people are unjustly treated, in this comparative sense, depends on whether they are treated *differently* from other people. Thus it may be unfair if, in a distribution of resources, some people are denied their share. Fairness may require that, if certain goods are given to some, they should be given to all.

Another kind of justice is *non-comparative*. Whether people are unjustly treated, in this other sense, depends only on facts about them. It is irrelevant whether others are treated differently. Thus, if we treated no one as they deserved, this treatment would be unjust in the non-comparative sense. But, if we treated everyone equally unjustly, there would be no comparative injustice.[7]

It can be hard to distinguish these two kinds of justice, and there are difficult questions about the relation between them.[8] One point should be mentioned here. Non-comparative justice may require us to produce equality. Perhaps, if everyone were equally deserving, we should make everyone equally well off. But such equality would be merely the effect of giving people what they deserved. Only comparative justice makes equality our aim.

When I said that, in my examples, no one deserves to be better off than others, I did not mean that everyone is equally deserving. I meant that, in these cases, questions of desert do not arise. It is only comparative justice with which we are here concerned.

There is another relevant distinction. In some cases, justice is *purely procedural*. It requires only that we act in a certain way. For example, when some good cannot be divided, we may be required to conduct a lottery, which gives everyone an equal chance to receive this good. In other cases, justice is in part *substantive*. Here too, justice may require a certain kind of procedure; but there is a separate criterion of what the outcome ought

[7] Cf. Joel Feinberg, 'Noncomparative Justice', *Philosophical Review*, 83 (1974).

[8] Cf. Philip Montague, 'Comparative and Non-comparative Justice', *Philosophical Quarterly*, 30 (1980).

to be. One example would be the claim that people should be given equal shares.[9]

We can now redescribe our two kinds of Egalitarianism. On the Telic View, inequality is bad; on the Deontic View, it is unjust.

It may be objected that, when inequality is unjust, it is, for that reason, bad. But this does not undermine this way of drawing our distinction. On the Deontic View, injustice is a special kind of badness, one that necessarily involves wrong-doing. What is unjust, and therefore bad, is not strictly the state of affairs, but the way in which it was produced.

There is one kind of case which most clearly separates these two views: those in which some inequality cannot be avoided. For Deontic Egalitarians, if nothing can be done, there can be no injustice. In Rawls's words, if some situation 'is unalterable . . . the question of justice does not arise.'[10]

Consider, for example, the inequality in our natural endowments. Some of us are born more talented or healthier than others, or are more fortunate in other ways. If we are Deontic Egalitarians, we shall not believe that such inequality is in itself bad. We might agree that, if we *could* distribute talents, it would be unjust or unfair to distribute them unequally. But, except when there are bad effects, we shall see nothing to regret in the inequalities produced by the random shuffling of our genes. Many Telic Egalitarians take a different view. They believe that, even when such inequality is unavoidable, it is in itself bad.[11]

[9] There is an intermediate case. Justice may require a certain outcome, but only because this avoids a procedural flaw. One such flaw is partiality. Suppose that we have to distribute certain publicly owned goods. If we could easily divide these goods, others might be rightly suspicious if we gave to different people unequal shares. That might involve favouritism, or wrongful discrimination. We may thus believe that, to avoid these flaws, we should distribute these goods equally.

How does this view differ from a view that requires equality for substantive reasons? One difference is this. Suppose that we have manifestly tried to distribute equally, but our procedure has innocently failed. If we aimed for equality only to avoid the taint of partiality or discrimination, there would be no case for correcting the result. (For discussions of these points, see Robert Goodin, 'Egalitarianism, Fetishistic and Otherwise', *Ethics*, 98 (1987); and Lawrence Sager and Lewis Kornhauser, 'Just Lotteries', *Social Science Information* (Sage, London, Newbury Park and New Delhi, Vol 27, 1988).)

[10] John Rawls, *A Theory of Justice*, (Cambridge: Harvard University Press, 1971), page 291.

[11] There is now a complication. Those who hold this second view do not merely think that such inequality is bad. They often speak of natural injustice. On their view, it is unjust or unfair that some people are born less able, or less healthy, than others. Similarly, it is unfair if nature bestows on some richer resources. Talk of unfairness here is sometimes claimed to make no sense. I believe that it does make sense. But, even on this view, our distinction stands. According to Telic Egalitarians, it is the state of affairs which is bad, or unjust; but Deontic Egalitarians are concerned only with what we ought to do.

These views differ in several other ways. The Telic View, for example, is likely to have wider scope. If we believe that inequality is in itself bad, we may think it bad whoever the people are between whom it holds. It may seem to make no difference whether these people are in the same or different communities. We may also think it irrelevant what the respects are in which some people are worse off than others: whether they have less income, or worse health, or are less fortunate in other ways. *Any* inequality, if undeserved and unchosen, we may think bad. Nor, third, will it seem to make a difference how such inequality arose. That is implied by the very notion of intrinsic badness. When we ask whether some state is in itself bad, it is irrelevant how it came about.

If we are Deontic Egalitarians, our view may have none of these features. Though there are many versions of the Deontic View, one large group are broadly contractarian. Such views often appeal to the idea of reciprocity, or mutual benefit. On some views of this kind, when goods are co-operatively produced, and no one has special claims, all the contributors should get equal shares. There are here two restrictions. First, what is shared are only the fruits of co-operation. Nothing is said about other goods, such as those that come from nature. Second, the distribution covers only those who produce these goods. Those who cannot contribute, such as the handicapped, or children, or future generations, have no claims.[12]

Other views of this kind are less restrictive. They may cover all the members of the same community, and all kinds of good. But they still exclude outsiders. It is irrelevant that, in other communities, there are people who are much worse off. On such views, if there is inequality between people in different communities, this need not be anyone's concern. Since the greatest inequalities are on this global scale, this restriction has immense importance.

Consider next the question of causation. The Telic View naturally applies to all cases. On this view, we always have a reason to prevent or reduce inequality, if we can. If we are Deontic Egalitarians, we might think the same; but that is less likely. Since our view is not about the goodness of outcomes, it may cover only inequalities that result from acts, or only those that are intentionally produced. And it may tell us to be concerned only with the

[12] See, for example, David Gauthier, *Morals by Agreement* (Oxford: Oxford University Press, 1980), pages 18 and 268.

inequalities that we ourselves produce. On such a view, when we are responsible for some distribution, we ought to distribute equally. But, when no one is responsible, inequality is not unjust. In such cases, there is nothing morally amiss. We have no reason to remove such inequality, by redistribution. Here again, since this view has narrower scope, this can make a great practical difference.

4

Let us now consider two objections to the Telic View.

On the widest version of this view, *any* inequality is bad. It is bad, for example, that some people are sighted and others are blind. We would therefore have a moral reason to take single eyes from the sighted and give them to the blind. That conclusion may seem horrific.

Such a reaction is, I believe, mistaken. To set aside some irrelevant complications, we can imagine a simplified example. Suppose that, after some genetic change, children are henceforth born as twins, one of whom is always blind. And suppose that, as a universal policy, operations are performed after every birth, in which one eye from the sighted twin is transplanted into its blind sibling. That would be non-voluntary redistribution, since new-born babies cannot give consent. But I am inclined to believe that such a policy would be justified.

Some people would reject this policy, believing that it violates the rights of the sighted twins. But that belief provides no ground for rejecting the Telic View. As pluralists, Telic Egalitarians could agree that the State should not redistribute organs. Since they do not believe equality to be the only value, they could agree that, in this example, some other principle has greater weight, or is overriding. Their belief is only that, if we all had one eye, this would be *in one way* better than if half of us had two eyes and the other half had none. Far from being horrific, that belief is clearly true. If we all had one eye, that would be much better for all of the people who would otherwise be blind.[13]

A second objection is more serious. If inequality is bad, its disappearance must be in one way a change for the better, however this change occurs. Suppose that, in some natural disaster, those who are better off lose all their extra resources,

[13] Cf. Nozick, *op. cit.*, page 206 (though Nozick's target here is not the Principle of Equality but Rawls's Difference Principle).

and become as badly off as everyone else. Since this change would remove the inequality, it must be in one way welcome, on the Telic View. Though this disaster would be worse for some people, and better for no one, it must be, in one way, a change for the better. Similarly, it would be in one way an improvement if we destroyed the eyes of the sighted, not to benefit the blind, but only to make the sighted blind. These implications can be more plausibly regarded as monstrous, or absurd. The appeal to such examples we can call *the Levelling Down Objection*.[14]

It is worth repeating that, to criticize Egalitarians by appealing to this objection, it is not enough to claim that it would be *wrong* to produce equality by levelling down. Since they are pluralists, who do not care only about equality, Egalitarians could accept that claim. Our objection must be that, if we achieve equality by levelling down, there is *nothing* good about what we have done. Similarly, if some natural disaster makes everyone equally badly off, that is not in any way good news. These claims do contradict the Telic Egalitarian View.

I shall return to the Levelling Down Objection. The point to notice now is that, on a Deontic view, we avoid this objection. If we are Deontic Egalitarians, we do not believe that inequality is bad, so we are not forced to admit that, on our view, it would be in one way better if inequality were removed by levelling down. We may believe that we have a reason to remove inequality only *when*, and only *because*, our way of doing so benefits the people who are worse off. Or we may believe that, when some people are worse off than others, through no fault or choice of theirs, they have a special claim to be raised up to the level of the others, but they have no claim that others be brought down to their level.

Given these differences between the Telic and Deontic Views, it is important to decide which view, if either, we should accept. If we are impressed by the Levelling Down Objection, we may be tempted by the Deontic View. But, if we give up the Telic View, we may find it harder to justify some of our beliefs. If inequality is not in itself bad, we may find it harder to defend our view that we

[14] Such an objection is suggested, for example, in Joseph Raz, *The Morality of Freedom* (Oxford: Oxford University Press, 1986) Chapter 9, and Larry Temkin, *op. cit.* pages 247–8.

should often redistribute resources. And some of our beliefs might have to go. Reconsider the Divided World, in which the two possible states are these:

(1) Half at 100 Half at 200
(2) Everyone at 145

In outcome (1) there is inequality. But, since the two groups are unaware of each other's existence, this inequality was not deliberately produced, or maintained. Since this inequality does not involve wrong-doing, there is no injustice. On the Deontic View, there is nothing more to say. If we believe that (1) is worse, and because of the inequality, we must accept the Telic form of the Egalitarian View. We must claim that the inequality in (1) is in itself bad.

We might, however, give a different explanation. Rather than believing in equality, we might be especially concerned about those people who are worse off. That could be our reason for preferring (2).

Let us now consider this alternative.

5

In discussing his imagined case, Nagel writes:

> If one chose to move to the city, it would be an egalitarian decision. It is more urgent to benefit the second child . . . This urgency is not necessarily decisive. It may be outweighed by other considerations, for equality is not the only value. But it is a factor, and it depends on the worse off position of the second child. An improvement in his situation is more important than an equal or somewhat greater improvement in the situation of the first child.[15]

This passage contains the idea that equality has value. But it gives more prominence to another idea. It is more important, Nagel claims, to benefit the child who is worse off. That idea can lead us to a quite different view.

Consider first those people who are badly off: those who are suffering, or those whose basic needs have not been met. It is widely believed that we should give priority to helping such

[15] Nagel, *op. cit.* page 124.

people. This would be claimed even by Utilitarians, since, if people are badly off, they are likely to be easier to help.

Nagel, and others, make a stronger claim. On their view, it is more urgent to help these people even if they are *harder* to help. While Utilitarians claim that we should give these people priority when, and because, we can help them *more*, this view claims that we should give them priority, even when we can help them *less*.

Some people apply this view only to the two groups of the well off and the badly off.[16] But I shall consider a broader view, which applies to everyone. On what I shall call

The Priority View: Benefiting people matters more the worse off these people are.

For Utilitarians, the moral importance of each benefit depends only on how great this benefit would be. For *Prioritarians*, it also depends on how well off the person is to whom this benefit comes. We should not give equal weight to equal benefits, whoever receives them. Benefits to the worse off should be given more weight.[17] This priority is not, however, absolute. On this view, benefits to the worse off could be morally outweighed by sufficiently great benefits to the better off. If we ask what would be sufficient, there may not always be a precise answer. But there would be many cases in which the answer would be clear.[18]

On the Priority View, I have said, it is more important to benefit those who are worse off. But this claim does not, by itself, amount to a different view, since it would be made by all Egalitarians. If we believe that we should aim for equality, we shall think it more important to benefit those who are worse off, since such benefits reduce inequality. If *this* is why we give such benefits priority, we do not hold the Priority View. On this view, as I

[16] Cf. H. Frankfurt, *The Importance of What We Care About* (Cambridge: Cambridge University Press, 1988), Chapter 11, and Joseph Raz, *op. cit.* Chapter 9.

[17] Several other writers have suggested such a view. See, for example, Thomas Scalon, 'Nozick on Rights, Liberty, and Property', *Philosophy & Public Affairs*, 6 (1976), pages 6 to 10, Joseph Raz, *op. cit.*, Harry Frankfurt, 'Equality as a Moral Ideal', in *The Importance of What We Care About* (Cambridge: Cambridge University Press, 1988), David Wiggins, 'Claims of Need', in his *Needs, Values, Truth* (Oxford: Blackwell, 1987), Dennis McKerlie, 'Egalitarianism', *Dialogue*, 23 (1984), and 'Equality and Priority', *Utilitas*, 6 (1994).

[18] Like the belief in equality, the Priority View can take either Telic or Deontic forms. It can be a view about which outcomes would be better, or a view that is only about what we ought to do. But, for our purposes here, this difference does not matter.

define it here, we do *not* believe in equality. We do not think it in itself bad, or unjust, that some people are worse off than others. That is what makes this a distinctive view.

The Priority View can be easily misunderstood. On this view, if I am worse off than you, benefits to me matter more. Is this *because* I am worse off than you? In one sense, yes. But this has nothing to do with my relation to you.

It may help to use this analogy. People at higher altitudes find it harder to breathe. Is this because they are higher up than other people? In one sense, yes. But they would find it just as hard to breathe even if there were no other people who were lower down. In the same way, on the Priority View, benefits to the worse off matter more, but that is only because these people are at a lower *absolute* level. It is irrelevant that these people are worse off *than others*. Benefits to them would matter just as much even if there *were* no others who were better off.

The chief difference is, then, this. Egalitarians are concerned with *relativities*: with how each person's level compares with the level of other people. On the Priority View, we are concerned only with people's absolute levels. This is a fundamental structural difference. Because of this difference, there are several ways in which these views have different implications.

One example concerns scope. Telic Egalitarians may, I have said, give their view wide scope. They may believe that inequality is bad even when it holds between people who have no connections with each other. This may seem dubious. Why would it matter if, in some far off land, and quite unknown to me, there are other people who are better off than me?

On the Priority View, there is no ground for such doubts. This view naturally has universal scope. If it is more important to benefit one of two people, because this person is worse off, it is irrelevant whether these people are in the same community, or are aware of each other's existence. The greater urgency of benefiting this person does not depend on her relation to the other person, but only on her lower absolute level.

These views differ in other ways, which I have no space to discuss here. But I have described the kind of case in which these views most deeply disagree. These are the cases which raise the Levelling Down Objection. Egalitarians face this objection because they believe that inequality is in itself bad. If we accept the Priority View, we avoid this objection. On this view, except when it is bad for people, inequality does not matter.

6

Though equality and priority are different ideas, this distinction has been often overlooked.

One reason is that, especially in earlier centuries, Egalitarians have often fought battles in which this distinction did not arise. They were demanding legal or political equality, or attacking arbitrary privileges, or differences in status. These are not the kinds of good to which our distinction applies. And it is here that the demand for equality is most plausible.

Second, when Egalitarians considered other kinds of good, they often assumed that, if equality were achieved, this would either increase the sum of these goods, or would at least not reduce this sum. In either of these cases, equality and priority cannot conflict.

Third, even when a move to equality would reduce the total sum of benefits, Egalitarians often assumed that such a move would at least bring *some* benefits to the people who were worse off. In such cases, equality and priority could not deeply conflict. Egalitarians ignored the cases in which equality could not be achieved except by levelling down.

Since this distinction has been overlooked, some writers have made claims that are not really about equality, and would be better stated as claims about priority. For example, Nagel writes:

> To defend equality as a good in itself, one would have to argue that improvements in the lot of people lower on the scale of well-being took priority over greater improvements to those higher on the scale.[19]

In the example with which we began, Nagel similarly claims that it would be 'more urgent' to benefit the handicapped child. He then writes:

> This urgency is not necessarily decisive. It may be outweighed by other considerations, for equality is not the only value.[20]

These remarks suggest that, to the question 'Why is it more urgent to benefit this child?', Nagel would answer, 'Because this would reduce the inequality between these two children'. But I doubt that this is really Nagel's view. Would it be just as urgent to benefit the handicapped child, even if he had no sibling who was better off? I suspect that, on Nagel's view, it would. Nagel would

[19] *Reading Nozick*, edited by Jeffrey Paul (Oxford: Blackwell, 1981) page 203.
[20] *op. cit.* p. 124.

then, though using the language of equality, really be appealing to the Priority View.[21]

Consider next the idea of distribution according to need. Several writers argue that, when we are moved by this idea, our aim is to achieve equality. Thus Raphael writes:

> If the man with greater needs is given more than the man with lesser needs, the intended result is that each of them should have (or at least approach) the same level of satisfaction; the inequality of nature is corrected.[22]

When discussing the giving of extra resources to meet the needs of the ill, or handicapped, Norman similarly writes:

> the underlying idea is one of equality. The aim is that everybody should, as far as possible, have an equally worthwhile life.[23]

As before, if that were the aim, it could be as well achieved by levelling down. This cannot be what Norman means. He could avoid this implication by omitting the word 'equally', so that his claim became: 'the aim is that everybody should, as far as possible, have a worthwhile life.' With this revision, Norman could not claim that equality is the underlying idea. But that, I believe, would strengthen his position. Distribution according to need is better regarded as a form of the Priority View.[24]

What these writers claim about need, some have claimed about all kinds of distributive principle. For example, Ake writes:

> Justice in a society as a whole ought to be understood as a complete equality of the overall level of benefits and burdens of each member of that society.

The various principles of distributive justice, Ake claims, can all be interpreted as having as their aim 'to restore a situation of complete equality to the greatest degree possible'.[25] Some writers

[21] Similar remarks apply to section 117 of my *Reasons and Persons* (Oxford: Oxford University Press, 1984). For a later discussion of the choice between these views, see Nagel's *Equality and Partiality*, op. cit., Chapters 7 and 8.

[22] D.D. Raphael, *Justice and Liberty* (London: Athlone Press, 1980), page 10. Cf. page 49.

[23] Richard Norman, *Free and Equal* (Oxford: Oxford University Press, 1987), page 80.

[24] See, however, the excellent discussion in David Miller, 'Social Justice and the Principle of Need', in *The Frontiers of Political Theory*, ed. Michael Freeman and David Robertson (Brighton: Harvester Press, 1980).

[25] Christopher Ake, 'Justice as Equality', *Philosophy & Public Affairs*, 5 (1975), pages 71 and 77.

even make such claims about retributive justice. They argue that, by committing crimes, criminals make themselves better off than those who keep the law. The aim of punishment is to restore them to their previous level.

These writers, I believe, claim too much for equality. But there are some plausible views which are rightly expressed in egalitarian terms. For example, Cohen suggests that 'the right reading of egalitarianism' is that 'its purpose is to eliminate involuntary disadvantage'.[26] He means by this *comparative* disadvantage: being worse off than others. This is an essentially relational idea. Only equality could eliminate such disadvantage. Cohen's view could not be re-expressed in the language of priority. Similar assumptions underlie Rawls's view, whose complexity leads me to ignore it here.

Some Egalitarians are not moved by the Levelling Down Objection. For example, Ake writes

> What about the case of someone who suddenly comes into good fortune, perhaps entirely by his or her own efforts? Should additional burdens . . . be imposed on that person in order to restore equality and safeguard justice? . . . Why wouldn't it be just to impose any kind of additional burden whatsoever on him in order to restore the equality? The answer is that, strictly speaking, it would be . . .[27]

Ake admits that, on his view, it *would* be just to level down, by imposing burdens on this person. What he concedes is only that the claim of justice would here be overridden. Levelling down would be in one way good, or be something that we would have a moral reason to do. Similarly, Temkin writes:

> I, for one, believe that inequality is bad. But do I *really* think that there is some respect in which a world where only some are blind is worse than one where all are? Yes. Does this mean I think it would be better if we blinded everybody? No. Equality is not all that matters.[28]

Several other writers make such claims.[29]

[26] G.A.Cohen, 'On the Currency of Egalitarian Justice', *Ethics*, 99 (1989).

[27] *op cit.* page 73.

[28] *Inequality*, page 282.

[29] See for example Amartya Sen, *Inequality Reexamined* (Oxford: Oxford University Press, 1992), pages 92–3.

7

Since some writers are unmoved by the Levelling Down Objection, let us now reconsider that objection. Consider these alternatives:

(1) Everyone at some level	
(2) Some at this level	Others better off

In outcome (1) everyone is equally well off. In outcome (2), some people are better off, but in a way that is worse for no one. For Telic Egalitarians, the inequality in (2) is in itself bad. Could this make (2), all things considered, a worse outcome than (1)?

Some Egalitarians answer Yes. These people do not believe that the avoidance of inequality always matters most. But they regard inequality as a great evil. On their view, a move to inequality *can* make an outcome worse, even when this outcome would be better for everyone. Those who hold this view we can call *Strong Egalitarians.*

Others hold a different view. Since they believe that inequality is bad, they agree that outcome (2) is *in one way* worse than outcome (1). But they do not believe that (2) is worse all things considered. In a move from (1) to (2), some people would become better off. According to these Egalitarians, the loss of equality would be morally outweighed by the benefits to these people. (2) would be, on balance, better than (1). Those who hold this view we can call *Moderates.*

This version of Egalitarianism is often overlooked, or dismissed. People assume that, if we are Egalitarians, we must be against a move to inequality, even when this move would be bad for no one. If we regard such inequality as outweighed by the extra benefits, our view must, they assume, be trivial.[30]

That assumption is mistaken. If some change would increase inequality, but in a way that is worse for no one, the inequality must come from benefits to certain people. And there cannot be a *great* loss of equality unless these benefits are also great. Since these gains and losses would roughly march in step, there is room for Moderates to hold a significant position. They believe that, in all such cases, the gain in utility would outweigh the loss in equality.

[30] See, for example, Antony Flew, *The Politics of Procrustes* (Buffalo, New York: Prometheus, 1981), page 26, McKerlie, 'Egalitarianism', *op. cit.*, p. 232. See also Nozick, *op. cit.* p. 211.

That is consistent with the claim that, in many other cases, that would not be so. Moderates can claim that *some* gains in utility, even if *great*, would *not* outweigh some losses in equality. Consider, for example, these alternatives:

(1) All at 100
(4) Half at 100 Half at 200
(5) Half at 70 Half at 200.

Moderates believe that, compared with (1), (4) is better. But they might claim that (5) is worse. Since (5) would involve a much greater sum of benefits, that is not a trivial claim.

Return now to the Levelling Down Objection. Strong Egalitarians believe that, in some cases, a move towards inequality, even though it would be worse for no-one, would make the outcome worse.[31] This view may seem incredible. One of two outcomes *cannot* be worse, we may claim, if it would be worse for no one. To challenge Strong Egalitarians, it would be enough to defend this claim. To challenge Moderates, we must defend the stronger claim that, when inequality is worse for no one, it is not in any way bad.

Many of us would make this stronger claim. It is widely assumed that nothing can be bad if it is bad for no one. This we can call the *Person-affecting View.*

This view might be defended by an appeal to some account of the nature of morality, or moral reasoning. According to some writers, for example, to explain the impersonal sense in which one of two outcomes can be *worse* – or worse, period – we must appeal to claims about what would be worse *for* particular people. The Person-affecting View can also be supported by various kinds of contractualism.[32]

Egalitarians might reply by defending a different meta-ethical view. Or they might argue that, when the Person-affecting View is applied to certain other questions, it has unacceptable

[31] I am assuming here that inequality is not in itself bad for people. It is not bad for me if, unknown to me and without affecting me, there exist some other people who are better off than me. That assumption is implied, not only by hedonistic theories about well-being, but also by plausible versions both of desire-fulfilment theories, and of theories that appeal to what Scanlon calls *substantive goods.* For a contrary view, however, which would need a further discussion, see John Broome, *Weighing Goods* (Oxford: Blackwell, 1991) Chapter 9.

[32] Such as the view advanced in Thomas Scanlon's 'Contractualism and Utilitarianism', in ed. Amartya Sen and Bernard Williams, *Utilitarianism and Beyond* (Cambridge: Cambridge University Press, 1982).

implications, since it conflicts too sharply with some of our beliefs.[33] Since I have no space to discuss these questions here, I shall merely express an opinion. The Person-affecting View has, I believe, less plausibility than, and cannot be used to strengthen, the Levelling Down Objection.

8

I shall now summarise what I have claimed. According to Telic Egalitarians, it is in itself bad, or unfair, if some people are worse off than others through no fault or choice of theirs. Though this view is widely held, and can seem very plausible, it faces the Levelling Down Objection. This objection seems to me to have great force, but is not, I think, decisive.

Suppose that we began by being Telic Egalitarians, but we are convinced by this objection. We cannot believe that, if the removal of inequality would be bad for some people, and better for no one, this change would be in any way good. If we are to salvage something of our view, we then have two alternatives.

We might become Deontic Egalitarians. We might come to believe that, though we should sometimes aim for equality, that is not because we would thereby make the outcome better. We must then explain and defend our beliefs in some other way. And the resulting view may have narrower scope. For example, it may apply only to goods of certain kinds, such as those that are co-operatively produced, and it may apply only to inequality between members of the same community.

We may also have to abandon some of our beliefs. Reconsider the Divided World:

(1) Half at 100 Half at 200
(2) Everyone at 145

On the Deontic View, we cannot claim that it would be better if the situation changed from (1) to (2). This view is only about what people ought to do, and makes no comparisons between states of affairs.

Our alternative is to move to the Priority View. We could then

[33] See Temkin, *op. cit.*, Chapter 9. Another objection to the Person-affecting View comes from what I have called *the Non-Identity Problem* (in my *Reasons and Persons*), Chapter 16).

keep our belief about the Divided World. It is true that, in a change from (1) to (2), the better off would lose more than the worse off would gain. That is why, in utilitarian terms, (2) is worse than (1). But, on the Priority View, though the better off would lose more, the gains to the worse off count for more. Benefits to the worse off do more to make the outcome better. That could be why (1) is worse than (2).

The views that I have been discussing often coincide. But, as I have tried to show, they are quite different. They can support different beliefs, and policies, and they can be challenged and defended in different ways. Taxonomy, though unexciting, needs to be done. Until we have a clearer view of the alternatives, we cannot hope to decide which view is true, or is the best view.

All Souls College
Oxford OX1 4AL
England

[19]

Equality, Priority, and Time

KLEMENS KAPPEL
University of Copenhagen

The lifetime equality view (the view that it is good if people's lives on the whole are equally worth living) has recently been met with the objection that it does not rule out simultaneous inequality: two persons may lead equally good lives on the whole and yet there may at any time be great differences in their level of well-being. And simultaneous inequality, it is held, ought to be a concern of egalitarians. The paper discusses this and related objections to the lifetime equality view. It is argued that rather than leading to a revision of the lifetime equality view, these objections, if taken seriously, should make us account for our egalitarian concerns in terms of the priority view rather than the equality view. The priority view claims that there is a greater moral value to benefiting the worse off. Several versions of the priority view are also distinguished.

I. INTRODUCTION

Among egalitarians, most writers have taken for granted that egalitarian concern, however other details are to be understood, applies to persons' full lives. Thus, ignoring a number of details, egalitarians have claimed that it makes an outcome better if people are equally well off when their lives as wholes are compared, and they have assumed that this is all there is to egalitarian concern.[1] Let this be the *lifetime equality view*.

Recently, however, the lifetime equality view has been questioned, notably by Dennis McKerlie and Larry Temkin.[2] They suggest that the lifetime equality view leads to consequences that egalitarians should not welcome. The problem is that equality with respect to well-being in persons' whole lives is compatible with great inequality with respect to well-being at any particular moment.[3] Lifetime equality does not rule out *simultaneous inequality*. McKerlie and Temkin hold simultaneous inequality as such to be a bad feature of an outcome, and they think that egalitarians ought to revise or amend the lifetime equality view to accommodate this.

One main objective of this paper is to discuss the merits of these arguments. I shall suggest that the examples produced by McKerlie

[1] Rawls, when saying that inequality is justified only if the worst off group is better off in terms of primary goods than in any other possible distribution, assumes that the correct unit over which to compare distributions is typical life spans. So do writers like Ronald Dworkin and Thomas Nagel. For Dworkin's views, see his 'What is Equality? Part 1: Equality of Welfare', *Philosophy and Public Affairs*, x (1981). A statement of Nagel's views can be found in his 'Equality', *Mortal Questions*, Cambridge, 1979.

[2] See Dennis McKerlie, 'Equality and Time', *Ethics*, xcix (1989), 283–345, and Larry Temkin, *Inequality*, Oxford 1993, ch. 8.

[3] For simplicity I discuss only equality with respect to well-being. However, most of what is said below applies to other metrics of egalitarian concern, such as resources.

and Temkin which are intended to persuade us that simultaneous inequality as such is bad really trade on other features. Once they are removed there is in my view no convincing case that simultaneous inequality as such is bad. A second objective of the paper is to comment on the revisions of the lifetime equality view, and the additions to this view that McKerlie and Temkin propose. I shall suggest that they do not work, or lead to implausible results. This provides a further reason to be sceptical about the view that simultaneous inequality as such is bad.

A third aim of the paper goes further and is to suggest that what the examples offered by McKerlie and Temkin point to is that we should abandon equality as our basic way of accounting for egalitarian concern. Rather, what matters in our egalitarian judgement is giving priority to the worse off. This is the *priority view*. Briefly, and to be elaborated in a moment, the priority view claims that there is a greater moral value to benefiting the worse off.

Finally, the fourth aim of the paper is to open a discussion of a number of problems in the priority view which are similar to the problems of the equality view that initiates the discussion of the lifetime equality view. In this final part, however, more questions are raised than answered.

II. EQUALITY AND PRIORITY

Before turning to the objections to the lifetime equality view, I shall state the claims of the priority view and the equality view in slightly more detail. Following Parfit and others, we can distinguish two different positions accounting for egalitarian concern.[4] On one view, the *equality view*, equality as such has value. It is valuable that people are equal with respect to well-being or resources or whatever the metric of our egalitarian concern. Since equality has value we should often make life better for the worse off in order to promote this value. But obviously the particular value of equality can also be promoted by making life worse for the better off.[5] In the lifetime version of the equality view two further claims are made. One is that the relevant unit to compare is peoples' full lifetimes. The second is that when comparing people in their lifetimes, we should compare how good peoples' lives are as wholes. I shall later discuss other views which take life-

[4] Derek Parfit in his Lindley Lecture 'Equality or Priority?', University of Kansas, 1995. See also the recent discussion in Temkin, ch. 9.

[5] This description covers only what Parfit in 'Equality or Priority?' calls *Strong Egalitarianism*. Another position, *Weak Egalitarianism*, claims that we should never lower the level of well-being for the sake of equality, since the loss in utility will always outweigh the gain in equality. I shall be concerned only with the former view.

times as the relevant unit of time, but which deny that goodness of a life as a whole is the relevant unit to compare.

The priority view denies that equality as such has value. Instead it claims that benefiting the worse off is more important than benefiting the better off: if our efforts and the resultant increases in well-being are the same, then the value of benefiting someone is greater the worse off he or she is. In other words, the moral value of an increase in well-being depends not only on the size of the increase but also on the absolute level of well-being that the person in question has.[6] According to the priority view the best distributions of resources tend towards equality with respect to well-being, but as explained this is not because equality as such has value. The priority view comes in several versions depending on how we construe the notion of being worse off. One version may focus on those who are worse off in their lives as wholes, whereas other versions may focus on persons who at particular times are worse off, but not necessarily worse off in their lives as a whole. In section VIII I shall discuss several versions of the priority view.

One important motivation for accepting the priority view rather than the equality view is the *levelling down objection*. Suppose that in a population there is an unequal distribution of well-being. Suppose next that through a natural disaster everybody is made equally well off, and that each is now at a lower level. According to the equality view the situation has thereby improved in one respect, even if it may be on the whole worse. However, since everybody is worse off some writers find it hard to believe that things can have improved, even in one respect. It is hard to believe that equality could be valuable when its realization does not improve the life of anyone. This motivates the priority view, which allows us to say that the lot of the worse off should be improved but does not imply that we have a reason to make everybody worse off for the sake of equality.[7]

As I shall mention in this paper, another reason for preferring the priority view to the equality view is that certain objections to the lifetime equality view, if taken seriously, really indicate that we should abandon not only the view that the unit of egalitarian concern should be lifetimes, but also the view that equality as such is what matters in egalitarian concern. Instead, we should accept the priority view in a version that does not focus on lifetimes.

[6] The priority view should not be confused with the law of diminishing returns, i.e. the often true empirical fact that the more resources a person possesses, the less good more resources will do him. The priority view does not concern what brings more or less well-being about. It concerns the different issue of the moral value of making someone better off.

[7] For a detailed discussion of some of the basic assumptions behind one version of the levelling down objection, see Temkin, ch. 9.

III. CHANGING PLACES EGALITARIANISM

Return now to the lifetime equality view. According to this view it is good if people are equally well off when their lives as wholes are compared. McKerlie objects as follows to the lifetime equality view:

> It does not see any disvalue in inequality between parts of lives as long as the inequality is compensated for at earlier or later times so that there is no inequality between complete lives. This enables us to imagine a new kind of egalitarian society. It contains great inequality, with happier lives attached to certain social positions. But at a fixed time people change places and switch from a superior position to an inferior one or vice versa. One example would be a feudal society in which peasants and nobles exchange roles every ten years. The result is that people's lives as wholes are equally happy. Nevertheless during a given time period the society contains great inequality, and in one sense this always remains true.[8]

This is what McKerlie has named *changing places egalitarianism.* McKerlie and Temkin believe that there must be egalitarian reasons to object to changing places egalitarianism. And they locate what is bad about it in the great simultaneous inequality that changing places egalitarianism allows for. Simultaneous inequality as such is bad, they suggest. It is a bad thing if, while some prosper, others are at the same time badly off, even if everybody has an equally good life on the whole. And since the lifetime equality view does not recognize anything bad about simultaneous inequality, this view cannot be all there is to egalitarian concern.

As I said in the introduction, I shall question the claim that it is simultaneous inequality as such which is bad in the examples of changing places egalitarianism, on which McKerlie and Temkin rest their case. I suggest that there are other explanations of the badness of the particular cases of changing places egalitarianism that McKerlie and Temkin offer.[9]

[8] McKerlie, 479. As McKerlies goes on to say, this has obvious analogues in our society. Elderly members of our society may be much worse off than younger members without this necessarily giving rise to objectionable inequality, according to the lifetime equality view. Similarly, Temkin imagines Job1 and Job2: '[Job1] and his family are healthy and wealthy. He has the love and respect of all who know him. In addition, his plans are realised, his desires fulfilled, and he has complete inner peace. Job2, on the other hand, has led a wretched life. His health is miserable, his countenance disfigured. He has lost his loved ones. He is a penniless beggar who sleeps fitfully in the streets, and whose efforts and desires are constantly frustrated.' This will go on for forty years, after which 'the situations of Job1 and Job2 will be reversed during the second half of their lives, so that in fact the overall quality of their lives, taken as a whole, will be completely equal' (Temkin, pp. 235–6).

[9] There are two complications which should be set aside. First, it might be suggested that if a life contains both times of great suffering and times of great happiness, this life is worse than if the same total sum of benefits and burdens were distributed more evenly. Someone might suggest that what is objectionable about changing places egalitarianism is that it involves this unfortunate distribution of well-being. I suspect that

First consider the fact that the cases of changing places egalitarianism we are offered all involve suffering or having a very hard time. This may be quite important for our responses to the cases. Think of a case of changing places egalitarianism without suffering:

> In a society everybody has a life well worth living. During periods of ten years some members of the society have a kind of sabbatical during which they are much better off than others. Since everybody gets one sabbatical in their life, and since everybody not on sabbatical is equally well off, all members of society have lives that on the whole are equally good. However, there is no way that everybody can have their sabbatical at the same time without a substantial loss in well-being for everybody.

It seems much less convincing that this case of simultaneous inequality is objectionable. This suggest that what is objectionable about the examples that McKerlie and Temkin produces might not be simultaneous inequality after all, but the fact that they involve suffering.

Secondly, note that the examples involve fairly long periods of time. In McKerlie's feudal society peasants and nobles change places every ten years, whereas positions are switched only after forty years in Temkin's case of Job1 and Job2 (see n. 8). And the time span involved may be even longer in the real life example of elderly people who have a relatively poor quality of life compared to contemporary younger members of the society in which they live. But imagine now a case of changing places egalitarianism which involves both simultaneous inequality and relative hardships, but much smaller time-spans:

> In a small society all the male members survive by hard labour in the mines. They work day and night shifts. Thus, while the day shift suffers, the other part of the work force relaxes, and vice versa. On the whole, however, everybody has an equally good life.

This is changing places egalitarianism, and in extreme cases it may contain as much simultaneous inequality as for example the peasant and nobles society. There would be many reasons to improve working conditions in this case, but the persistent simultaneous inequality is not obviously a candidate among them.

I have now suggested that simultaneous inequality as such does not account for what is objectionable about the cases of changing places egalitarianism that are offered by McKerlie and Temkin. However, the cases of changing places egalitarianism raise other issues, and I shall return to those in sections V and VI. Before doing so, however, I shall

this line of reasoning is confused, and I shall ignore it in what follows. Secondly, there is an issue about individual responsibility. If simultaneous inequality is the product of peoples' free and rational choices, there may for that reason be nothing objectionable about it. Here I shall merely assume that we consider only cases of changing places egalitarianism in which facts about individual responsibility are not reasons for claiming that there is nothing objectionable about them.

in the next section comment on McKerlie's and Temkin's proposals as to how an egalitarian should accommodate the view that simultaneous inequality as such is bad. This, I suggest, will reveal further reasons to be sceptical of the view that simultaneous inequality is bad.

IV. CATERING FOR SIMULTANEOUS INEQUALITY

McKerlie and Temkin both believe that simultaneous inequality as such is bad. But they disagree about how best to revise the lifetime equality view on the basis of this. I shall now comment on their proposals. McKerlie suggests that we should revise the lifetime equality view so as to operate on different time spans. Equality should concern segments of people's lives rather than their whole lives.[10] Now, a segment of a person's life (say, a period of ten years) can be compared to segments of other persons' lives in several different ways. But since this revision of the lifetime equality view is motivated by a concern about simultaneous inequality, we should obviously compare *simultaneous* segments of peoples' lives. This is the *simultaneous segment view*, named and defended by McKerlie. On this view, it is regrettable if one person during a segment has a rather poor quality of life while others during the same segment prosper. It is irrelevant if the same person in previous or later segments is much better off than others. Levels of well-being during past or future segments are of no importance to judgements as to whether inequality in one particular segment is objectionable.[11]

There are several problems in the simultaneous segment view. The major problem, it seems to me, is that this revision of the lifetime equality view takes the wrong form. Comparison of well-being in segments of time is conceptually quite different from comparisons of well-being at particular times. An analogy may help here: comparing velocity of objects at particular times is conceptually quite different from comparing distances that objects have moved in a particular span of time. If, for some reason, we were interested in two objects moving at the same velocity at particular times it would be quite unhelpful to compare the distance they had moved during spans of time, unless, of course, either the spans of time were very short, or we knew that the objects moved at a constant speed. Similarly, unless we choose segments that are very short, or unless we can safely assume that people's quality of life remain constant within a segment, it seems that equality of total well-being within segments is irrelevant to simul-

[10] See McKerlie, ibid.

[11] This is of course not to deny that if one can now foresee that future segments will contain inequality, then one should now do something to avoid it, if possible.

taneous inequality. Or, to put the same point in another way, simultaneous segmental equality will allow for large simultaneous inequality within segments. If one really thinks that simultaneous inequality matters there seems to be no case for the simultaneous segments view.

A further point here is that McKerlie argues that simultaneous inequality as such is bad, but also admits that this view has consequences that are really hard to believe. Suppose that two persons plan to go to their respective dentists. As McKerlie asks: 'Are there serious egalitarian reasons for preferring two 10:30 appointments [with the dentist] to an appointment at 10 and an appointment at 11?'.[12] It is in part to avoid such undesirable implications that McKerlie goes on to suggest the simultaneous segments view. However, as explained, this line of reasoning seems mistaken. Simultaneous segmental equality is something completely different from simultaneous equality. To propose the simultaneous segments view is not a way of defending the belief that simultaneous inequality is bad; it is to accept a completely different view.

This suggests that if one finds simultaneous inequality as such to matter, one is not entitled to capture this in terms of segmental equality. But one might claim that there are several kinds of inequality that makes an outcome worse, and that simultaneous inequality is one of them. This calls not necessarily for a revision of the lifetime equality view, but merely for admitting concern for simultaneous inequality in our general egalitarian concern.

However, this way of capturing the alleged importance of simultaneous inequality also leads to implausible consequences. Simultaneous differences in well-being between two persons must somehow be aggregated over time to yield a total of simultaneous inequality between these two persons in a period of time. Otherwise there would even in principle be no way one could say that one option is better than another with respect to simultaneous inequality.

There seem to be two different possibilities as to how to aggregate simultaneous inequality between persons over time, and neither of them is attractive.[13] On one view, several instances of simultaneous inequality can cancel each other out in the following way. If one person for a period of time is better off than another in terms of well-being, but this second person later is better off than the first by the same measure of well-being, and for a similar period of time, then this may be said to have restored equality between the two. Let this be the view that *reverse simultaneous inequalities cancel each other out.*

[12] McKerlie, 483.

[13] These two ways of aggregating inequality are mentioned by both McKerlie and Temkin.

The problem in this view is the following. Suppose that I am better off than you during a period of time. During this period of simultaneous inequality I gain ten units of well-being more than you. To counterbalance this inequality there needs to be some other times during which you gain ten units of well-being more than me. Only such periods of simultaneous inequality can cancel out the first one. This, however, amounts to saying that there will be no uncompensated simultaneous inequality just in case we gain the same total amount of well-being in the period of time during which both of us exist. Hence, the view that several instances of simultaneous inequality tend to cancel each other out collapses into the lifetime equality view, in the case of people who live simultaneously and for an equally long time. But this allows for changing places egalitarianism.

Suppose then that we reject the view that reverse simultaneous inequalities cancel out. We could then make the opposite claim: rather than cancel out, reversed instances of simultaneous inequality make things worse. If I am better off than you for a period of time, and thereby gain ten units of well-being more than you, then a second period of time in which you gain ten units more than me in total will not counterbalance the badness of the first period of simultaneous inequality. Rather it will make the situation worse with respect to total simultaneous inequality. Let this be the view that *simultaneous inequalities add up*.[14] According to this view, nothing can serve to cancel out simultaneous inequality. My appointment with my dentist today makes me worse off than you, and this simultaneous inequality is not rectified by your similar appointment tomorrow. Only if everybody is simultaneously equally well off at all times can we avoid it completely. In my view, this makes it more difficult to believe that simultaneous inequality, if aggregated this way, is bad in itself. Should we really sacrifice, say, well-being, to diminish the prevalence of simultaneous inequality under these assumptions?

This all adds up to a further reason to be sceptical about the importance of simultaneous inequality. The reason is simply that there seems to be no attractive way to incorporate it into a general egalitarian view. The segment view suggested by McKerlie fails to capture the alleged importance of simultaneous inequality. It may then be suggested that simultaneous inequality is merely a feature of an outcome that tends to make it worse. However, it is not obvious how several instances of simultaneous inequality should be aggregated. One proposal for aggregation leads to a view that allows for changing places egalitarianism. Another makes it harder to believe that simultaneous inequal-

[14] This is the view that McKerlie seems to accept, at least when applied to equality between simultaneous segments.

ity is important after all. This should make us all the more interested in alternative ways of accounting for what is objectionable about the cases of changing places egalitarianism produced by McKerlie and Temkin. I now turn to that issue.

V. SUFFERING

Above I have argued that examples of changing places egalitarianism intended to show that simultaneous inequality is bad depend on the fact that they involve suffering. As soon as the assumption of suffering is removed from these examples, it is much less clear that anything objectionable remains. Therefore these examples do not show that simultaneous inequality as such is bad. However, examples involving suffering may in other ways be used in objections against the lifetime equality view. I shall now discuss such objections. Consider this example:

> In a casualty department in a hospital two persons, A and B, are in pain. Unfortunately, only one of them can be given pain relieving treatment, since there is a shortage of the particular painkiller needed. A's pain is much stronger than B's pain. However, B did in the past suffer from pains similar to those he has now, and these pains went on for a considerably longer time and were not treated. On the whole B is therefore worse off than A in terms of well-being, and in that sense B has suffered more pain. Suppose finally that whoever we treat, the benefit will be the same because A's stronger pain will be only partially relieved, whereas B's pain will be fully relieved.[15]

At least some would think that in this case A ought to be treated, since A's pain is worse. However, according to the lifetime equality view there are egalitarian reasons for treating B rather than A, since B is worse off on the whole. And as the example is construed, the defender of the lifetime equality view cannot appeal to the greater benefit being produced by treating stronger rather than weaker pains. Since A's greater pain cannot be relieved fully, whereas B's lesser pain can, the benefit will be the same whichever we treat. Therefore, on the lifetime equality view, one is likely to judge that B rather than A ought to be treated. As I said, at least some would think that this consequence of the lifetime equality view is hard to believe.[16]

I shall refer to this as *the severe suffering objection*. The thrust of the objection is the observation that severe suffering like strong pain exerts a special claim upon us. I suggest that to capture the contention

[15] This is elaborated from an example of Derek Parfit's; see his 'Comments', *Ethics*, xcvi (1986). And the particular version presented here has been helped by Parfit's comments on an earlier draft of this paper.

[16] McKerlie also discusses this example as an argument against the lifetime equality view, and it is part of the reason that he finds a revision of that view plausible.

that A should be treated with an egalitarian view, we should not revise the lifetime equality view. Rather, we should reject both the assumption that egalitarian concern operates over lifetimes, *and* the assumption that equality as such has value. In this section I shall mostly be concerned with the latter claim.

Is there any reasonable version of the equality view that would explain why we should treat A whose pains are stronger rather than B? Someone might appeal to simultaneous inequality between A and B and claim that the reason we should treat A rather than B is that A is simultaneously worse off than B.[17] This view, however, seems absurd, or at least it would have rather unwelcome consequences. Imagine that we have to choose between two outcomes. In one, two patients both suffer severe pain, but at different times. In the other they suffer the same pain but at the same time. Assume that in our judgement we consider only the suffering of the two patients and no other people. In neither of the outcomes can we relieve the pains. If the claim for pain relief is generated only or even partly from simultaneous inequality it seems that we have a reason to choose the outcome where they suffer simultaneously. This might be so, even if they are slightly worse off in the outcome where they suffer at the same time. This, I suggest, is particularly hard to believe in the cases of pain.[18] In addition to this there are the previously considered problems with the view that simultaneous inequality as such is bad. Therefore, it just seems implausible that the claim generated in cases of persons in pain has anything to do with simultaneous inequality. And if this is true, it is doubtful if there is any version of the equality view that could explain our judgement that A's stronger pain ought to be treated rather than B's weaker, despite B's suffering more pain on the whole.

I shall now suggest why what the severe suffering objection is about is much better explained by the priority view. First, as just mentioned, the importance of relieving someone's pain has to do with his or her suffering, not with others' not being in pain. This suggests that the moral claim arising from the relief of pain has a central feature in common with the priority view: it concerns how badly off people are in terms of pain. It does not, like the equality view, concern a relation between people's levels of well-being or pain.

Secondly, the relief of pain has diminishing importance. As the example indicates, even if the benefits are the same, and our efforts are the same, we should relieve the stronger pain. In other words, the less someone is in pain, the less important is pain relief, even if our efforts and the size of the benefit in terms of well-being remain the

[17] McKerlie's view indeed seems to be that it is because of the simultaneous inequality between A and B that A rather than B ought to receive the treatment.

[18] This argument is of course just another instance of the levelling down argument.

same. Again this is not because of the relation between those in more and less pain, but because of the claim that the stronger pain exerts. This again suggests the priority view rather than the equality view.

Consider thirdly that what has here been said about pain might be generalized to other sorts of suffering, and to well-being in general. If we accept that what has here been exemplified with pain really applies to well-being in general, we have arrived at a version of the priority view.

Notice finally that the severe suffering objection suggests a version of the priority view that does not focus only on the total amount of well-being in a lifetime. The lifetime priority view would tell us to treat B rather than A, since B in his life as a whole is worse off than A. Rather, the version of the priority view that we are approaching suggests that what matter are episodes of severe pain, or at least that they have a special importance. What is suggested, then, by the severe suffering objection, and the discussion of simultaneous inequality above, is that the lifetime equality view is implausible for two reasons: egalitarian concern should not focus on lifetimes, and egalitarian concern should not be captured in terms of equality. Rather, a version of the priority view that does not take persons' lives as wholes as the relevant unit to focus on now seems more plausible. Later I shall discuss versions of the priority view that are intended to capture this.[19]

Before concluding this section it may be worth mentioning briefly certain other issues that the severe suffering objection could be thought to raise. It might be suggested that the objection relies on the fact that much of B's pain is in the past at the time a decision whether to treat A or B must be made. This could be held to be what justifies that we treat A rather than B. Past pain doesn't count at all, or doesn't count with the same weight as present or future pain. If this were right it might be that we should after all not reject equality in favour of priority. But suppose the temporal order in the example is reversed. We must either partially treat A's more severe pain now or B's lesser pain now, while knowing that B, in the future, will suffer more pain, and for that reason B will be overall worse off. It seems that if we choose in the original case to treat A rather than B, so we shall in this modified case. Thus, our judgement that A's greater pain rather than B's lesser pain should be treated does not seem to rely on a greater concern about the future than the past.

[19] One is obviously not forced by the severe suffering objection to give up the lifetime equality view. One might simply posit that the relief of suffering (or prioritarian concerns more generally) and lifetime equality are two separate issues, both of which are to be given some weight. Or, one might insist that simultaneous inequality as such is bad, besides lifetime inequality being bad. My only general objection against these proposals is that they might be less economical than merely assuming the priority view.

Similarly, the severe suffering objection does not rely on the assumption that we should do more to relieve present suffering at the cost of temporally distant suffering (be it in the future or in the past). Suppose that we must now choose to relieve either A's strong pain now or B's lesser pain in the future. Or the other way around: we must choose to relieve either B's lesser pain now or A's stronger pain in the future. Assume in both cases that B is overall worse off than A. I think that in both cases we would find it fair to relieve A's stronger pains. These considerations show that the kind of concern involved in the severe suffering objection in a clear sense exhibits temporal neutrality. What is at stake in this example is something different from attitudes to time. However, I shall now turn to a different case in which attitudes to time might be involved.

VI. EQUALITY AND TEMPORAL DISTANCES

Earlier I claimed that one of the crucial features in the examples of changing places egalitarianism offered by McKerlie and Temkin is that they employ a fairly large temporal span between the suffering or hardship that someone undergoes and the benefits which, according to the lifetime equality view, restore lifetime equality. My suggestion was that once we imagine examples in which there are no such large temporal spans it is much less clear that anything objectionable remains in cases of changing places egalitarianism, even if they still involve simultaneous inequality to the same degree. Hence, it is not clear that simultaneous inequality as such is bad.

However, examples employing large temporal spans may in a different way be used in an objection against the lifetime equality view. This is what I shall discuss in the present section. Consider this example:

> A had a happy childhood, whereas B endured a miserable one. Luckily for B, this left no enduring traces. It does not disadvantage B now, and has not adversely affected him in other parts of his life after childhood. Therefore, apart from their childhoods, the lives of A and B do not differ in morally significant ways. Now, many years later, they are equally in need of a place in a residential home for elderly people. However, there is only room for one of them, so a choice must be made. Each of them would benefit equally from being admitted, and they would now be equally well off without admission to the residential home.

According to the lifetime equality view, there is an egalitarian reason for preferring B for the vacancy at the residential home, and from the example it is clear that there is no countervailing reason to prefer A to B. Consequently, if we accept the lifetime equality view we should all things considered choose B rather than A. This could be so even if A now were slightly more in need of the care given at the home, since a slightly greater need of A could be outweighed by the worse childhood that B endured.

This judgement will, at least to some, seem wrong. On their view, A and B each have an equally strong claim to be admitted to the residential home. How can something, it might be asked, which does not matter in the least to B now be what decides the issue? As in the previous section, I shall for the sake of argument assume that an egalitarian would accept this judgement. That is, I shall assume that a reasonable egalitarian view should remain indifferent between A and B. However, were the hardships of B more recent, it would be more reasonable to let this fact influence our egalitarian judgements as to who should be admitted to the residential home.

A view of this kind might be implicated in what was discussed earlier in connection with cases of changing places egalitarianism. As noticed, our responses to these cases seem to be somewhat dependent on the temporal span between times people are badly off and the times they are better off. Large temporal spans seem to give rise to some objectionable features in changing places egalitarianism which are not preserved when the temporal distances are made shorter.

I shall now ask how one might best give an egalitarian account of this view, i.e. how best to give egalitarian reasons for giving less or no weight to inequalities in the distant past. Are these egalitarian reasons best to be explained invoking the equality view or the priority view? I shall argue that the priority view is a much more plausible candidate. That is, if we accept a general egalitarian view according to which hardships in the distant past count for less or not at all, we should do so on the basis of the priority view rather than the equality view. The reason is that the priority view much better than the equality view combines with reasons for caring less about the distant past. I shall argue for this first by discussing four proposals that all assume the equality view in some version, but nevertheless endeavour to explain why past inequalities matter less or not at all. I shall argue that none of them is plausible.

The first version of the equality view is this. Suppose we decide that lifetime equality is irrelevant, and that simultaneous inequality is bad. And suppose we further decide that instances of simultaneous inequality add up rather than cancel out (as discussed in section IV). According to this view, there is a plain egalitarian reason to remain indifferent between A and B. Even if we choose B in preference to A this will not counterbalance the badness of the simultaneous inequality that arose because of B's suffering during his childhood. And whichever we choose for the vacancy at the residential home, the same amount of simultaneous inequality will be brought about. Therefore, there is no egalitarian reason to prefer A to B.

While this view might have some appeal, it is hard to believe for the reasons given in section IV. A further problem in this context is that the view will not be sensitive to temporal distances. It will make no difference whether the simultaneous inequality between B and A is in the distant past, the recent past, or the future. This makes it a less plausible explanation of why egalitarians should remain indifferent between A and B in the present example.

A second option is this. Someone might suggest that there is a *temporal asymmetry* with respect to equality. Only future inequality matters.[20] At any time we should aim at there being no inequality in the future. This would explain why B's unfortunate childhood would be irrelevant to the decision about who should be admitted to the residential home.

This view might seem to be in line with the general view that the future matters more than the past. Suppose that you wake up in a hospital bed, not knowing which is true about you: either you are the patient who underwent a long, painful operation yesterday (which you do not remember), or you are the patient who will undergo a short and not so painful operation later today. There is no question that you would be greatly relieved to learn that you had the greater suffering yesterday and no operation to undergo later, rather than no long-lasting, painful operation yesterday but a short operation later today.[21] We have, it seems, a strong bias towards the future. We prefer, so to speak, greater pain in the past to lesser pain in the future.

The problem, however, is whether this general line of reasoning supports a temporal asymmetry with respect to the negative value of inequality. Recall that the alleged negative value of inequality does not in itself consist of individual's being better or worse off. Rather, the negative value of inequality as such is supposed to be based on a mere relation between people's levels of well-being. Therefore, it is hard to see why a temporal asymmetry in rational individuals' concern for their well-being in the past and the future should be a ground for accepting a similar asymmetry with respect to the value of equality.

A third proposal is this. It has been held that we should discount the value of goods for *temporal distance*, or more exactly for certain psychological relations that roughly correlate to temporal distances.[22] Briefly, the view I have in mind says that future and past harms and

[20] McKerlie discusses this option.

[21] This is Parfit's example. See *Reasons and Persons*, pp. 165–6.

[22] Notice that the appeal to temporal asymmetry is quite different from the appeal to temporal distance. Temporal distance is neutral with respect to whether something is in the past or in the future; only the temporal distance as such matters, irrespective of direction (or, as explained, certain psychological relations that correlate with temporal distance matter, irrespective of whether they extend into the future or into the past). Temporal asymmetry, on the other hand, concerns only whether something is in the past

benefit rationally matter less to me now in so far as I am less psychologically related to my future and past selves that will experience or has experienced these harms and benefits.[23] Someone might now suggest that we should similarly discount the negative value of inequality. Since the time when B was worse off than A is now in the distant past, this past inequality matters less.

This proposal, however, suffers from the same defect as did the previous one. Even if we accepted discounting past or future benefits and burdens this would not provide a reason for discounting the value of past or future equality, since equality as such is a relation between persons, not something to which we might be more or less psychologically related. Of course, I may now more or less vividly remember inequalities that I experienced in the past. But this is beside the point. According to the equality view, there are forms of inequality which are bad in themselves, apart from any consciousness of them, or psychological relations to experiences of them that we may have. Therefore, the general view on discounting past and future benefits and burdens does not justify discounting the value of past and future equality.

My fourth and final suggestion is this. Suppose that we accept the general reasons for discounting past and future benefits and burdens. We can then assume that there is at each time t, and for each person P, a discounted sum of well-being, which is the aggregation of all past and future benefits and burdens (or levels of well-being) in P's life, the value of these discounted in correlation to the strengths of psychological relations that hold between P at t and the time the benefits and burdens occurred in P's life. Or, as we might say more briefly, we can ascribe discounted and time-relative sums of well-being to persons at particular times. A version of the lifetime equality view could claim that equality with respect to these discounted and time-relative sums of well-being is what matters.[24]

This view would explain why we could, on egalitarian grounds, remain indifferent as to whether A or B ought to be admitted to the residential home: the discount rate could be so steep that the hardships of B's childhood did no longer count at all. In that case, there would be no egalitarian reason to prefer one to the other.

or in the future; how distant in the past or the future something is located is irrelevant. Obviously, the views could be combined.

[23] This view, or a view very much like it, has recently been defended by Jeff McMahan. See Jeff McMahan, 'Preferences, Death, and the Ethics of Killing', *Preferences*, ed. C. Fehige and U. Wessels, forthcoming. This view need not deny that numerical identity as such matters for egoistic concern. Numerical identity as such might be a necessary but not sufficient condition for selfish concern. Thus the view under discussion differs from that defended by Parfit in *Reasons and Persons*.

[24] Notice that this view still takes the lifetimes of people to be relevant, but it denies that the goodness of peoples' lives as wholes should be compared.

I suggest that we should not accept this view. Even if we do accept the general reasons for discounting, we should not take the further step and accept that simultaneous inequality with respect to discounted and time-relative sums of well-being is bad in itself (as the view claims).[25] My reason for this suggestion is that reflection upon this kind of simultaneous inequality fails to reveal it as an attractive feature of outcomes. Consider the following case:

C and D are born on the same day. Eighty years later they die the same day. When their whole lives are considered C and D have lives well worth living, and they are equally well of with regard to total well-being. Hence, there is no lifetime inequality between C and D. However, C has a miserable life for the first 10 years of her life, and D has an equally miserable time for the last ten years. Apart from these two periods of times, C and D have equally good lives. Assume finally that both have lives that are well worth living.

Assume for simplicity that we discount past and future well-being with a constant ratio per year. On this assumption there will be one point in time without simultaneous inequality between C and D with respect to discounted sums of well-being. This is when C and D are celebrating their fortieth birthday. At all other times there will be simultaneous inequality between C and D with respect to discounted sums of well-being. There is nothing formally dubious about this. The problem is that it is hard to believe that this kind of simultaneous inequality is bad. Consider C and D when they celebrate their thirtieth birthday. They are equally well off at that time with respect to well-being, and have been so for twenty years, and they will be so for another forty. There is no ordinary simultaneous inequality between them, just as there is no lifetime inequality, and both of them have a life well worth living. However, there is still simultaneous inequality with respect to discounted sums of well-being, and this simultaneous inequality has been present for years.[26] But can we really believe that this sort of inequality matters, that it is bad in itself?

Let me conclude this section by summing up. I have argued that there is no egalitarian position based on the equality view which allows for an explanation of the importance of temporal distances. Either the considered versions of the equality view fails to connect with the more fundamental reasons relating to temporal asymmetries

[25] As this remark implies, the view under consideration shares a feature with the view that simultaneous inequality as such is bad. Where that view tells us to compare time-relative levels of well-being, the present view tells us to compare time-relative and discounted sums of well-being. Hence, they agree that simultaneous inequality with respect to some time-relative notion is bad. However, the two views disagree about what sort of simultaneous inequality matters.

[26] This again raises the question about how to aggregate across time simultaneous inequality with respect to time-relative and discounted sums of well-being between persons. Do these instances of simultaneous inequality add up, or do they rather cancel out?

or to temporal distances, or the view is simply too hard to believe. I shall now turn to the priority view.

VII. PRIORITY AND TEMPORAL DISTANCES

Let me start by making two almost trivial points. First, if we accept the severe suffering objection and the intuitions regarding temporal distances, we should obviously reject the lifetime priority view. The second point relates to the discussion of reasons for giving less weight to inequalities in the distant past, either because of temporal asymmetries or because of temporal distances. It is understandable why the priority view should connect with the view that we should be more concerned about future suffering than about past, or the view that we should discount for diminished psychological relations (or temporal distances). However, as I shall now discuss, from the fact that the connection between the priority view and temporal asymmetries and reasons for discounting past and future well-being is intelligible, it does not follow that we should accept those combinations of views.

Suppose we are convinced by Parfit's example that it is egoistically rational for persons to care only about their futures; the past as such does not count. This means that in her egoistic concern, what matters to a person is, at any time, that the rest of her life be as good as possible, not that her life as a whole be as good as possible.[27] If we accept this view, we might consider a revision of the priority view which ignores past well-being and suffering for the simple reason that for individual agents it is egoistically rational to ignore the past. A version of this future directed priority view would claim that at each time we should give priority to those who, when the rest of their lives are considered as a whole, are worse off.[28] However, although this view is more plausible than the future directed equality view, it is certainly not obvious that we should accept it, even if we generally accept the priority view as the basis of our egalitarian concern, and accept that only the future matters in egoistic concern.

I shall mention just two examples. Suppose one person suffered immensely in the recent past, but is now completely unaffected by this. Another person did not suffer at all in the past, and is now just as well off in terms of well-being as the first one. We can benefit either of them equally much today, but unfortunately only one of them. Which of these persons should we rather help? It is hard to believe that the fact

[27] In addition to what is clear from Parfit's example, Warren Quinn has shown how these two notions are importantly different in his paper 'The Puzzle of the Self-Torturer', *Morality and Action*, Cambridge, 1993.

[28] There are other versions of future directed priority views, i.e. ones that do not focus on total amounts of well-being.

that one of them suffered in the recent past should not be given any weight at all, even if these persons in their egoistic concern care only about their future. This, however, is still consistent with a version of temporal asymmetry. We might say that in our prioritarian concern we should give *more* weight to the future than to the past. That is, other things being equal it is more important to benefit those who will be worse off in the future than those who were equally badly off in the past. Consider then another example. Suppose that we today can benefit either of two persons equally much. Today they are equally well off. But while one of them suffered one week ago, the other will suffer the same in a week. Does this provide a reason for helping one rather than the other today? I guess that many on reflection would deny this. That is, even if we accept that future directedness is rational with respect to egoistic concern, we might deny that this temporal asymmetry should be extended also to the priority view.[29]

Let us then turn to the issue of discounting. As explained above, we might accept the discounting of past and future benefits and burdens in correlation to diminished psychological relations across time. According to this view there is for each person at each time a discounted sum of well-being. Just as a proponent of the equality view could suggest that inequality with respect to this discounted sum has negative value, so the proponent of the priority view could suggest that we should give priority to those whose discounted and time-relative sums of well-being are smallest. This would explain the intuitions that hardships in the distant past count for less in egalitarian concern, or not at all, as discussed in the residential home example.

According to this version of the priority view, if our efforts and the benefits produced are the same, and if two persons are now equally well off, and were so in the past, and will be so in the future, we should nonetheless help one person rather than another, *merely* because one person more vividly remembers some past hardship they both had to endure, or because one person is better able to relate to future burdens that both must bear. However, it seems difficult to accept that it could be more important to help one person rather than another merely because this person has richer psychological relations to some past or future hardship that both endured or must endure. The priority view holds that there is an added moral value to benefiting the worse off, even if the benefit is the same. This view has a great appeal. But giving priority to those who are worse off with respect to discounted sums of well-being has no similar appeal: it is simply difficult to believe that

[29] Another problem concerns the possible instability of the future directed priority view. Assume that we should at any time do most for those who from that time are worse off. This view may be hard to implement in a policy because it gives us new aims all the time. McKerlie discusses a similar objection to the future directed equality view.

mere differences in psychological relations can be a reason for benefiting one person rather than another.

It might be replied that we find it difficult to believe that mere differences in psychological relations can have the importance assumed in this version of the priority view *because* we believe that numerical identity rather than psychological relations is what matters for morality and for self-interest.[30] However, when we realize that numerical identity cannot be what matters, and realize that certain psychological relations are what matter, we should make certain corresponding revisions of other beliefs of ours.[31] Among the beliefs that should be revised is the belief that mere differences in psychological relations cannot be a reason for benefiting one person rather than another.

However forceful this reply may be, I remain unconvinced. Suppose that we initially believe that identity is what matters for the priority view. More exactly, suppose that our starting point is that the priority view should take lifetimes, or total amounts of well-being in lives, as the unit to focus on. Suppose next that we become convinced by Parfit's arguments that numerical identity as such cannot have the significance that we used to believe it to have. Suppose also that we become convinced that certain psychological relations should take on some of the importance that we used to assign to numerical identity. In that case I would find it reasonable to suggest that we should accept that mere differences in persons' psychological relations to their past and future can be a reason for benefiting one rather than another. And I would find it more reasonable to accept the version of the priority view that focuses on discounted and time-relative sums of well-being.

But this is not the line of reasoning that is presented here. The argument here *starts* with a rejection of the view that the priority view should operate over lifetimes, i.e. with the assumption that a version of the priority that operates over lifetimes is implausible. This is very much like saying that numerical identity is not what matters for the priority view. But here we do not say so for metaphysical reasons, but merely because the lifetime priority view does not fit our considered judgements. But if this is our starting point, then the subsequent admission that numerical identity for metaphysical reasons cannot be what matters for self-interest and morality will *not* be a reason to revise the priority view so that it instead operates on sums of well-being which are time-relative and discounted for degrees of psychological relations. When we already assume that the lifetime priority is wrong, then the admission that identity as such is not what matters is *not* a reason for assuming a version of the priority view in which certain psychological relations takes the place of identity.

30 Nils Holtug suggested this excellent reply.

31 This is what Parfit argues in *Reasons and Persons*, pt. iii.

There are further problems in the version of the priority view that operates on discounted and time-relative sums of well-being. Sometimes memories of past suffering or hardships are not themselves painful or distressing. Similarly, though this is perhaps more rare, the anticipation of future pains or suffering may itself not be distressing. Consider now these *neutral psychological connections*, as we may call them. We might improve someone's time-relative and discounted sum of well-being *merely* by weakening her neutral psychological connections, for example by erasing her neutral memories of past suffering, or by interfering with her neutral anticipation of future pain. Therefore, according to the version of the priority view that takes discounted sums of well-being as its focus we can morally improve a situation by weakening neutral connections to past or future burdens. This seems absurd, and it comes close to admitting that things can improve in one way, even if no-one thereby becomes better off. But, as will be remembered from section II, one important reason to accept the priority view rather than the equality view is that the equality view faces the levelling down objection. According to the equality view, situations may improve with respect to equality without anyone becoming better off in any respect. This leads some people to deny that equality as such can be valuable. The priority view avoids that problem. However, what we have seen now is that a version of the priority view that takes discounted sums of well-being as the relevant unit to focus on faces a similar problem.

Of course there is a very attenuated sense in which someone might be said to be better off when a time-relative and discounted sum of well-being has increased. But it is hard to accept that this is what counts. This indicates that there is a problem in reconciling the basic idea in the priority view with this way of including discounting in the priority view. This in turn suggests that even if we accept discounting for diminished psychological relations, and accept the priority view as the basis for egalitarian concern, we should not accept priority with respect to discounted and time-relative sums of well-being.

Let me here sum up the previous discussion of the priority view and temporal asymmetry and temporal distances. The view that there is a temporal asymmetry in our rational concern about our well-being such that only the future matters, or such that the future matters more, can be combined with the priority view in an intelligible way. However, even if we accept this temporal asymmetry *and* have a presumption in favour of some version of the priority view, we might not accept the combination of these two. Similarly, if we believe that past and future benefits and burdens should be discounted for temporal distance (that is for psychological relations that roughly correlate with temporal distance), then it is understandable why one might want to revise the priority view correspondingly. However, as I suggested above, we then get a version of the priority that is hard to believe, since it seems to recommend benefiting some rather than others for the wrong reasons.

There are further questions as to the most plausible version of the priority view. Suppose we are confronted with the following choice:

> If we do nothing A and B will be in pain today. Since their pains are equally strong, and other relevant factors are the same, their levels of well-being today will be equal, if we do nothing. Tomorrow neither A nor B will be in pain, and their levels of well-being will again be equal, but at a higher level, again on the assumption that we do nothing. Imagine finally that what we can do is this: either we provide pain relief only for B today, or we increase only A's level of well-being tomorrow. In either case the increase in well-being and our efforts will be the same.

I suspect that most people who are attracted to the priority view would say that we ought to relieve the pain of B today rather than bring about an increase in A's well-being tomorrow. I shall now discuss what this example indicates about the priority view.[32]

Two initial remarks are evident. Firstly, if we share the response that I have described we should obviously reject versions of the priority view which operate over total sums of well-being in persons' lives. Secondly, and more importantly, the case described obviously has nothing to do with temporal assymmetries, or with discounting for diminished psychological relations. The timespan is too short for discounting for diminished psychological relations to matter, and temporal assymmetries could not explain why treating B's pain today is more important than providing a similar increase in A's well-being tomorrow.

There are, however, two different responses to these cases that one could make on behalf of the priority view. One response is to claim that the priority view should exhibit a particular kind of *distribution sensitivity*, which is to be explained as follows. For any person there is a well-being/time function mapping this person's well-being against time. On the view that the goodness of peoples' lives on the whole is what matters, what should guide our distribution of benefits and burdens is, so to speak, the area under the curve. However, what the examples just considered suggest is that we should be concerned also about the shape of the curve. According to the intuitions evoked by the examples, we should give some weight to changing the shape of a

[32] Another example is this. Suppose that C will suffer a decline in well-being tomorrow. Either we prevent this and thereby bring about a certain increase in well-being. Or we produce just as great an increase in well-being some days later in C's life. Again, those sympathetic to the priority view will, I think, believe that we should prevent C's decline in well-being rather than merely increase his level of well-being by the same measure, even when our efforts and the effects in terms of well-being are the same.

person's well-being/time function such that its minima become larger. Increasing the value of the well-being/time function where it is low is morally more valuable than increasing it where it is higher.[33] Notice that this response still assumes that lifetimes are the correct unit for the priority view to operate over. But it denies that the goodness of a life as a whole is ultimately relevant for the priority view.

Whatever its possible intuitive plausibility, this view does have counter-intuitive consequences. In some cases it will be morally better if the low values of a person's well-being/time curve are increased, even if this has as a result that the person's overall sum of well-being is decreased. Suppose that we can either prevent C's suffering tomorrow and thereby increase his well-being by a certain amount, or by the same effort produce a somewhat greater increase in well-being at some later time when he is not suffering at all. On the distribution sensitive version of the priority view it could be the case that we should prevent C's suffering tomorrow, even if C on the whole is better off if we rather increase his level of well-being more later. In other words, one could rationally prefer a life with a greater but more unevenly distributed total sum of well-being to a life with a smaller but more evenly distributed sum of well-being, despite the fact that it might be *morally* better if one live the life with lesser but more evenly distributed total well-being.

This shows that the priority view in the distribution sensitive version faces a problem that shares some features with the levelling down objection – the objection to the equality view from which the priority view borrows much of its plausibility. In one version, the levelling down objection claims that the equality view is implausible since it implies that things can morally improve in one respect despite no-one's being better off in any respect. In a way, the distribution sensitive version of the priority view shares this feature. By avoiding times of low well-being, one can make the world a morally better place, even if no person in her life as a whole is better off. The analogy to the levelling down objection is limited, however, since there are, after all, beneficiaries when we increase the minima of the well-being/time curve at the cost of goodness of a life as a whole: the worse off at particular times become better off, that is, the worse off *persons-at-times* become better off.[34] But it remains the case that no person in her life as a whole may be better off, and yet the world has become a morally better place.

[33] One should not confuse this with diminishing marginal utility. To claim that the priority view must be sensitive to distributions of well-being is a general claim about the moral value of well-being. It is not a claim about what brings more or less well-being about.

[34] This reply was suggested by Derek Parfit.

The severity of this problem of the distribution sensitive version of the priority view depends on what we take the moral importance to be of the notions of persons-at-times and of persons in their full lifetimes. What the discussion shows is that there is a tension in accepting distribution sensitivity *and* insisting that persons in their full lifetimes are the basic units of moral concern. But as soon as we give some weight to persons-at-times, as if they were separate entities to be subjected to moral concern, the problem disappears.

This leads to the second response on the behalf of the priority view that I wish merely to mention. Rather than accepting that persons' lifetimes are the relevant units, but accepting distribution sensitivity within that, we could claim that the priority view should take persons-at-times at the relevant unit. This view claims that goodness of life as a whole is what matters, but it denies that this should be viewed over people's full lifetimes. It is goodness of the lives of persons-at-times that matters. Sometimes, when persons-at-times become better off, this may come at the cost that persons, in their lives as a whole, become worse off. But this need not worry us, since we now generally claim that persons' lives as wholes are the wrong unit to focus on. This version avoids some of the problems mentioned above. But, needless to say, it remains to be discussed more fully.[35]

[35] Numerous people have commented on earlier versions of this paper. I would in particular like to thank Derek Parfit for his extensive written comments, and Roger Crisp for suggesting many improvements. Also, I would like to thank Ingmar Persson, Dennis McKerlie, Nils Holtug, Kasper Lippert Rasmussen, Peter Sandøe, Govert den Hartogh and his workparty on practical philosophy at Amsterdam University, and audiences at Utrecht University for comments and suggestions on earlier versions of this paper.

[20]

Equality

Dennis McKerlie

Egalitarianism is powerfully represented in recent moral and political philosophy. John Rawls, Ronald Dworkin, and Thomas Nagel have defended views that count as egalitarian. An egalitarian theory might be based on an appeal to equality itself as a moral ideal. In this article I consider views that call for substantive equality in the conditions of people's lives: not just political equality, or equality in the sense of having the same set of basic rights, but equality with respect to the opportunities open to them, or the resources available to them, or in the quality of their lives themselves. For the issues I discuss it is not necessary to decide in which of these ways (or in some other way) our lives should be equal. But for the sake of simplicity I will sometimes assume that the view requires equality in the quality of lives, and that the quality of a person's life is a matter of that person's level of welfare or well-being.

Equality is a relationship between different people. There is equality when they are equally supplied with resources, or equally happy. Whether there is equality or inequality, and how much there is, will depend on the overall distribution of whatever is valued. A moral view based on the value of equality will give us the goal of creating this relationship in outcomes. It will tell us to aim at outcomes with the overall pattern of distribution that maximizes equality or minimizes inequality.

A view might give us this aim because it sees equality as a property that makes an outcome better. It is simpler to explain this view in terms of the disvalue of inequality. The view counts inequality as making an outcome worse: the more inequality there is, the worse—other things being equal—the outcome is.[1] Since inequality

1. The ceteris paribus clause could be activated in two different ways. One possibility is that the badness of the inequality is outweighed by the positive value of some other feature of the outcome—for example, an increase in the overall well-being of people in general—so that the outcome is on balance good. On the other hand, some egalitarians think that the presence of certain other factors can simply cancel the badness

Ethics 106 (January 1996): 274–296

is a relationship between people, the most important and distinctive feature of the view is that it supposes that there can be value in relationships between lives, not just in the content of lives. In other words, the view denies that the only things that can have value are benefits and harms experienced by individuals, or gains and losses in the quality of their lives. It supposes that the existence of a certain relationship between the quality of two lives—inequality—can be bad, just as a decline in the quality of one of the lives would be a bad thing. Since the view says that inequality makes an outcome worse, and it tells us to aim at the best outcome, it is appropriate to call it the teleological equality view.

This is not the only way of explaining why we should aim at equality. Some egalitarians think that inequality is not bad, but unfair. Suppose that I have goods to distribute, and I distribute them unequally. These egalitarians would not claim that the inequality between lives that I create is bad in the way that suffering or deprivation inside one life is bad. They would say instead that the existence of the inequality shows that I have treated the people concerned unequally. And they think that at least some ways of treating people unequally count as treating them unfairly. Their view could be called a deontological equality view. It links inequality to unfairness rather than to the badness of an outcome, and it explains the unfairness in terms of people being treated unfairly by other individuals or by social institutions. The basic claim is that we have a duty to treat people fairly and that some ways of bringing about inequality violate that duty.[2]

It is possible to believe both that inequality is a property that makes outcomes worse and that we have a duty to treat people equally in certain respects. Consequently, the distinction between a teleological concern with equality and a deontological concern with equality, as I have drawn it, is not exhaustive. In this article I will be concerned with versions of the deontological view that would not say that inequality is bad. The claim that inequality is bad in itself is controversial, and one of the attractions of a deontological equality view is that it might enable us to explain the concern for equality without making that claim.

The general distinction between teleological and deontological moral views has been drawn in many different ways, and different features have been proposed as the distinguishing mark of a deonto-

of the inequality. For example, they might believe that inequalities that match people's desert, or inequalities that result from the free choices of the people concerned, are not bad at all.

2. The distinction between the teleological equality view and the deontological equality view is discussed in Derek Parfit's "On Giving Priority to the Worse Off" (unpublished manuscript) and in Larry Temkin's book *Inequality* (Oxford: Oxford University Press, 1993), p. 11.

logical view. Many different things could be meant by describing a concern for equality as "deontological." Minimally, the name could be given to *any* view that holds that we have duties with respect to reducing inequality or not bringing about inequality. Characterized in this minimal way a deontological equality view might say that I have a duty not to cause inequality myself and in addition a duty to eliminate inequality however that inequality may have been caused. In effect this view holds that the concern for equality is fully agent neutral. I will discuss a more specific version of a deontological concern for equality. It suggests that the concern for equality is based on an agent-relative duty not to treat people unequally ourselves. This view has the advantage of contrasting sharply with the teleological concern for equality. More important, I think it would be difficult to explain a purely agent-neutral duty with respect to equality without claiming that inequality itself is bad.[3]

Principles of equality are usually defended as part of a theory of justice. Inequality is not merely regrettable, but unjust. It is not clear what it means to treat a value as part of justice, but it does at least imply that the value of equality is especially weighty and important. If equality is a matter of justice, we would expect it to outweigh other values that are not themselves included in justice.

Understandably, egalitarians welcome the conclusion that equality is an especially important value. But it does create a problem for their view. If we consider the teleological equality view, inequality cannot be the only feature that makes an outcome worse. We could create perfect equality by giving everyone the same extremely low standard of well-being, but we would consider this to be a very bad outcome. We care about improving the quality of people's lives as well as equalizing the quality of their lives, so we must add a principle that will express this concern to the principle of equality. But if we think that the principle of equality always outweighs the other principle—either because only equality is included in justice, or because the other principle is also a concern of justice but the theory of justice gives priority

3. This brief description of the deontological equality view leaves many questions about it unanswered. I am not using 'deontological' in the same way as John Rawls, for whom any moral view that treats a distribution of good things across different people as being itself good counts as deontological (John Rawls, *A Theory of Justice* [Cambridge, Mass.: Harvard University Press, 1971], pp. 25, 30). The view that I have described should also be distinguished from a view that says that only inequality caused by individuals, or social institutions, is bad or objectionable. That view seems to be a restricted version of the teleological concern for equality. I have not tried to explain what it means to treat people equally. But I think that treating people equally requires that my actions should have equal effects for the better or for the worse on the people that they influence, not that my actions must be designed to create an overall equality between the people concerned in terms of welfare or quality of life.

to equality—we will be led to conclusions that are intuitively implausible. Any sacrifice in people's standard of living that reduces inequality would be justified.

One response to this problem is for egalitarians to move away from caring for equality as such. For example, Rawls wants his egalitarian distributive principle to have absolute priority over utilitarian considerations. So he formulates the principle in terms of promoting the interests of the worst-off group rather than achieving equality. In effect he bases egalitarianism on the idea of priority for the worst off rather than on the idea of equality.

I will assume that the teleological equality principle does not always outweigh other moral principles. This could be so if equality is a matter of justice, but the theory of justice is pluralist and contains other principles, and the theory does not give equality lexical priority over the other values. Alternatively, the connection between equality and justice might not be as strong as egalitarians have typically thought. Perhaps, although inequality is always bad, it is not always unjust. It might only be extreme inequalities, or inequalities that have certain kinds of causes, that count as injustices.[4]

Sometimes people say that they will not accept equality as a value unless they are given convincing arguments in favor of it. Many egalitarians have attempted to argue for the importance of equality. But whether we interpret the principle of equality teleologically or deontologically, it is not obvious that it is right to expect or demand some further argument for it. The teleological version of the principle claims that inequality is itself something that is bad. Suppose that the claim is true. If it is true, there is no reason to think that we will be able to explain the badness of inequality in terms of the badness of other things, or in terms of some other value that is not a matter of badness, any more than we can explain why suffering is bad in that way. There is no obvious reason for saying that the claim that inequality is bad must be supported by an argument while the claim that suffering is bad does not require that support.

I am not suggesting that it is pointless to argue for equality in the ways that egalitarians have done. Some philosophers accept metaethical views—for example, contractualism—which mandate that any first-order moral claim must be supported by a special kind of argument, and they apply their metaethical view to equality. In this article

4. The problem of balancing equality against other values is worse for the deontological equality view. If inequality is merely bad, it is easy to understand how its badness might be outweighed by the goodness of other things—e.g., by the positive value of gains in welfare for better-off people. But the deontological view objects to inequality because it represents unfair treatment. It is harder to see how gains for better-off people could justify treating other people unfairly in order to achieve those gains.

I will not be concerned with contractualism, or similarly ambitious metaethical theories. Others might say that if moral claims are not supported by arguments then it is only reasonable to accept them if they cohere in a strong way with our considered judgments. And they might add that the claim that inequality itself is bad does not find strong support in our considered judgments, unlike the claim that suffering is bad. So the first claim does require support by arguments while the second claim does not. I will not attempt to measure the strength of the support for equality in our considered judgments. But in considering this question it is important to distinguish between the view that equality is a dominant value and the view that equality is at least one valuable thing. It might be easier to find support for the second view among ordinary people's considered judgments. Perhaps one reason that philosophical egalitarians have been so prone to argue for equality is that they are attempting to defend the strong position that equality is a dominant value.[5]

One argument for equality has been influential. It points out that serious inequality often directly results from, or at least is strongly influenced by, differences in people's natural abilities and the social positions into which they are born. For a given individual, that person's genetic endowment and initial social position are things that she is not in any way responsible for. Being born with a certain range of abilities, for example, is a matter of luck. The argument claims that it is a bad thing if these morally arbitrary factors influence the course of people's lives. The way to prevent them from having a significant influence is to create social institutions and policies that will replace the inequality that would otherwise have resulted by equality. The argument justifies distributive equality as a way of neutralizing factors which should not be allowed to influence people's lives.

The argument is persuasive. It draws on ideas that, for many people, explain their commitment to egalitarianism. Accounts of egalitarianism often focus on the sheer misfortune of being born with disadvantages like family poverty or mental disability and on the profound consequences they have on the rest of people's lives. One source of the argument is Rawls's claim that the system of natural liberty,

5. Philosophical egalitarians usually propose very strong versions of egalitarianism. The strength of an egalitarian view depends on two things: the content of its egalitarian principle (e.g., a principle that calls for strict equality is stronger than a principle that only tells us to eliminate extreme inequalities), and how it weighs the egalitarian principle against other principles (a view that gives strict priority to equality is stronger than a view which sometimes allows utilitarian reasons to outweigh equality). Egalitarians also typically agree that moral theories should fit the considered moral judgments of ordinary people. This is true, in different ways, of Rawls, Dworkin, and Nagel. Perhaps they have not done enough to show that their strong egalitarian theories satisfy this condition.

and the liberal interpretation of his principles of justice, should be rejected because they allow distributive shares to be influenced by factors like the natural distribution of talents.[6] His own theory of justice, by contrast, tries to mitigate the effects of the "genetic lottery." Rawls himself seems to put relatively little weight on this argument compared to his contractualist argument for his principles of justice. But it has been taken up by other egalitarian writers, some of whom suggest that it is stronger than the contractualist argument.[7]

The argument has also been strongly criticized.[8] I will suggest that it does not provide independent support for the value of equality. If the argument is intended to derive the importance of equality from some quite distinct value—the value of preventing morally arbitrary factors from determining the quality of people's lives—it is unconvincing.

Suppose that a society has a competitive economy. How well people do is strongly influenced by their natural ability, their family background, and good and bad luck of other kinds. Also suppose that there is considerable variation between people in all these respects, but that coincidentally the factors cancel one another out and the result is distributive equality. Different people end up doing equally well, but it is still true that how well each individual does is causally influenced by these factors. If the argument is correct and our basic objection is to factors like these ones changing the course of people's lives, we should find the result objectionable. The result is equality, but in this special case equality does not mean that the influence of morally arbitrary factors has been eliminated. So we should find this particular equal distribution as objectionable as inequality would be. But I think that people with egalitarian sympathies would not find it objectionable. It seems that they care more about achieving equality itself than eliminating the effects of morally arbitrary factors.

Another example makes the same point. Suppose that the world contained only one person and that she used her native intelligence to greatly improve her life. Would we be troubled by the fact that these advantages are gained thanks to her morally arbitrary natural ability? Would we wish instead that her intelligence had not had any

6. Rawls, sec. 12. Thomas Nagel discusses egalitarianism with special reference to inequality resulting from these factors in *What Does It All Mean?* (Oxford: Oxford University Press, 1987), chap. 8, and *Equality and Partiality* (Oxford: Oxford University Press, 1991), chap. 10.

7. Brian Barry, *Theories of Justice* (Berkeley: University of California Press, 1989), chap. 27; Will Kymlicka, *Contemporary Moral Philosophy* (Oxford: Oxford University Press, 1990), chap. 2, sec. 2.

8. Most effectively by Robert Nozick, *Anarchy, State, and Utopia* (New York: Basic, 1974), pp. 213–27.

influence on her life at all, for the better or for the worse? Surely not. The example involves one life being influenced by morally arbitrary factors, but we do not object because it does not involve inequality between different people.

The examples show that our fundamental objection is not to the specified factors having consequences for the quality of people's lives. We object when the factors are responsible for differences in the quality of different lives. But this is just to say that we object to *inequalities* that are caused by these factors. The argument does not derive the importance of equality from some other value. To be persuasive, it must start from the idea that a certain kind of inequality, or inequality with a certain kind of cause, is objectionable.

This does not mean that the argument is useless. If we are trying to persuade someone without egalitarian sympathies, it might be best to point to cases of deep inequality directly rooted in factors totally beyond the control of the people concerned. If a person does not feel that this kind of inequality is objectionable, they probably cannot be brought to care about inequality at all. But we should not misunderstand the nature of the argument. It does not show us that we value an equal distribution for the deeper reason that it eradicates the influence of what is morally arbitrary. The argument presupposes that we do care about, or can be brought to care about, a certain kind of inequality. It expresses this concern, but it does not explain or justify it.

Assuming that we are willing to accept equality as a value, should we understand it teleologically or deontologically? I will discuss the two views in turn, focusing on the most serious problem for each. The deontological view faces a problem about the scope of its principle of equality, about the kinds of inequality that it can count as objectionable. I will argue that it is difficult to find a satisfying solution to the problem. The teleological view faces a problem about its central claim, that equality itself has value. This issue is complicated, but I will suggest that the problem is not a good reason for abandoning the teleological view.

The deontological view objects to inequality when it results from the activity of some agent, where the agent is subject to a duty to treat people equally and what the agent does violates that duty. But inequality can occur independently of the agency of any individual or institution. The deontological equality view cannot object to these inequalities. If they are inequalities that egalitarians would want to object to, this is a reason to understand the principle teleologically.

One relevant case is inequality between people who do not stand in any significant relation to one another. Imagine two people living on separate desert islands, one bountiful and the other barren. There will be a great difference in the quality of their lives. But the difference is not the result of any kind of treatment by individuals, institutions, or societies, so it cannot result from unfair treatment. The deontological view cannot criticize this inequality.[9]

The first case is not decisive, since many egalitarians are willing to say that this kind of inequality is not objectionable. The second example deals with people living under common institutions in the same society. A natural difference between them might directly cause a significant difference in their lives. For example, one person might suffer from birth from a painful and incurable disease that prevents him from living happily. It might be possible to reduce the inequality by compensating the person with the disease in other ways. But actions by individuals or institutions do not play any part in causing the inequality. So the deontological view cannot draw the conclusion that it should be reduced. It is harder to claim that this kind of inequality is unobjectionable, that nothing should be done to redress the difference between the two lives.

Nevertheless, some egalitarians would say that this inequality is not objectionable.[10] They might suggest that their theories, and distributive justice itself, are primarily or exclusively concerned with inequality that is a consequence of the social and economic system of a particular society. Here we reach a kind of inequality that egalitarians must object to—the inequalities generated in a competitive economic system. The inequality may be based in part on differences in the natural ability of the people concerned, but in this case—unlike the previous example—institutions, laws, and conventions do have a role in explaining why people end up with different shares of resources. However, it is questionable whether the deontological equality view can object to even this kind of inequality.

Nozick explains the difficulty.[11] To apply the deontological equality principle, we must be able to identify an agent responsible for the differential, unfair treatment. One possibility is that the agent is the state or the social and economic system itself. Society treats some people unfairly by distributing different shares of resources to different people.

Nozick thinks that this answer is unconvincing. He claims that in a competitive, free enterprise economic system there is no centralized agent that can be literally said to distribute resources. People end up

9. Nozick uses the example to criticize Rawls (Nozick, pp. 185–86). He thinks it is implausible to hold both that there is no claim to equality in this example and that a claim to equality is created when social cooperation is introduced. Egalitarians who agree with Rawls about the two cases might decide that the deontological principle of equality provides a better explanation of their view than the claim that cooperation is a necessary condition for the application of egalitarian principles.

10. Nagel, *Equality and Partiality*, p. 107.

11. Nozick, p. 223.

with a certain share of resources as a result of many different choices made by many different people and institutions. Moreover, it is not enough, in order to apply the deontological principle, to find someone or something to accuse of failing in the duty to treat people equally. The inequality must also result from positive action, rather than passive inaction, on the part of that agent.[12] Society does not act to prevent inequalities from arising as a result of individual choices, and it does not act to eliminate the inequalities once they are present. But Nozick claims that the failure to act to overturn the result of individual choices is not the same thing as positive action.

Nozick's way of describing the generation of inequality in a competitive economy is disputable, but many egalitarians themselves describe it in the same way. Rawls says that the fault of the system of natural liberty is that it *permits* distributive shares to be influenced by morally arbitrary factors like natural ability.[13] Nagel says that a free enterprise system *allows* rewards to depend on talent and consequently *permits* serious inequality to develop.[14] Neither writer thinks it is important to insist on saying that society actively brings about the inequality.

The other possibility is to make individuals the relevant agents. By making their choices within the context of a competitive economy, they create inequality and fail in their duty to treat other people fairly. Egalitarian writers prefer to blame the system rather than the individuals who live under it, but the possibility is worth considering. Nozick argues against it with an example. When I buy a ticket at one cinema rather than another I create an inequality between the cinema owners, but I do not act wrongly even if I have no serious reason for making the choice in the way I do. The example is persuasive, but perhaps partly because my choice creates a trivial inequality. If I met two people in serious need, and I had enough money to divide it between them and help them both, it is not obvious that it would be acceptable to instead give all the money to one of them. At least people with egalitarian sympathies are not committed to thinking that this choice could not be criticized.

Nagel seems especially conscious of the difficulty. He discusses inequality in terms of its fairness or unfairness. And he thinks that the basic egalitarian concern is equality of treatment.[15] So his view at

12. If the deontological principle of equality is like other deontological principles the issue of whether the agent intended to produce inequality, or merely forsaw that it would result, might also be important.

13. Rawls, p. 72.

14. Nagel, *Equality and Partiality*, pp. 100–101.

15. Ibid., pp. 106–7. Nagel distinguishes this idea from the "pure priority view" that he describes in chap. 7.

least resembles what I have called the deontological equality view. However, Nagel is uncomfortable about simply saying that the individual choices that cumulatively create the inequality are themselves wrong. And he also respects Nozick's point that a free enterprise society does not distribute unequal shares of resources to its citizens in the same way that a parent might give out unequal pieces of pie to her children.

Nagel begins with the idea that what is objectionable is not inequality itself but people being treated unequally, and he recognizes that there is a difficulty about identifying an agent that treats people unequally in the case of this kind of inequality. His solution is to suggest that in the special case of the state it is just as objectionable to permit these inequalities to develop as it would be to directly distribute resources in an unequal way. A society—unlike an individual—has a negative moral responsibility for the inequality that it allows.[16] With respect to the social and economic framework there is no morally important distinction between what the state does and what the state permits. If the state allows rewards to depend on talent it is responsible for the resulting inequality, since the inequality would not have occurred if the state had egalitarian institutions instead. Allowing inequality counts as a choice, and it needs to be justified.

Few would disagree with Nagel's conclusion that noninterference on the part of the state requires justification just as interference does. But if we understand the value of equality in the deontological way, we will not think that allowing inequality requires the same kind of justification as bringing about inequality. In the case of a different deontological principle allowing someone to die does indeed require justification, but not the same justification that killing would require. In fact, if we think that inequality has moral importance *only* in the deontological way, it should not be difficult to justify permitting it. Inequality that is merely permitted and not produced by an agent will not be in any way bad.

Clearly Nagel's view about negative responsibility is meant to be stronger than this point about justification.[17] He really believes that permitting inequality is morally equivalent to deliberately creating it.

16. Ibid., pp. 84, 99–102, 107–8. Rawls also seems to assimilate societies that actively distribute goods on the basis of natural differences between people ("aristocratic and caste societies") and societies that merely allow these differences to influence distribution (Rawls, p. 102).

17. In his discussion Nagel also points out that a laissez-faire economic system cannot be justified on the ground that noninterference on the part of the state is "natural" (Nagel, *Equality and Partiality*, pp. 100–101). But this does not help us to understand how the deontological concern for equality can be applied to the state's noninterference.

A society that permits shares to depend on natural ability is morally equivalent to a society that directly distributes resources to its members and distributes them unequally in accordance with natural ability. Nagel says that the society that permits inequality fails to treat people as equals. And he says that if social mechanisms are involved in the causation of a benefit then its unequal distribution is a form of unequal treatment and so is unfair.

But if we understand the value of equality in the deontological way, why should we agree with these stronger claims? This sense of negative responsibility—the moral equivalence of permitting and producing inequality—does not follow from Nagel's point that the inequality could have been prevented if different institutions were in place. Allowing inequality will be failing to treat people as equals if treating them as equals requires taking positive action to provide them with equal shares. But someone who accepts the deontological view has no reason to agree that treating people as equals does involve that, or to agree that *permitting* inequality is a way of *treating* people unequally.

In explaining why he applies negative responsibility to the state and not to individuals Nagel says that the state does not have a life of its own.[18] In other writings Nagel has suggested that deontological moral reasons are rooted in the personal perspective of agency.[19] Apparently the state does not have a personal point of view. This might be given as a reason for applying the strong version of negative responsibility to a state that permits inequality. But its persuasiveness would depend on accepting Nagel's explanation of the source of deontological reasons. Someone who thought that deontological reasons were ultimately explained by the content of the moral claims of the victim, rather than by the personal point of view of the agent, would have no reason to accept the conclusion about inequality. Moreover, Nagel says that the strong version of negative responsibility does not apply to the state and the rights of individual citizens.[20] We would need to explain why the fact that the state does not have a life of its own makes permitting and producing inequality morally equivalent without also undermining the importance of the distinction—in the case of the state—between violating a right and not preventing its violation.[21]

18. Ibid., p. 101.

19. Thomas Nagel, *The View from Nowhere* (Oxford: Oxford University Press, 1986), pp. 175–85.

20. Nagel, *Equality and Partiality*, pp. 166–68.

21. My argument assumes that Nagel accepts something like the deontological principle of equality that I have described, and then tries to solve the problem about merely permitted inequality by arguing that a strong doctrine of negative responsibility applies to the state although it does not apply to individuals. If Nagel is using a different

I am not arguing that the inequalities generated in a free enterprise economy are unobjectionable. It is crucial for egalitarianism to insist that they be eliminated or reduced. But when egalitarians explain their objection, it seems better to appeal to the teleological principle of equality rather than the deontological principle. Applying the deontological principle to this case requires criticizing the particular choices made by individuals, or arguing that society produces these inequalities rather than merely permitting them to arise, or agreeing with Nagel that the state has a special negative responsibility for inequality that it permits. Each of these proposals has difficulties, and that makes it difficult to confidently draw the conclusion that the inequality is objectionable.[22] The teleological view has no difficulty in reaching that conclusion. It would also condemn the inequality in my first two examples, but not all egalitarians will see that as a disadvantage.

The most important issue about the teleological principle of equality is whether it is reasonable to care about a relation between lives. It asks us to see as having value something that is in an important sense divorced from benefits for people. Achieving equality does not necessarily mean improving the quality of any life. This will be true regardless of how the quality of a life is understood: whether it is explained in terms of the satisfaction of the desires of the person who lives it, or in terms of that person's happiness, or in terms of the life containing objectively valuable experiences and activities. Some would claim that equality is not the kind of thing that can be valuable.[23] They

basic principle my criticism would not apply. For example, he might value equality in the teleological way and also believe that it is only inequality that is caused by, or could have been prevented by, social institutions that should count as bad. But this suggestion would have difficulty accounting for the importance to Nagel of the special way in which negative responsibility is supposed to apply to the social and economic framework.

22. This is only true of the kind of deontological view that I have discussed: a view that does not claim that inequality is bad in itself, and thinks that the moral concern with equality involves an agent-relative reason not to cause or bring about inequality. It is possible that even a deontological view of this kind will tell one agent to respond in some way to the inequality wrongfully caused by others, even if it does not claim that inequality is a bad thing in itself. But this possibility will not help to defend Nagel's position. His view is not that the state should act to reduce inequality because that inequality was wrongfully caused by individuals; rather he thinks that the state has a negative responsibility for permitting inequality even if the individual actions that produced the inequality did not themselves violate the deontological principle.

23. Hastings Rashdall (*The Theory of Good and Evil*, vol. 1 [Oxford: Clarendon, 1907], pp. 266–67) points out that equality is a distribution of what is good between different people. Equality is not the good of any one person or of people collectively. Rashdall thinks that a distribution is too abstract to count as being good itself. Larry Temkin (Temkin, chap. 9) emphasizes the importance of this objection to equality. He calls the view about the good that motivates the objection "the slogan," and he makes a serious and sustained attempt to undermine it. I will comment on some of the ideas in his answer to the objection.

think that value must be connected to improvements in at least some lives, not to a mere relationship between the quality of different lives.[24]

The general issue of whether equality itself can have value leads to a particular difficulty for egalitarians. The principle of equality says that something is gained if we reduce inequality in a way that harms some people and helps no one. It judges that the unequal outcome would be improved if the better-off person were simply reduced to the level of the worse-off person. That is a claim that nonegalitarians find unacceptable, and many egalitarians themselves are embarrassed by.

The principle of equality will be part of a combined moral view containing another principle concerned with improving, not equalizing, lives. For the sake of simplicity, suppose that the second principle is the principle of utility. It says that the outcome will be worse if the better-off person is simply reduced to the level of the worse-off person without anyone being benefited. So it is possible that the combined view will say that, all things considered, the change is a change for the worse. Derek Parfit has suggested that there might be a version of the combined view which is significantly egalitarian (it would at least sometimes choose an outcome better in terms of equality but worse in terms of utility) but which nevertheless concludes in every case of this sort that the utilitarian reasons against leveling the better off down to the level of the worse off outweigh the egalitarian reasons for doing it.[25] This version of the combined view would never recommend leveling down, all things considered.

Resolving the general issue about the value of equality seems to take priority over deciding whether the combined view can avoid leveling down, all things considered. Even if the combined view never draws that conclusion, this would presumably be explained by how it balances egalitarian reasons against utilitarian reasons, not by the account it gives of egalitarian reasons themselves. The combined view would still contain a principle saying that equality has value in itself, so it would not be acceptable to those who deny that this is possible. On the other hand, suppose we were willing to agree that equality did have value. Then we might not reject the combined view even if it did sometimes call for leveling down. Now we can understand why

24. The strongest challenge to the teleological principle of equality does not come from the metaethical view that 'good' in "equality is (a) good" must *mean* "good for someone." It comes from the first-order ethical view that the only things that can have intrinsic value are things which make the lives of people better. We might agree that an improvement in the quality of a person's life is good, not merely good for the person in question. The difficult question is whether, when we understand what equality is, we think that it too has this kind of value.

25. Parfit, "On Giving Priority to the Worse Off."

the view reaches that conclusion. There is a value—the value of a decrease in inequality—to weigh against the harm done to people, and there is no reason to think that this value would always be less important than the harm.[26]

Some egalitarians would be willing to change the principle of equality to avoid both the general issue about the value of equality and the particular problem about leveling down. One possibility is to base an egalitarian theory on the idea of helping people who are badly off. We might think that benefits for them take priority over benefits for better-off people. If we hold this view we are not attributing value to equality. Instead we believe that improvements in the lives of badly off people have more value than similar improvements in the lives of better-off people. We would not reduce the welfare of the better off when this would not benefit the badly off. We give priority to helping the badly off, but when we cannot help them we should benefit, not harm, the better off.[27]

This response to the problems explicitly abandons the principle of equality. The resulting moral view might deserve to be called egalitarian, but not because it cares about equality for its own sake. A second response attempts to modify rather than abandon the principle.

The principle of equality counts all inequality as objectionable. One way to modify the principle is to restrict it, to say that only certain kinds of inequality are objectionable. The restriction relevant to the problems claims that inequality is only objectionable if it harms, or disadvantages, the people who are worse off under the inequality. Inequality that does not harm those who are worse off is not wrong. The test of whether inequality disadvantages the worse off is whether eliminating the inequality would benefit them.[28]

26. Parfit discusses ideas that would give the issue about leveling down priority over the question about the value of equality ("On Giving Priority to the Worse Off"). People who feel uncertain about the value of equality might nevertheless confidently judge that leveling down is wrong, all things considered. According to some views about the role of considered judgments in supporting moral theories, this judgment can legitimately be used to draw a conclusion about the principle of equality: the principle should be rejected unless the combined view can avoid leveling down. However, I think it is possible that they are certain that leveling down is wrong because, at some level, they are taking it for granted that equality is not important for its own sake. If this is the explanation of their considered judgment, it would not justify drawing a conclusion against the principle of equality.

27. As I have already suggested, I think Rawls's difference principle is best understood as a version of the priority view that gives absolute priority to the interests of the very-worst-off social and economic class. The egalitarian view that Nagel explains in *Equality and Partiality*, chap. 7, is also a version of the priority view.

28. This revision of the principle of equality is not equivalent to Rawls's difference principle. The difference principle gives a special role to the very-worst-off group: inequality is objectionable if it disadvantages them. The revised principle of equality

Clearly this response avoids the problem about leveling down. Leveling down ends inequality by lowering the better off to the level of the worse off, without benefiting the people who are (or were) worse off. But if an inequality can be eliminated without benefiting the worse off, it is not an inequality that disadvantages them. The revised principle of equality will say that the inequality is not objectionable, so it will not tell us to eliminate it.

The response does not avoid the general question about the value of equality. It restricts the principle of equality, but it does not change the fundamental idea that the principle expresses. Only inequality that disadvantages the worse off is counted as objectionable, but if we consider this kind of inequality, the principle of equality objects to the difference between the lives of the people concerned. The objection is supposed to be concerned with the relationship between the lives, not the content of the lives.

This leads to the most serious problem for the suggestion. Why should inequality only be bad when it disadvantages those who are worse off under the inequality? If we object to inequality that satisfies the condition but not to other kinds of inequality it may not be the revised principle of equality that explains our judgments. The principle criticizes inequality because it thinks that inequality is bad in itself. But our real objection may be to the fact that the badly off are being deprived of benefits, benefits that they would receive if the inequality were eliminated. If this is what we think, then our conclusion is explained by some version of the priority view, not by the revised principle of equality. We care most about helping the badly off, so we are willing to reduce the level of the better off in order to benefit them. This change will also end the difference between their lives, but we are interested in helping the badly off, not in eliminating that difference. That is why we only eliminate the difference when it *does* benefit the badly off. Arguably this attempt at revising the principle of equality also ends by abandoning it.

The problems for the principle of equality might prompt egalitarians to switch from the teleological equality view to the deontological view. The deontological view seems to avoid the general issue about

---

does not single that group out for special treatment. It tells us to minimize objectionable inequality, where an inequality is objectionable if it disadvantages those who are worse off under the inequality. Groups other than the very-worst-off group might be disadvantaged by an inequality. So minimizing objectionable inequality will not always maximize the prospects of the very worst off. Even if we decide that inequality is only objectionable if it disadvantages the very-worst-off group, the principle of minimizing objectionable inequality is still not the same as the difference principle. The difference principle tells us to increase inequality when this would benefit the worst off. The revised principle of equality would not count this inequality as objectionable, but it does not require us to produce the inequality.

equality. It does not say that inequality itself should be regarded as something bad. It says that some actions that produce inequality violate the duty to treat people fairly.

The deontological view only escapes the problem if it is reasonable to believe in a duty to treat people equally without also thinking that inequality is bad in itself. If the existence of the duty could be explained only by the badness of the resulting inequality then the deontological equality view would presuppose the teleological equality view. But I think we can understand acknowledging the duty without explaining it in that way. We might object to treating people in different and unequal ways just because we think that this is an unfair way to treat them. The differential treatment will cause a difference between their lives, but we are not compelled to think that the difference is itself bad.

It is not obvious that the deontological view avoids leveling down.[29] The issue concerns actions that create inequalities which benefit, or at least do not harm, the people who are worse off under the inequalities. Since the deontological view does not suppose that inequality is bad, it might be thought that we can just stipulate that these actions are not wrong. However, the deontological view objects to inequality when it results from differential treatment that is unfair. In order to show that this kind of inequality is not objectionable we must be able to explain why the differential treatment that creates it is not unfair. In these cases the treatment does not harm the people who turn out to be worse off under the inequality. But that does not guarantee that it is fair. The deontological equality view counts actions as wrong because they treat different people in different ways, not because they cause harm to individuals. If we assume that the action must be fair because it does not harm those who are worse off, it might seem that our concern is really with harms and benefits for the people who are badly off, not with the fairness of their treatment compared to the treatment of those who are better off. Again the danger is that our conclusion will really be explained by the priority view, not the deontological equality view.

Other egalitarians would choose to defend the teleological equality view. They think equality can be said to have value even when it is clearly distinguished from improvements in the content of any lives. They are suggesting that something can be good without making anyone's life better, without contributing to the good of any person. Someone who finds this general position about value persuasive would also not be convinced that leveling down is a fatal problem for the combined view. They would be prepared to admit that the combined

29. When Nagel discusses a version of the deontological equality view he suggests that it might sometimes tell us to level down (Nagel, *Equality and Partiality*, p. 107).

view might sometimes tell us to harm some people without benefiting others.[30]

Egalitarians will be anxious to make this position persuasive. One way of supporting it is to find examples of things that we count as good that are similarly unrelated to benefits for people. With these examples as analogies, it will be easier to believe that equality can have value.

For example, we might compare the principle of equality to the principle of average utility. Unlike the principle of total utility, there is a sense in which the principle of average utility cares about distribution. It is concerned with how happiness is divided between lives, although not in the same way as the principle of equality. Like the principle of equality, the principle of average utility can be said to attribute value to an overall property of an outcome: the ratio between the total amount of happiness and the number of people.

This is revealed when the principle of average utility is applied to questions about population. It would prefer an outcome in the distant future in which a few people live very happy lives to an alternative outcome in which many more people live less happy lives. The two outcomes contain different people. In saying that the first outcome is better than the second, the principle is making a claim of value that is independent of benefits for particular people. No individual is better off in the first outcome than in the second or lives a better life in the first outcome than in the second. It might seem that the principle of average utility, like the principle of equality, counts as being valuable something that is distinct from the goods of particular people.

There are strong reasons for thinking that the principle of average utility gives the wrong answers to questions about population size.[31] But the principle is not rejected because it has the formal features described in the previous paragraph. So it still might be useful as an analogy. If the principle of equality shares those formal features it should also not be automatically rejected.

However, the analogy is not close enough. As the population example shows, the principle of average utility does value something that is independent of benefits for particular, identifiable people. But in another sense this principle is concerned with what is good for people. In explaining why the first outcome is better, it is natural to say that it is better because under it people in general would be better off, even if no specific individuals are better off in the first outcome

30. This is Temkin's view. He does not think it is crucial for the plausibility of egalitarianism to show that the combined view would never recommend leveling down, all things considered.

31. Derek Parfit, *Reasons and Persons* (Oxford: Oxford University Press, 1984), pp. 419–22.

than in the second (I do not mean to underestimate the difficulty of explaining exactly what this claim means). The principle of average utility preserves a connection between what is good and a better quality of life for people. This is not true of the principle of equality. It says that an equal outcome is better, in one respect, than an unequal outcome in which everyone is much better off. There is no respect in which the equal outcome is better for specific individuals or for people in general. The principle of equality breaks more radically with the view that what is good must involve improvements in people's lives. From the fact that the claim of value made by the principle of average utility is not unacceptable in principle it does not follow that the same is true of the claim made by the principle of equality.[32]

Temkin answers the objection to equality by first diagnosing the view about the good that it expresses. He suggests that it asserts a general relationship between what is good for people and what makes outcomes better. It claims that the only factors relevant to the goodness of outcomes are factors that are also relevant to what is good for people. Temkin identifies a view about what is good for a person with a view about what would be in that person's self-interest. He concludes that the objection makes the goodness of outcomes a direct function of the self-interest of individuals. He replies to the objection by arguing in detail that there is no view about what is good that is plausible *both* as a theory of what makes outcomes good *and* as a theory of what is in the self-interest of individuals.[33]

I suspect that this strategy will not lead to a satisfying answer to the objection. In the end the objection counts against equality because it makes a claim about the goodness of outcomes. The objection will seem persuasive as long as we think it is reasonable to locate the factors that make outcomes good *inside* people's lives—for example, as long as we think that an outcome can be made better only in virtue of some valuable activity that an individual performs or some valuable state that a person is in. A view of the good that recognizes objectively valuable activities and states of people may not be persuasive as a theory of self-interest, if we think that self-interest must be explained in terms of the satisfaction of desires or happiness (although it might be acceptable as an account of when a person's life is made better, understood in terms of the amount of good that the life contains).

32. Temkin points out that the objection against equality can be isolated from these examples by restricting the objection to cases where the identities of people are not affected by our choices (Temkin, pp. 255–56). I think there is another problem with using the examples to defend equality: the most general and persuasive version of the objection can accommodate the examples.

33. Ibid., chap. 9, secs. 6–9.

However, the objection to equality will have force as long as it does seem persuasive as a theory of the goodness of outcomes.[34]

In fact, Temkin considers a view of this sort which he calls the Objective List Theory of the goodness of outcomes. He suggests that once we have distinguished the factors that make outcomes better from the satisfaction of desires and happiness there is no reason why equality itself cannot be such a factor. He compares equality to moral ideals and values that ask us to respect freedom, autonomy, or rights. In respecting those ideals we might act in ways that do not further the self-interest of anyone, but we do bring about a better outcome.[35] Perhaps the same is true of equality.

However, I think that these analogies also do not help to defend equality. In the case of respecting autonomy, the gain in autonomy may not be in the person's self-interest, but it seems that something valuable is added to the content of the person's life. In other cases respecting rights might not involve adding anything good or valuable to lives, apart from the issue of self-interest. But the ideals that Temkin mentions are often interpreted deontologically rather than teleologically. When we respect someone's rights, our basic moral reason for doing so is not that our action will lead to the best overall outcome (if the outcome does count as better it is because the deontological moral claim expressed by the right has been respected). By contrast, the teleological equality view seems to make its basic moral claim about the goodness of outcomes. The equal outcome is better in its own

34. There is another way of using a theory of the good to answer the objection against equality. It might be claimed that equality does make people's lives better—that the relation of equality holding between my life and someone else's life counts as a good added to my life, in the way that the relation of truth holding between one of my beliefs and the facts might make my life better. Temkin considers the possibility of defending equality by appealing to an Objective List Theory of what is in a person's self-interest, with equality as one of the items on the list, although he does not himself endorse this defense (Temkin, pp. 273–75). I think that it is more persuasive to make this claim about equality if we distinguish between self-interest and the value contained in a life and say that equality makes a person's life better even though it does not contribute to that person's self-interest. Nevertheless, I think the claim is unconvincing. My own life is not made worse by the mere existence of another person whose happiness is greater than my own. John Broome does think that inequality makes lives worse (John Broome, *Weighing Goods* [Oxford: Blackwell, 1991], chap. 12). He suggests that this is so because inequality involves unjust treatment, and being treated unjustly itself counts as a harm for the person who is treated unjustly (Broome, pp. 181–82, 192–99). In the cases being considered (where the inequality does not harm the people who are worse off in other ways, apart from the existence of the inequality itself) it seems to me more reasonable to say that the treatment is unjust despite the fact that it does not harm anyone or make anyone's life worse.

35. Temkin, pp. 275–76.

right, not because some deontological obligation has been fulfilled.[36] So these analogies should not persuade us that equality can have value even though it does not involve lives being made better in any sense.

The analogies with other goods and values do not provide a strong defense of the principle of equality. It might be better to think about the principle directly and try to weaken the force of the objections against it.

The basic criticism is that the principle asserts a value that is too far removed from individuals and their goods. As I have explained, the principle says that equality is a good despite the fact that it does not benefit anyone. But the extent of its separation from considerations of what is good for people should not be exaggerated. The principle attributes value to equality between people, not other things. And it says that people should be equal with respect to the good. So it is not comparable to the view that beautiful things have a value that is completely independent of any possibility of their being understood and enjoyed by people (or other intelligent observers). The principle of equality is concerned with the human good, even if equality does not itself make anyone's life better.

The principle attributes value to an overall property of an outcome, not to an aspect of people's lives. But that property is just the distribution of what is good between different individuals. Attributing value to this property of outcomes is not incompatible with saying that the case for equality is essentially a response to the moral claims of individuals.[37] The property of outcomes has value because people have a moral claim to an equal enjoyment of the good.[38] So the princi-

36. Ross's discussion is helpful (W. D. Ross, *Foundations of Ethics* [Oxford: Clarendon, 1939], pp. 285–89), although he considers the value of a distribution in accordance with merit, not an equal distribution. Ross thinks that such a distribution is a good. Because it involves a relation between people he calls it a situational good. Situational goods are distinguished from personal goods—personal goods are activities and experiences that would be included in the goods of people. Ross thinks that the goodness of a distribution in accordance with merit explains our duty to produce it, and not the other way around. He argues that we would value this distribution even if it were produced by nature and not by human agency. I think that defenders of the teleological principle of equality should explain it in the way that Ross explains his distributive principle.

37. Temkin proposes this way of understanding equality. He specifies principles of equality in terms of the complaints that individuals might have against inequality (Temkin, chap. 2). He says that the ultimate concern of egalitarians is satisfying moral claims held by individuals, where those claims are concerned with how those fare with respect to other individuals (Temkin, p. 200).

38. A different example will show why it seems helpful to talk about individual claims. Suppose that I have made a promise to a particular person, and I now see that keeping my promise will hurt that person's interests. I might nevertheless believe that

ple of equality can be understood in terms of the moral claims of individuals, even if it does not contribute to the good of individuals.

Relating equality to individual claims might have another advantage. It can explain an asymmetry in how we think about the value of equality. Inequality is a difference between two lives. If the moral principle about equality simply says that such a difference is a bad thing, the objection to inequality will not have more to do with the position of the person who is worse off than with the position of the person who is better off. It would be like the aesthetic objection some might feel to a painting of two people of very different heights. The difference in heights might be a blemish, but the blemish does not have more to do with the shortness of the short person than with the tallness of the tall person. Arguably it is different in the case of inequality. There the objection does have more to do with the position of the worse-off person than with the position of the better-off person. The asymmetry can be explained (assuming that it does exist) if we can say that the inequality violates a claim of the worse-off person in a way that it does not violate a claim of the better-off person.[39]

Appealing to the moral claims of individuals to explain the value of equality might seem odd given the nature of the principle of equality that I am discussing. It will sometimes tell us to lower everyone's level of welfare, including the people who were worse off under the inequality. The problem is not that this conclusion is absurd, but that it might seem questionable whether it can be explained in terms of the moral claims of individuals. According to this view, an inequality is objectionable because it violates a claim of the worse-off person. That person has less than a fair share. Can we satisfy that person's claim by giving him even *less* than he already had?

---

there is a moral reason for me to keep my promise. The moral reason is not related to the person's self-interest, or to making that person's life better in some other way. Nevertheless, it is not an impersonal moral reason. Keeping the promise satisfies a moral claim held by that individual, even if it does not further that person's good.

39. Despite Temkin's use of the language of individual complaints, I am not sure that he would agree that the asymmetry I have described is an important feature in our judgments about inequality or that individual claims can provide some significant further explanation of why inequality is wrong. He discusses (Temkin, pp. 45–46) the view that since equality itself is symmetrical we could attribute a complaint against inequality to the people who are better off under the inequality, or to *us* when we think about the inequality and find it objectionable, as well as to the people who are worse off under the inequality. Used in this extended way the language of complaints seems to be just a way of signaling that an inequality is objectionable. Claims understood in this way would not explain why the inequality is objectionable, and they would not be able to account for the asymmetry. John Broome also appeals to the moral claims of individuals in discussing the value of equality (Broome, pp. 192–200). Broome thinks that individual claims can help us to understand how inequality itself can count as a harm to individuals. I think it is better to use individual moral claims to explain how inequality can be wrong even though it does not harm individuals.

However, I think that this objection can be answered. The worse-off person has less than a fair share because he has less than someone else. His claim is to have a fair share, but determining what is a fair share will depend on comparisons with the shares of other people. The principle of equality says that we should use the idea of equality in making these comparisons. So it can make sense to say that we are satisfying the claim of the person who is worse off to have a fair share, even if we give him less than he had when the inequality existed. The outcome that results will be fairer, even if it is worse in other respects.[40]

I think that these points about the principle of equality do weaken the force of the objections against it. Equality is a value concerned with the human good, and it is a response to the moral claims of individuals. It expresses a view about the best way to divide what is good between different people. If there are any moral claims that are fundamentally concerned with the distribution of the good across different lives, it seems that they must be separate from considerations about promoting or increasing people's goods. When we understand the work that the principle is supposed to do, there is less reason to reject it because equality might not benefit anyone. To reject it for that reason is in effect to claim that the distribution of the good cannot itself be a basic moral concern. Why should egalitarians, or people who find this principle or some other distributive principle at least persuasive, simply grant that claim?

The explanation of the value of equality in terms of individual claims, and the idea that the claims are claims to a *fair* share, might suggest that this view is really a particular kind of deontological concern with equality. Perhaps individual claims do not explain why an equal outcome is better than an unequal outcome, they explain why we should care about equality independently of the goodness of outcomes. I will not argue that the view must be interpreted teleologically according to all of the different ways of drawing the distinction between teleological and deontological equality views. What is important for the argument of this article is that it does not match the deontological view that I have described, a view involving an agent-relative reason not to cause inequality ourselves. So there is no reason to suppose that the claims are only relevant when some person or institution acts to create inequality. Inequality that is wholly explained by natural

40. The explanation of equality in terms of claims has another advantage. In the case of many moral claims, we think that it is possible for the person with the claim to resign it or not to insist on it. If inequality is objectionable because it violates a claim of the people who are worse off, it will not be objectionable if they choose to relinquish their claim. And we might suspect that this would happen when achieving equality would make everyone worse off. So there might be another reason why the principle of equality would not lead to leveling down.

causes might also be opposed to the relationship that should exist between different lives, and it would count as violating individual claims as I understand them.

I have discussed several issues about equality. I began by distinguishing between the teleological equality view and the deontological equality view. The deontological interpretation of equality has difficulty with the scope of its egalitarian principle. On the other hand, the teleological interpretation seems initially to be more vulnerable to the most fundamental objections against thinking that equality can matter for its own sake. I have argued that these objections can be answered, or at least that we can come to see why they are not decisive reasons against the teleological principle of equality.

[21]

LARRY S. TEMKIN

Inequality

I. Introduction

The notion of equality has long been among the most potent of human ideals, and it continues to play a prominent role in political argument. Views about equality inform much of the debate about such wide-ranging issues as racism, sexism, obligations to the poor or handicapped, relations between the developed and underdeveloped countries, and the justification of competing political, economic and ideological systems.

Most discussions of equality have focused on two questions: Is equality really desirable? And, what kind of inequality do we want to avoid (that is, insofar as we are egalitarians, do we want equality of opportunity, or primary goods, or need satisfaction, or welfare, or what)? These are important questions. But I shall be asking a third question. When is one situation *worse* than another with respect to inequality? It is only by addressing this question, I believe, that one can begin to understand the nature and complexity of the notion of inequality.

Of course, in some cases the answer to my question can easily be given. We know, for instance, that among equally deserving people, the inequality in a situation would be worse if the gaps between the better- and worse-off people were large, than if they were small. Consider, however, a situation where many are better off, and a few are worse off. How would the inequality in such a situation compare to the inequality in a situation

I would like to thank the Oxford University Moral Philosophy Reading Group for their useful suggestions concerning this article. I am also grateful to Larry Lohmann, Thomas Nagel, William Rued, James Ward Smith, and the Editors of *Philosophy & Public Affairs* for their helpful comments. Amartya Sen and Derek Parfit deserve special thanks for their insightful suggestions and penetrating criticisms. This work has been generously supported by the National Humanities Center, for which I am most grateful.

where a few are better off and many are worse off? How would both of these compare to a situation where the better-off and worse-off groups were equal in size? It is with questions such as these that I shall be concerned. As we shall see, they are complicated questions, and ones to which several plausible but conflicting answers might be given.

II. Preliminary Comments

I shall mainly discuss inequality of *welfare*, but this does not affect my arguments. Analogous arguments could be made in terms of inequality of opportunity, or primary goods, or need satisfaction, or whatever.

For each of my examples, where some people are better off than others, I shall assume that the better- and worse-off are equally skilled, hard-working, morally worthy, and so forth. While it would not be necessary, I shall also assume that the better-off are not responsible for the plight of the worse-off, either directly, through exploitation, or indirectly, through unwillingness to share their good fortune. This would be so, for instance, if the inequality were due to irremediable differences in health. Since my concern is with what we might say about these examples with respect to inequality, I make these assumptions in order to insure that our judgments about them are as free as possible from the disturbing influence of our other moral beliefs and ideals.[1]

Although there may be many differences between them, as I shall use the term, an *egalitarian* is any person who attaches *some* value to equality *itself* (that is, any person that cares *at all* about equality, over and above the extent it promotes other ideals). So, equality needn't be the only ideal the egalitarian values, or even the ideal she values most. Still, throughout this article, when I consider what one might say about a situation with respect to inequality, I shall be considering what an egalitarian would say about it *if* equality *were* the only ideal she valued.

1. One might think that if one assumes people are equally deserving, the concern for equality reduces to a special instance of the more general concern for proportional justice—the concern that each person receives what she deserves, or that there ought to be a proportion between doing well and faring well. But although equally deserving people will be equal if they each get what they deserve, on reflection I think it is apparent that the concern for equality is distinct in both its foundations and implications from the concern for proportional justice. In any event, even if our concern for equality *were* a special instance of our concern for proportional justice, this would not affect the importance of the issues I address or the arguments I present.

There is, I believe, an intimate connection between people's views about equality and certain of their views about justice and fairness. In particular, I believe egalitarians have the deep and (for them) compelling view that it is a bad thing—unjust and unfair—for some to be worse off than others through no fault of their own. Unfortunately, as we shall see, it is one matter to note that egalitarians have such a view, and quite another to unpack what it involves.[2]

In this article, I assume that egalitarians should care about natural, and not merely social, inequality (cf. note 2). However, even if one only cared about social inequality, most of the same questions and considerations would apply. (My own view is that most concern about social inequality must ultimately ride piggyback on concern about natural inequality. It is difficult to see why social inequality would be intrinsically bad, if natural inequality is not.)

I am concerned with inequality from a *moral* perspective. This is why my question is not "When is there *more* inequality in one situation than another?" but rather, "When is one situation *worse* than another with respect to inequality?" There is, I believe, good reason to allow for the possibility that in some cases one situation might be *worse* than another with respect to inequality, though in a certain objective sense it has *less* inequality.[3]

2. A word about the notion of injustice. Some people think that the notion of injustice can be applied to a situation only if something could be done to prevent or improve that situation. They regard the concept of natural injustice as bogus. I believe it is a natural injustice that some people are born blind while others are not. However, the difference between these positions is, at least for some writers, largely terminological. Even if one wants to say, with Rawls or Nozick, that an *injustice* has been suffered only where there is a perpetrator of the injustice, we can still recognize that a situation is such that if someone *had* deliberately brought it about she would have been perpetrating an injustice. This tells us something about the situation. It tells us that if the situation were such that we *could* do something to improve it, we *should*, or, more accurately, that there would at least be some reason to do so. (Rawls would, I think, agree with this, though perhaps Nozick would not.)

3. One reason I say this is I think the egalitarian is not committed to the view that deserved inequalities—if there are any—are as bad as undeserved ones. In fact, I think deserved inequalities are not bad *at all*. Rather, what is objectionable is some being worse off than others *through no fault of their own*. Thus, while objectively there may be *more* inequality in one situation than another, that needn't be *worse* if the greater inequality is deserved, but the lesser inequality is not. (This is another reason why I make the assumption that people are equally skilled, hardworking, morally worthy, and so forth.)

It is, of course, extremely difficult to decide when people are worse off than others through no fault of their own. Some think this is nearly always the case, others, almost never the

Finally, although I think most of the arguments that have been offered against equality can be refuted, let me emphasize that this article is neither a defense, of, nor an attack on, the ideal of equality. I do not address the question of whether one *should* care about inequality, or the question of *how much* one should care about inequality. It seems to me that until one understands the notion of inequality, these questions are premature.

III. Individual Complaints

My main aim is to consider our judgments about how situations compare with respect to inequality. However, there is another, more particular kind of judgment it will help to consider first. This kind of judgment is about how bad the inequality in a situation is from the standpoint of particular individuals in that situation.

Such judgments can be made using the terminology of "complaints." Thus, for any situation where some people are better off than others, we can say that the best-off have nothing to complain about, while the worst-off have the most to complain about. (Here, and in what follows, I often drop the locution "with respect to inequality." Henceforth all references to complaints are to be understood as complaints with respect to inequality, unless stated otherwise.)

To say that the best-off have nothing to complain about is in no way to impugn their moral sensibilities. They may be just as concerned about the inequality in their world as anyone else. Nor is it to deny that, insofar as one is concerned about inequality, one might have a complaint *about* them being as well off as they are. It is only to recognize that, since they are at least as well off as every other member of their world, *they* have nothing to complain about. Similarly, to say that the worst-off have a complaint is not to claim that they will in fact complain (they may not). It is only to recognize that it is a bad thing (unjust or unfair) for them to be worse off than the other members of their world through no fault of their own.

For any situation, then, in which some people are worse off than others,

case. Fortunately, one need not decide this issue in order to recognize that only undeserved inequalities are bad.

Another reason for couching my question in normative rather than purely quantitative terms is presented obliquely in Section V.

two questions arise. Who is it that has a complaint, and, how should we compare the seriousness of different people's complaints? To the first question there seem to be two natural but competing answers, neither of which can easily be dismissed.

According to the first answer, only those people who are *worse off than the average* have a complaint. This answer might be defended as follows. In a world of *n* equally deserving people, the fairest distribution would be for each person to receive one *nth* of the total, since among equally deserving people, a fair share is an *equal* share. Those who receive less than one *nth* of the total would thus have a complaint, since they are receiving *less* than their fair share. Moreover, they are the only people who have a complaint, since those who receive one *nth* or more of the total are *already* receiving their fair share or *more* than their fair share. But, in a world of *n* people, one *nth* of the total welfare is the average level of welfare. Hence, all and only those people below the average have a complaint.

This line of reasoning often expresses itself in our thinking. Thus, whatever someone is complaining about, the lament "Why me?!" will tend to meet with a sympathetic response if she is below average in the relevant respect, but ring hollow otherwise. So, for instance, given that the second best-off person is *already* much better off than the average member of her world, most people would tend to think that such a person has *nothing* to complain about with respect to her level of welfare.

There is another natural answer to the question of who has a complaint: *all but the very best-off* have complaints. A defense of this answer may be made with the aid of diagram 1.[4]

In A, we may judge that q has a complaint, because, among equally deserving people, we think it is a bad thing (unjust or unfair) for one person to be at q's level, while another is at p's. In B, for instance, it may seem that q would have just as much to complain about as in A, since she is no better off than she was in A, and since p is no worse off than she was in A. True, in B there is another person who is as bad off relative

4. In the following discussion, it will be easier if I assign numbers to the levels of welfare represented in my diagram. This naturally suggests that the bottoms in the diagram represent the zero-level of welfare—the point at which life ceases to be worth living. While presumably there *is* such a zero-level, there are large disagreements about where that level is. Let me note, therefore, that neither here, nor elsewhere in the article do my results depend upon our being able to determine the zero-level precisely. In this respect my work differs significantly from that of most economists.

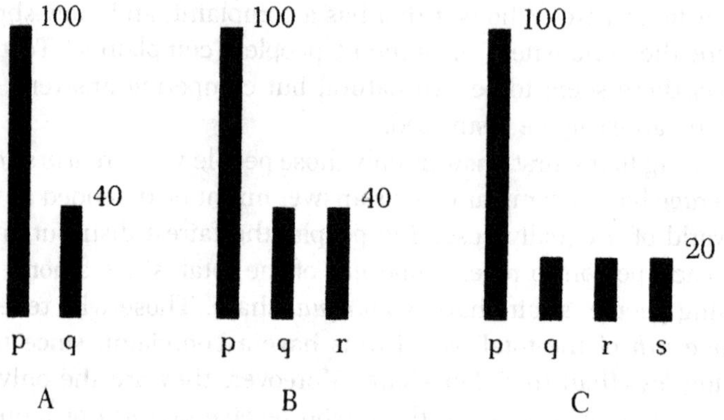

DIAGRAM 1

to p as q is. However, that doesn't lessen the injustice of q's being worse off than p—it only makes it the case that instead of there being *one* instance of injustice there are two!

Consider C. From q's perspective, it seems the inequality in C might appear worse than the inequality in A, since p is just as well off as she was in A, while q is worse off. More particularly, it appears that q's complaint would be larger in C than in A, since it is worse to be at level 20 while another is at level 100, than it is to be at level 40 while another is at level 100. Again, the presence in C of r and s may not seem to lessen the injustice of q's being worse off than p through no fault of her own; their presence only makes it the case that instead of there being one person with a larger complaint than q had in A, there are three. But note, in C, q fares better relative to the average than she did in A. Hence, the view that q would have a larger complaint indicates that we determine q's complaint by comparing her to p, and not to the average.

Extended, such reasoning yields the conclusion that all but the very best-off have a complaint. And on reflection, it is clear that this, too, often expresses itself in our thinking.[5] Applied to our earlier example, this reasoning suggests that even if we admit that *relatively speaking* the second best-off person has "nothing" to complain about, when we focus on the individual comparison between the best-off person and that person,

5. For example, most people who earn $10,000 a year are better off than the vast majority of people alive today, people who have ever lived, and the other living organisms. But this doesn't stop us from comparing such people to the relatively few who are much better off than they, and thinking they have a complaint with respect to inequality.

it will appear to be unfair or unjust for the one to be worse off than the other through no fault of her own. So, even the second best-off person will have a complaint; though her complaint may be small, both in absolute terms, and relative to the complaints of others. Thus, on the question of *who* has a complaint, there appear to be two plausible answers: those below the average, and all but the very best-off.

Let us next consider the question of how we assess the seriousness, or size, of someone's complaint. To this question there seem to be three plausible answers. The first two parallel the division in our thinking about who has a complaint. Thus we might think that the size of someone's complaint will depend upon how she compares to either *the average member of her world*, or *the best-off member of her world*.

These two ways of regarding the size of someone's complaint correspond to two natural ways of viewing an unequal world: as a deviation from the situation which would have obtained if the welfare had been distributed equally, and as a deviation from the situation in which each is as well off as the best-off person. On both views it would be natural to determine the size of someone's complaint by comparing her level to the level at which she would cease to have a complaint. On the first view, this would be the average level of her world—the level she would be at if fate had treated each person equally. On the second view, it would be the level of the best-off person—the level at which she would no longer be worse off than another.

There is a third way of measuring the size of someone's complaint. This way accepts the view that all but the very best-off have a complaint, but contends that the size of someone's complaint depends not on how she fares relative to the best-off person, but on how she fares relative to *all of the other people who are better off than she.*

This view might be defended as follows. It is bad for someone to be worse off than another through no fault of her own. This is why any person who is in such a position will have a complaint. But if it is bad to be worse off than one person through no fault of your own, it should be even worse to be worse off than two such people. And, in general, the more people there are who are better off than someone (and the larger the gap between them), the more that person should have to complain about with respect to inequality. Therefore, to determine the size of someone's complaint, one must compare her level to those of *all* people who are better off than she, and not only to the level of the very best-off person.

Although this third way of regarding the size of someone's complaint may seem less natural than the first two, it does not seem less plausible. Indeed, it is arguable that this position captures certain of the most plausible features of the first two views, while avoiding their most implausible features.

Let us summarize the argument so far. Our notion of inequality allows us to focus on particular individuals, and make judgments about whether or not, and the extent to which, they have a complaint with respect to inequality. There is, however, a division in our thinking concerning who has a complaint, and how we determine the magnitude of a complaint. Specifically, one might plausibly maintain that only people below the average have a complaint, and the size of their complaint depends upon (1) how they fare relative to the average (henceforth, I shall call this the *relative to the average* view of complaints). Alternatively, one may claim that all but the best-off have a complaint, and the size of their complaint depends either upon (2) how they fare relative to the best-off person (henceforth, the *relative to the best-off person* view of complaints), or upon (3) how they fare relative to all those better off than they (henceforth, the *relative to all those better off* view of complaints).

With respect to how we actually measure the size of someone's complaint on the views given above, I shall make the simplifying assumption that we merely subtract her level of welfare from the level of the average person, or the best-off person, or the levels of all those better off than she. I believe that the figures thus arrived at should be differentially weighted to reflect the view that inequality matters more at low levels than at high levels. But for my purposes here, we can ignore this complication. Doing so will not affect the main conclusion I shall reach.

IV. The Sequence

We are now in a position to consider our most general judgments about how situations compare with respect to inequality. In order to explore the reasoning underlying and influencing such judgments, I shall be looking at a group of artificially simple worlds, which I shall refer to as the *Sequence.* This consists of 999 outcomes, or *worlds,* each of which contains two groups of people, the better-off and the worse-off. In each world the total size of the population is 1,000, but the ratio between the two groups steadily changes. In the first world there are 999 people better off and one person worse off, in the second 998 better off and two worse off, and so on. By the end world one person is better off and 999 are worse off. The first, middle, and last worlds of the Sequence are represented below.

THE SEQUENCE

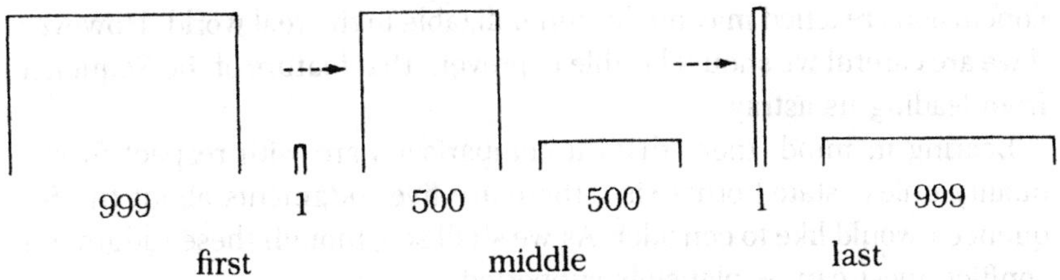

DIAGRAM 2

How do the worlds of the Sequence compare with respect to inequality? Notice, I am *not* asking how they compare "all things considered."[6] These are very different questions. Consider the worlds shown in diagram 3. We may think that "all things considered" A is better off than B, since everybody in A is better off than everybody in B, and since the inequality in A is fairly slight. Nevertheless, *with respect to inequality*, B is better than A. In B there is perfect equality. In A there is not.[7]

DIAGRAM 3

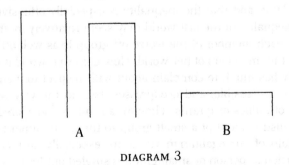

6. In this respect, then, as the reader may be aware, I am addressing a different question than the one Derek Parfit addresses when he employs similar diagrams in his work on future generations. Parfit is concerned with overall judgments of better and worse. I am concerned only with inequality.

7. Throughout this article, when I say that x is better than y with respect to inequality, I mean that the inequality in x is not as bad as the inequality in y.

A word about my terminology. Where philosophers talk in terms of *equality*, most econ-

I emphasize this point because the Sequence is getting progressively worse in terms of both total and average utility. This is an unavoidable feature of the Sequence, but one which should not mislead us as long as we bear in mind the question we are interested in. Similarly, there is a danger in looking at such neatly divided worlds in that some of the conclusions reached may not be generalizable to the real world. However, if we are careful we should be able to prevent that feature of the Sequence from leading us astray.

Bearing in mind, then, that all comparisons are with respect to inequality unless stated otherwise, there are five judgments about the Sequence I would like to consider. As we shall see, though these judgments conflict, most can be plausibly supported.

V. Better and Better

When one first considers the Sequence, one might judge that the worlds are getting *better and better*, partly, because as the Sequence progresses, it appears to be less and less the case that anyone is being especially victimized by the situation. In the *first* world, for instance, it is as if the entire burden of the inequality in that world is borne by the one, lone member of the worse-off group. Given that that person is worse off than *every other* member of her world, it may seem both that she has a very large complaint, and that the inequality is especially offensive. By contrast, the inequality in the last world may seem relatively inoffensive. In *that* world, each member of the worse-off group is as well off as *all* but *one* of the other members of her world. Hence, in that world it may seem as if nobody has much to complain about with respect to inequality.

This view is plausible, and it expresses itself in the way we react to the actions of bullies or tyrants. Though, all things considered, we may prefer the mistreatment of a small group to the mistreatment of a large group, certain of our egalitarian views are especially attuned to cases where a particular person or small group is singled out for harmful treatment.[8] Indeed, I think it is the singling out in this way of an individual

omists talk in terms of *inequality*. The issues are the same, but not the words. On this point, I think the economists' terminology is more perspicuous, and for the most part I follow their practice. However, in a concession to more standard philosophical usage, I refer below to several principles of *equality*.

8. Whether we would actually prefer the mistreatment of a large group to that of a small

or small group that is the paradigm of a harmful discrimination which is grossly unjust and unfair (and not merely bad). Having seen that certain elements of our thinking support the "better and better" ordering, let us next try to get clearer about what those elements might involve.

One principle which might seem relevant here is the *maximin* principle of justice. This principle states that a society's political, social, and economic institutions are just if they maximize the average level of the worst-off group. In one form or another, many philosophers have come to advocate a maximin principle of justice, and one can see why. There is strong appeal to the view that just as it would be right for a mother to devote most of her effort and resources to her neediest child, so it would be right for a society to devote most of *its* effort and resources to its neediest members. This view is captured by the maximin principle which, in essence, maintains that it would be unjust for society to benefit the "haves," if instead it could benefit the "have-nots."

Now, strictly speaking, a Rawlsian version of the maximin principle is not relevant to our discussion. The main reason for this is simply that it has been offered as a principle of *justice, not* as a principle of equality. According to Rawls, a society's principles would be better, and not *unjust*, if they were altered to improve the lot of the worst-off group. This is so even if in order to effect a small improvement, the lot of the better-off had to be improved immensely. Clearly, then, the maximin principle of justice is *not* a plausible principle of *equality*, for whether or not such an alteration in a society's institutions would make that society more *just*, it would certainly *not* make it better with respect to inequality.

Still, while Rawls's maximin principle may not itself be relevant to our discussion, the spirit of that principle *is* relevant. This is because the same basic concern for the worst-off people which supports a maximin principle of justice would also support a maximin principle of equality.

---

group, all things considered, would depend upon what the mistreatment consisted of. All things considered, we must be glad that Hitler only ordered the mass murder of the Jews, Gypsies, and homosexuals, and not of all the occupied peoples. However, it might have been better if he had made all of the occupied peoples shave their heads and wear yellow armbands, instead of just dehumanizing the Jews in such a manner.

Unfortunately, this example is not pure. Partly, we may think it would have been better if more people had had to shave their heads and wear yellow armbands because we may think that would have lessened the humiliation accompanying those practices. My position is that even if this were *not* so, it still would have been better with respect to certain important elements of the notion of inequality if Hitler had made more people shave their heads and wear yellow armbands.

One version of the maximin principle of equality might be stated as follows: How bad a world is with respect to inequality will depend upon how bad the worst-off group in that world fares with respect to inequality, so, if the level of complaint of the worst-off group is larger in one of two worlds, that is the world which is worse. If the level of complaint of the worst-off group is the same in both worlds, then that world will be better whose worst-off group is smallest; if the two worst-off groups are the same size, then the next worst-off groups are similarly compared, and so forth.

Notice the second clause of this principle comes into play *only* if the worst-off groups fare the same in two worlds. This is important, for depending upon which view of complaints one adopts, the magnitude of the worst-off group's complaint may decrease as the size of the worst-off group increases. In such a case the first clause of the maximin principle would tell us that the situation was improving, and the second clause would not apply. Intuitively, then, the maximin principle of equality would first have us maximize the relative position of the worst-off group, and then minimize the number of people in that group, as long as we were not thereby increasing the complaint of the remaining members of the worst-off group. It would then have us do the same thing for the next worst-off group (as long as this did not increase the complaints of the worst-off group), and so on, until all of the groups were as well off and as small as they could be.

We can now see one reason why the "better and better" ordering seems plausible. In accordance with certain plausible positions, the members of the worst-off group have less and less to complain about as the Sequence progresses. This is true on both the "relative to all those better off" and the "relative to the average" views of complaints (since as the ratio between the better- and worse-off groups decreases, the members of the worst-off group fare better and better with respect both to the number of people who are better off than they [by a certain amount], and to the average level of people in their world). Therefore, insofar as we accept a maximin principle of equality, there will be reason to think that the Sequence is getting better and better.

It is worth noting that the advocate of the maximin principle of equality is concerned not with the sum total of complaints, but with the *distribution* of those complaints. Specifically, he wants the inequality to be distributed in such a way that the "load" which each member of the worst-off group has to "bear" is as small as possible.

III *Inequality*

There is another line of thought which might support the "better and better" ordering. In the earlier worlds, the inequality may seem particularly offensive, as there seems to be virtually nothing gained by it. If a redistribution of the sources of welfare took place, the better-off group would hardly lose anything, and the worse-off group would gain tremendously. Moreover, the unavoidability of the inequality in those worlds may do nothing to lessen the feeling that it is so pointless and unnecessary. We may still feel that the worse-off have been especially unfortunate; and we still fully recognize that a situation of complete equality would have obtained *if only* each better-off person had received a tiny bit less welfare, and *if only* the "extra table scraps" of welfare had gone to the worse-off people. In the middle worlds, on the other hand, a redistribution of the sources of welfare would be "costly." Many would have to lose a lot to achieve complete equality. In the end worlds, redistribution would involve a tremendous loss in the quality of life for some, with virtually no gain to those thus "benefited." Therefore, the inequality might seem least offensive in the end worlds of the Sequence, as in those worlds the "cost" of the inequality might seem smallest, and the "gain" highest.

This position might be summed up as follows. Whether or not anything could be done about it, it will offend us as egalitarians for some to be badly off, while others are well off. But from one perspective, at least, we will be most offended if just a few are badly off, while the vast majority are well off, since the inequality then seems particularly gratuitous. In accordance with this way of thinking, it will seem that the Sequence is getting better and better.

VI. Worse and Worse

I would next like to suggest that certain elements of our thinking support the judgment that the Sequence is getting, not better and better, but *worse and worse*. One principle which can yield this ordering is what I shall call the *additive* principle of equality. According to this principle, the inequality in a world is measured by summing up each of the complaints that the individuals in that world have; where the larger that sum is, the worse the world is. This kind of principle involves two natural and plausible assumptions: (1) given any two situations, the best situation with respect to some factor f will be the one in which the *most* f obtains if f is something desirable (for example, pleasure, happiness, equality), and the one in which the *least* f obtains if f is something undesirable (for example, pain, misery, inequality); and (2) to determine how much f obtains in a situation one needs only to determine the magnitude of the individual instances of f which obtain and then add them together. Since this kind of principle is—understandably enough—associated with utilitarianism, let me note that where people usually disagree with utilitarianism is *not* with its claim that the best world with respect to utility is the one where the sum total of utilities is greatest, but with its claim that total utility is all that matters.

So, like the maximin principle, an additive principle of equality represents certain plausible positions. It captures the view that it is bad for one person to be in such a position that she has a complaint, and the corresponding view that it should be even worse if, in addition to the first person with her complaint, there is a second person who has a complaint.

As with the maximin principle, the additive principle does not *itself* yield an ordering of the Sequence. However, when combined with the "relative to the best-off person" view of complaints, it supports the judgment that the Sequence is getting worse and worse. After all, on that view, more and more people will have a complaint of a certain constant amount as the Sequence progresses, and according to the additive principle, the more people there are with a given amount to complain about, the worse the situation is with respect to inequality.

There is another view which supports the "worse and worse" ordering. Since the main elements of this view have already been examined, I can be brief. Earlier I noted how the maximin principle of equality could be combined with either of two plausible views about complaints to support the "better and better" ordering. However, when combined with the "relative to the best-off person" view, the maximin principle will yield the "worse and worse" ordering. This is because on this view of complaints, the worst-off groups fare the same throughout the Sequence, and according to the most plausible version of the maximin principle of equality, if the worst-off groups fare the same in two worlds, that world will be best whose worst-off group is smallest.[9] Thus, there is a second set of plausible views which combine to support the "worse and worse" ordering.

There is yet another position which would support the "worse and

9. I say more in Section IX about why I think the most plausible version of the maximin principle has the feature in question.

worse" ordering. One might arrive at this position by reasoning as follows. Despite its appeal, the maximin principle is less plausible when applied to more realistic worlds where people are spread out over a continuum of welfare levels. No matter what level is chosen to separate the worst-off group from the rest of society, it seems implausible to contend that we should be genuinely concerned with the complaints of the people at that level, but shouldn't be concerned at all (except in the case of ties) with the complaints of those who are just above that level. More generally, while it may seem reasonable to be more concerned about those with large complaints than those with small complaints, it seems implausible to contend that the complaints of the one group matter, but those of the other do not (except in the case of ties). Thus, an additive principle might seem preferable to a maximin principle, insofar as it is concerned with the complaints of *all* those who have a complaint, and not just with the complaints of some arbitrarily selected worst-off group. Yet a maximin principle might seem preferable to an additive principle, insofar as it is concerned with the *distribution* rather than merely the sum total of complaints. This suggests that a principle which plausibly combined these two elements would have great appeal.

Here is one such principle: We measure the inequality in a world by adding together people's complaints, after first attaching extra weight to them in such a way that the larger someone's complaint is, the more weight is attached to it. Let us call this the *weighted additive* principle of equality. Like the additive principle and the maximin principle, the weighted additive principle appears to be a plausible principle of equality. Combined with the "relative to the best-off person" view of complaints, such a principle supports the "worse and worse" ordering. This is because no matter how the individual complaints are weighted, since the complaints are non-negative, $n + 1$ weighted complaints will always be larger than n weighted complaints.

VII. First Worse, then Better

We have seen that certain plausible positions support the "better and better" ordering, and that others support the "worse and worse" ordering. Still others support the judgment that the Sequence *first gets worse, then gets better*.

It is easy to be drawn to such an ordering by reasoning as follows. In the first world of the Sequence everyone is perfectly equal except, regrettably, for one, isolated individual. As there is just an ever so slight deviation from absolute equality, that world may seem nearly perfect with respect to inequality. In the second world, there are two people who are not at the level of everyone else. The deviation from a state of absolute equality has become more pronounced; hence that world may seem worse than the first. As the Sequence progresses this reasoning continues for a while. The deviations from absolute equality become larger, and as they do so, the Sequence appears to be worsening. After the midpoint, however, the deviations from absolute equality begin to get smaller. By the end world, there is once again just an ever so slight deviation from absolute equality. Everyone is equal except, regrettably, for one, isolated individual. Like the first world, therefore, the last world may appear to be almost perfect with respect to inequality. In sum, since it seems almost tautological that the less a situation deviates from absolute equality the better it is with respect to inequality, it seems natural and plausible to judge that the Sequence first gets worse, then gets better.[10]

There is another line of thought which supports the "worse, then better" ordering. In the first world, only *one* person has a complaint, so as large as this complaint may be, this world's inequality may not seem too bad. Specifically, it may not seem as bad as the inequality in the second, where it may seem that *two* people have *almost* as much to complain about as the one person had in the first world. And these two may not seem *nearly* as bad as the inequality in the middle world. In the middle world, it may seem *both* that a large number have a complaint (half of the population), *and* that the magnitude of their complaints will be quite large (they are, after all, worse off than half of the population through no fault of their own). In the last world, on the other hand, the situation may seem analogous to, though the reverse of, the one obtaining in the first. Although almost everyone has *something* to complain about, it may seem that the size of their complaints will be virtually negligible. Hence, as with the first world, the inequality may not seem too bad.

This reasoning involves two by now familiar elements: the view that the size of someone's complaint depends upon how she fares relative to all those better off than she, and the additive principle of equality. Ac-

10. Implicit in such reasoning is an intuitive notion about how to measure the deviation from a state of absolute equality. On examination, this involves measuring the deviation from the median.

cording to the "relative to all those better off" view of complaints, the size of individual complaints will decrease as the Sequence progresses, as there will be fewer and fewer better off than those who have a complaint. According to the additive principle, how bad a world is depends upon both the magnitude *and* the number of complaints. Combined, these views support the judgment that, with respect to inequality, the middle worlds, where a fairly large number have fairly large complaints, will be worse than either the initial worlds, where just a few have very large complaints, or the end worlds, where many have very small complaints. Indeed, on the view assumed here, according to which the size of someone's complaint is measured by summing up the difference between her level of welfare and that of each person better off than she, it is a simple task to verify that the combination of views I have been discussing will support the judgment that the Sequence first gets worse, then gets better.

Similar reasoning would lead one to expect that the additive principle would also support the "worse, then better" ordering when combined with the "relative to the average" view of complaints. As the Sequence progresses, the situation changes from there being a few much worse off than the average, to there being many only a little worse off than the average. So, combining the two views in question, it might seem that the middle worlds, where a fairly large number have fairly large complaints, will be worse than either the initial worlds, where just a few have very large complaints, or the end worlds, where many have very small complaints. And it is easy to verify that the "worse, then better" ordering is yielded by these views, if one makes the assumption that how bad it is for someone to deviate from the average can be measured by taking the difference between her level of welfare and that of the average person.[11]

I believe that the *weighted* additive principle, in any plausible version, would also support the "worse, then better" ordering when combined with either the "relative to all those better off" or the "relative to the

11. This is slightly oversimplified. On the "relative to the average" view of complaints, it might seem that deviations *above* the average should be regarded as bad, as well as deviations *below* the average. Using the terminology of *complaints*, it might be said that while those above the average do not have a complaint, *we* have a complaint about them being as well off as they are. Correspondingly, on the additive principle, we would measure the inequality in a world by adding up all of the complaints of those below the average, and all of *our* complaints about those above the average.

While this complicates the overall view, it would not change the ordering, or the basic reasoning underlying it.

average" view of complaints. But while the additive principle would combine with the "relative to all those better off" and the "relative to the average" views of complaints to yield an ordering of the Sequence corresponding to a *symmetrical* curve—like A of diagram 4—the weighted additive principle would combine with those views to yield a *skewed* curve—like B of diagram 4. The extent to which the curve would be skewed would depend on the exact weighting system employed by the weighted additive principle. If larger complaints receive lots of extra weight it will be greatly skewed; if they only receive a little extra weight it will only be slightly skewed.[12]

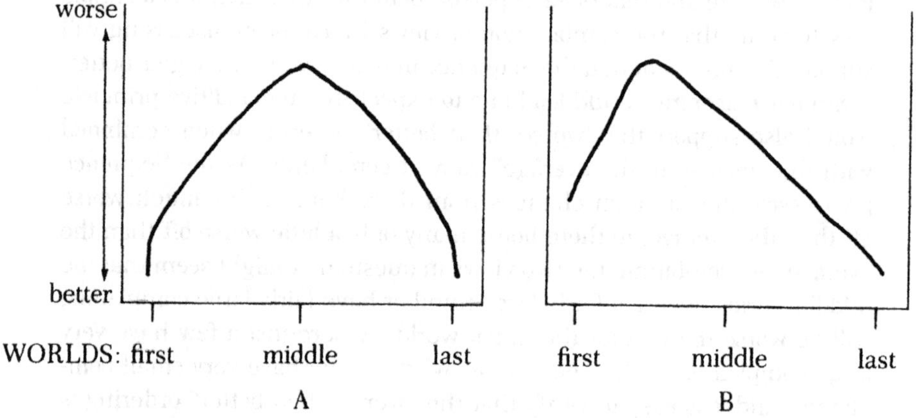

DIAGRAM 4

VIII. First Better, then Worse

By now it may seem that there are *bound* to be several plausible positions supporting the judgment that the Sequence *first gets better, then gets worse.* If there *are* such elements, however, I am not aware of them.

IX. All Equivalent

Do any plausible views support the judgment that the worlds of the Sequence are *all equivalent*? One principle which would support this

12. If the system of weighting were extreme enough, the weighted additive principle would actually support the "better and better" ordering when combined with the two views

ordering when combined with the "relative to the best-off person" view of complaints would be a maximin principle of equality which lacked the tie-breaking clause that if the worst-off groups in two worlds fare the same, then that world will be best whose worst-off group is smallest.

I believe that the version of the maximin principle with the tie-breaking clause is more plausible than the version without it. However, it might be charged that it is the latter version which actually captures our *maximin* views, and that the former version just represents an ad hoc attempt to reconcile our maximin views with certain other views. I think this charge cannot be sustained, and shall briefly suggest why.

The claim that we are especially concerned about the worst-off group is misleading insofar as it suggests a concern on our part about the *group itself* as opposed to the *members* of that group. We do not have a special concern for some real or abstract entity "the worst-off group." What we have, is a special concern for the worst-off members of our world, and it is *this* concern which a maximin principle expresses.[13] But surely, insofar as we are especially concerned with the worst-off members of our world, we would want to raise as many of them as possible above their present level, as long as by doing that we were not increasing the complaints of the others. This suggests that of the two competing versions, it is indeed the one first considered which accurately expresses our maximin views. Thus, one must look elsewhere to support the judgment that the worlds of the Sequence are all equivalent.

There is one sense in which it could be plausibly claimed that the worlds are equivalent. *If* the "relative to the best-off person" view of complaints is adopted, and *if* the worlds of the Sequence represented societies whose principles and institutions were responsible for the size of the gaps between the better- and worse-off groups, but not the *number* of people in those groups, then one might regard each of those societies as equivalent in that with respect to inequality the principles and institutions governing those worlds might seem equally unjust. Such a po-

of complaints in question. However, such a system of weighting would implausibly imply that to be worse off than 999 people is not merely *somewhat* worse than being worse off than 998, it is at least twice as bad. Given any plausible system of weighting, the "worse, then better" ordering results.

13. This claim may be controversial. It reflects the *individualistic* position that individuals are the proper objects of moral concern. This position opposes the *holistic* or *global* one which maintains that groups or societies may also be the proper objects of moral concern. Unfortunately, these issues are complicated and cannot be dealt with in this article.

sition would express the view that there is a sense in which two societies can be equally unjust, though in other respects one may be worse than the other. This is analogous to the view that two judges who accepted bribes in all of their cases might be equally corrupt, even if one tried fewer cases.

Thus, given certain assumptions, I think the "all equivalent" ordering would be plausible. However, let me add that the sense in which this is so is one where *social* justice is concerned. Where natural or cosmic justice is concerned, I am not aware of any independently plausible view which would support this ordering.[14]

X. Conclusion

I have examined the question of how we judge one situation to be worse than another with respect to inequality. I have suggested that a number of plausible positions might influence our egalitarian judgments. Specifically, I have suggested that the additive, weighted additive, and maximin principles could each be combined with the relative to the average, the relative to the best-off person, and the relative to all those better off views of complaints, to yield a judgment about how good or bad a situation is with respect to inequality. I have also suggested that in accordance with certain other plausible views, we might judge a situation's inequality in terms of either how gratuitous it appears to be, or how much it deviates from a state of absolute equality. Finally, I have suggested that we may judge how good or bad a society is with respect to inequality in terms of the principles and institutions of that society responsible for the inequality.

I have focused on situations where the levels of the better- and worse-off groups remain the same, but the ratios between those groups vary. It is important to bear in mind that many positions yielding the same judgment about these cases nevertheless represent different (combinations of) views. Hence, in more complicated cases, their judgments will often diverge.[15]

14. In this respect this ordering differs from the others discussed. Those orderings could be supported whether one's concern was natural *or* social justice.

15. By the same token, positions which diverge in their judgments about the Sequence may concur in their judgments about other cases. But whether the positions agree or not, they represent views with significantly different foundations, and hence, with significantly different implications for some cases.

Among the questions that still need to be explored are: What happens to our egalitarian judgments when (a) the better- and worse-off are spread out among a number of groups? (b) the levels between the better- and worse-off groups vary? and (c) the total number of people in the societies vary, and not merely the relative numbers of the better- and worse-off groups? These questions concern important elements of the egalitarian's thinking. However, their answers do not affect my main conclusions, as the arguments presented here are largely independent of the issues in question.[16]

In addition to the theoretical arguments presented, there are other, perhaps less significant, factors which support the conclusions I have reached. Of these, let me briefly note two.

First, there are many cases where people disagree as to which of two situations is worse with respect to inequality. Moreover, even among those who agree that one situation is better than another, there is often disagreement about the degree to which this is so. Such disagreements support my claims, as they can be explained by the complexity of the notion of inequality, and people's focusing on different elements of that notion in their (naive) judgments.

Second, economists have widely advocated a number of statistical measures of inequality, for example, the *range*, the *relative mean deviation*, the *gini coefficient*, the *variance*, the *coefficient of variation*, and the *standard deviation of the logarithm*. Unfortunately, they have mainly focused on these measures' judgments, but have not adequately pursued *why* these measures have the plausibility they do. On examination, these measures can be seen to give (rough) expression to, and derive their plausibility from, some of the positions I have argued for. Specifically, I believe the range gives expression to the relative to the best-off person view of complaints and the maximin principle of equality; the relative mean deviation gives expression to the relative to the average view of complaints and the additive principle of equality; the gini coefficient gives expression to the relative to all those better off view of complaints and the additive principle of equality; and the variance, the coefficient of variation, and the standard deviation of the logarithm each give expression, in their own way, to the relative to the average view of complaints and the weighted additive principle. Thus, between them, the standard

16. These issues are addressed in my book *Inequality* (forthcoming, Oxford University Press).

statistical measures give expression to each of the three principles of equality, as well as each of the three ways of measuring complaints. Hence, the extent to which these measures have been regarded as plausible, despite its being recognized that they face numerous shortcomings and often conflict, is further independent support for the different elements underlying them.[17]

In suggesting that many different positions underlie and influence egalitarian judgments, I am *not* suggesting that each of these positions is equally appealing, much less that everyone will find them so. But I do think that each represents certain plausible views which cannot easily be dismissed. Because these views often conflict, it may be possible, for each of the positions discussed, to construct examples where the judgment yielded by that position seems implausible. This does *not* show that the various positions are not plausible, nor does it show that they are not involved in people's egalitarian judgments. What it shows is that each position does not *itself* underlie each such judgment.

One conclusion this article suggests which the *non*-egalitarian might readily embrace can be put as follows. Upon examination, the notion of inequality turns out to involve a hodgepodge of different and often conflicting positions. Moreover, and more importantly, many of these positions are fundamentally incompatible, resting as they do on contrary views. It simply cannot be true, for instance, *both* that everybody but the best-off person has a complaint, *and* that only those below the average have a complaint. Nor can it be true that the size of someone's complaint should be measured by comparing her to the average, *and* by comparing her to the best-off person, *and* by comparing her to all those better off than she. The notion of inequality may thus be largely inconsistent and severely limited. While it may permit certain rather trivial judgments such as the judgment that an equal world is better than an unequal one, and that "other things equal" large gaps between people are worse than small ones, in many (and perhaps most) realistic cases, one cannot compare situations with respect to inequality.

Understandably, the egalitarian might try to resist this conclusion. She might contend that each of the positions presented in this article rep-

17. Again, I refer the reader to my forthcoming book on inequality, where a defense of these claims is provided. For a presentation and discussion of the statistical measures themselves, see, for example, J. E. Meade's *The Just Economy* (London: Allen and Unwin, 1976); also A. K. Sen's *On Economic Inequality* (Oxford: Clarendon Press, 1973).

resents a different *aspect* of the notion of inequality, and she might insist that what the conflict between these aspects illustrates is just how complex and multifaceted that notion truly is. What we need, it might be claimed, is to arrive at a measure of our notion of inequality which accurately captures each of the aspects involved in that notion, according them each their due weight. Such a measure would give us a way of accurately comparing many, though perhaps not all, situations with respect to inequality.

It appears, then, that once we see what the notion of inequality involves, we may come to think it is largely inconsistent and severely limited. Alternatively, we may come to think that it is complex, multifaceted, and partially incomplete. Either way, many of our common-sense egalitarian judgments will have to be revised. Few moral ideals have been more widely discussed, yet less well understood, than the notion of inequality.

Name Index

9781138711242

9781138711242